TRIGONOMETRY:

AN INFORMAL APPROACH

BY

STANLEY COTTER

PREFACE

The initial approach to teaching trigonometry has long been a source of division between two camps. As a senior at Bayside High School, Long Island, New York in 1947, I was fortunate to be taught the subject by a master, Isidore Dressler. Twenty-eight years later I was teaching from his textbook, TRIGONOMETRY[1] It begins, as most texts of that time, with a chapter on "Trigonometry of the Right Triangle". For many, the right triangle definitions of the trigonometric ratios serve as helpful models, not only in measurement problems, but also in later functional analysis.

However, after several years of teaching the calculus, I felt the need to introduce the trigonometric ratios as functions of real numbers, requiring an earlier presentation of the circle and its measurements. This text was written and published at Foothill Community College, Los Alto Hills, California in 1979, revised and updated in 1981, and successfully used in Trigonometry sections for over a decade. It was during these years that hand held calculators were replacing slide rules, and these tools were integrated into the current text.

In the present text, right triangle definitions of the ratios find there place in Chapter 8 on "Applications and the Laws of Triangles". Instructors using this text and preferring the traditional approach can start with section 8-1 and move back to Chapter 1 for continuity of development.

In a sense, this text serves as homage to Isidore Dressler, who not only was a model for me as a teacher, but also a stimulus to my efforts as an author. As he said in his own preface: "Each reader is urged to note the simplicity of language and the informal style. Students should have little or no difficuly reading and understanding the text. Note further the organization and exposition of the materials in a self-teaching order, meaningfully and pedagogically developed." I could think of no better way to honor his memory than to follow this dictum.

A more detailed exposition of the objectives of this text is provided by the author as a guide for instructors considering its adoption. This guide and a file containing solutions to even numbered problems may be accessed by contacting stancotter@yahoo.com and specifying "re: trig student guide" in the subject area.

[1]Trigonometry , Isidore Dressler and Barnett Rich, Amsco School Publications, Inc.,New York, 1975

TABLE OF CONTENTS

TABLE OF CONTENTS

TABLE OF CONTENTS

TABLE OF CONTENTS

TABLE OF CONTENTS

CHAPTER 1

THE CIRCLE, ARCS AND CENTRAL ANGLES,
DEGREE AND RADIAN MEASUREMENT

1.1. THE CIRCLE AND ITS MEASUREMENT

As a branch of mathematics, trigonometry was born in the pursuit of astro-
nomical measurements. The Ancient Greeks conceived of the heavens as an example
of the most perfect of geometric forms, the circle. To this day, the study of
trigonometry (literally translated as "triangle measurement") is most accessible
through analyzing the circle and its properties. A circle is defined as a set
of all points in a plane that are equidistant from a fixed point called the center. Figure 1-1A illustrates the definition. Point O is the center of the circle. P is a point on the circle. A circle is.drawn when the fixed point of a
compass is positioned at O and moved from P through Q and back to P in a clockwise revolution. Line segment $\overline{OP}$ is a radius of the circle. r is the measure
of the radius. Every such line segment drawn from the center to a point on the
circle has the same measure, r . We say that radii of the same circle are equal.
The word "radius" shall refer to either a line segment from the center to a point
on the circle, or to a number that measures the length of all such line segments.

Figure 1-1A

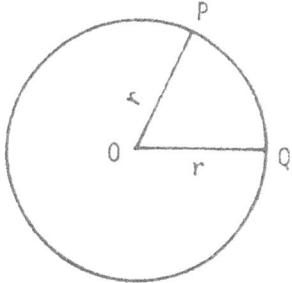

CIRCUMFERENCE

The circumference of a circle with radius r is given by the formula C = 2πr . This measure is the limit of the perimeters of the inscribed regular polygons as illustrated in Figure 1-1B. As the number of sides of the regular polygons increase, the perimeters give a closer and closer approximation to the perimeter (circumference) of the circle. This limiting notion is used as the definition of the circumference and is incorporated into the formula C = 2πr . From this formula we find that $\pi = \frac{C}{2r}$, which can be interpreted as the ratio of the circumference to the diameter (the diameter is twice the radius). This quantity is the same for all circles.

The Greek letter π ("pi") symbolizes an irrational number that has a nonterminating, nonrepeating decimal form: π = 3.1415927... .
In computations with π , the two place decimal approximation 3.14 is frequently used. Another common approximation is the rational number $3\frac{1}{7}$ or $\frac{22}{7}$. A comparison of these approximations will show that $3.14 < \pi < \frac{22}{7}$.

Figure 1-1B

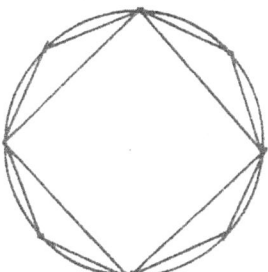

MEASURES OF ARC, REVOLUTIONS

Subdivisions of the circumference are known as arcs. Simple fractional parts of the circumference are usually expressed as multiples of π .

EXAMPLE: The radius of a circle is 3 inches. Find the measures of arcs that are one-half, one-third, three-fourths, and two-fifths of the circumference respectively.

SOLUTION: Since r = 3 inches, C = $2\pi(3)$ = 6π inches. Let s_1, s_2, s_3, s_4 be the required arcs. Then $s_1 = \frac{1}{2}(6\pi) = 3\pi$ inches, $s_2 = \frac{1}{3}(6\pi) = 2\pi$ inches, $s_3 = \frac{3}{4}(6\pi) = \frac{9\pi}{2}$ inches, $s_4 = \frac{2}{5}(6\pi) = \frac{12\pi}{5}$ inches.

EXAMPLE: The circumference of a circle is divided into six equal parts. If the radius is 5 centimeters (cm), find the measure of each part.

SOLUTION: Since r = 5 cm, C = $2\pi(5)$ = 10π cm. Therefore each part is $\frac{5\pi}{3}$ cm .

Arc measures can exceed the circumference. When they do it is convenient to use units of measure called <u>revolutions</u>. One revolution is equivalent to the circumference of the circle, so the number of revolutions required to generate an arc is the ratio of the arc to the circumference. If s is the measure of arc and C is the circumference, then revolutions = $\frac{s}{C}$.

EXAMPLE: Find the measure of arc generated by 3.2 revolutions about a circle with radius 14.3 . Leave answer in terms of π .

SOLUTION: The circumference, C = $2\pi(14.3)$ = 28.6π units. Let s be the arc generated. Then s = $(3.2)(28.6\pi)$ = 91.52π units.

Revolutions imply movement around the circle or, as in the case of a bicycle wheel on pavement, the transformation of circular motion to linear motion.

EXAMPLE: Find the straight-line distance traveled by a bicycle with 24 inch diameter wheels if the wheels make 19 revolutions. Express the answer to the nearest foot.

SOLUTION: Since the diameter is twice the radius, C = $2\pi r$ = πd where d is the diameter. C = 24π inches = $\frac{24\pi}{12}$ feet = 2π feet. Let s = distance traveled. Then s = $19(2\pi)$ = 38π feet = 119 feet to the nearest foot.

An arc that measures less than the circumference may be expressed as a <u>partial revolution</u>. Numerically it expresses the ratio of the arc to the circumference.

EXAMPLE: Find the arc generated by a partial revolution of .54 in a circle with radius 7 cm . Express the answer to the nearest centimeter.

SOLUTION: $C = 2\pi(7 \text{ cm}) = 14\pi$ cm . Let s be the arc. Then $s = (.54)(14\pi \text{ cm})$. To the nearest centimeter $s = 24$ cm .

EXAMPLE: How many revolutions does a wheel make in moving a straight-line distance of one mile (5280 feet) if the diameter of the wheel is 32 inches. Express the answer to the nearest hundredth of a revolution.

SOLUTION: $C = \pi d = 32\pi$ inches $= \dfrac{32\pi}{12}$ feet $= \dfrac{8\pi}{3}$ feet = 1 revolution.

Let x be the number of revolutions in one mile. Then by proportion,

$$\frac{x}{1 \text{ revolution}} = \frac{5280 \text{ feet}}{\frac{8\pi}{3} \text{ feet}} . \quad x = (5280)(\frac{3}{8\pi}) = 630.25 \text{ revolutions}.$$

EXERCISE SET 1-1.

1) Find the perimeters of the square and octagon in Figure 1-1B if the radius of the circle is 2 units. Round off answers to hundredths.*

2) Find the radius of a circle if the circumference is

    a) 12   b) 5.86   c) 1   d) 0.357

3) Find the radius of a circle if the circumference is

    a) 146   b) 17.8   c) π d) 0.075

* In this exercise set and those that follow, the approximation for π is the value displayed by the π key on a scientific electronic calculator, accurate to seven decimal places (3.1415927). Answers to the odd numbered problems are found in the rear of the text. All numerical answers are either expressed in exact form or rounded to two decimal places, unless exceptions are noted. The student is expected to have access to an electronic calculator.

4) What multiple of π must the radius be if the circumference is exactly $14\pi^2$?

5) What is the percentage of error involved in using 3.14 as an approximation to π ?

6) What is the percentage of error involved in using $3\frac{1}{7}$ as an approximation to π ?

7) What is the diameter of a circle with a circumference of 37.95 meters?

8) The circumferences of two circles are in the ratio of 3:2 . What is the ratio of their radii?

9) Combine the following into single fractions and find their decimal equivalents to two decimal places with the help of an electronic calculator:

a) $\frac{2\pi}{3} + \frac{3\pi}{4}$ b) $\frac{7\pi}{6} - \frac{3\pi}{4}$ c) $\frac{8\pi}{3} + \frac{11\pi}{6}$ d) $2\pi - \frac{4\pi}{3}$

10) Draw a circle with r = 5 . Divide the circumference into 8 equal parts as in Figure 1-1C. What is the length of each arc? Complete the labeling of the division points in a counterclockwise direction according to their distance from 0 .

Figure 1-1C

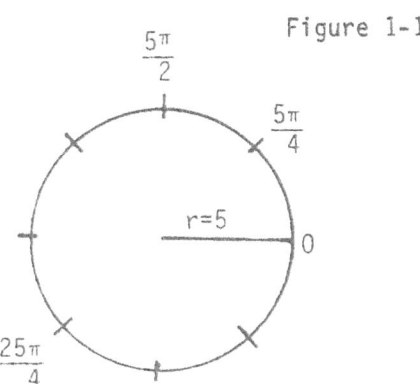

11) Follow the same directions as in exercise 10, dividing the circle into 12 equal parts.

12) How do the units on the circumference in exercise 10 change if the circle has a radius of 7 ?

13) How do the units on the circumference in exercise 11 change if the circle has a radius of 1 ?

14) Referring to Figure 1-1C, how many counterclockwise revolutions are indicated by a division point of $\frac{75\pi}{4}$ on the same circle of radius 5 ?

15) Find the arc generated by 3.7 revolutions on a circle of radius 9 . Leave answer in terms of π .

16) Find the arc generated in 8 hours by the minute hand of an electric clock if it measures 6 inches from the clock center.

17) A child pedals her tricycle 175 feet on a sidewalk. The pedals are attached to the front wheel which measures 50 inches in circumference. If the rear wheels have diameters that are one-third of the diameter of the front wheel, how many revolutions do they make in the distance covered?

18) How far forward does a bicycle travel in one revolution of the pedal gear if the gear ratio is 8:3 and the rear wheel has a diameter of 27 inches? Express your answer to the nearest hundredth of a foot.

19) How fast must the child in problem 17 pedal in order to cover the distance in 20 seconds. Express your answer in pedal revolutions per second.

20) How many pedal revolutions per mile are required of the bicycle in problem 18 in order to maintain a speed of 15 miles per hour?

1-2. STANDARD POSITION. QUADRANTS. THE UNIT CIRCLE.

Circles are fixed for arc measurements in trigonometry by placing them in what is called <u>standard position</u>. By this convention, the initial point or zero point of an arc is placed to the right of all others on the circumference. Positive arcs are measured in a counterclockwise direction from zero and nega- tive arcs are measured in a clockwise direction. Figure 1-2A shows a positive arc, s_1 , that is equal to one-third of the circumference and a negative arc, s_2 , that measures one-fourth of the circumference.

Figure 1-2A

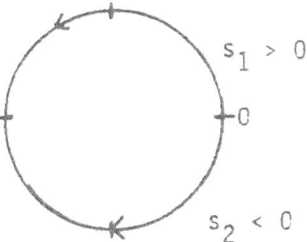

Another convention that is frequently useful is to divide the circumference into quadrants as in Figure 1-2B. These are labeled in Roman numerals in the positive counterclockwise direction and are fixed in their location.

Figure 1-2B

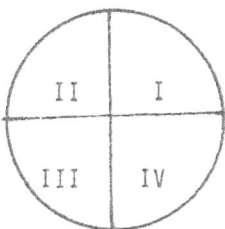

To determine the quadrant in which an arc terminates, division points be- tween quadrants are labeled in either positive or negative quartile measures according to whether the arc is positive or negative:

Figure 1-2C

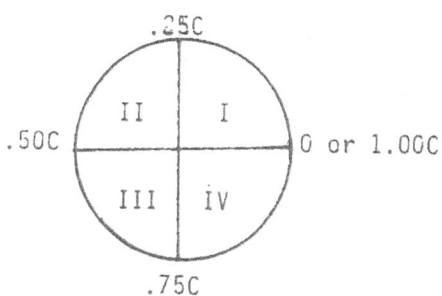 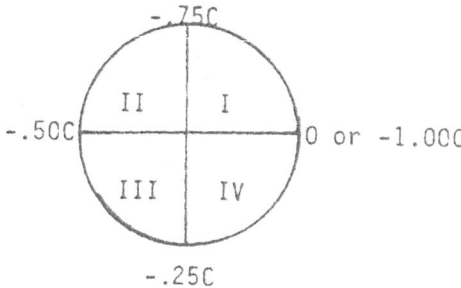

EXAMPLE: The circumference of a circle measures 8.7 cm . In what quadrant do arcs measuring 5.2 cm and -3.8 cm terminate if they are placed in standard position?

SOLUTION: An arc of 5.2 cm is equivalent to a partial revolution of $\frac{5.2}{8.7}$ = .60 rounded. Since .50C < .60C < .75C , this arc terminates in quadrant III. Similarly, -3.8 terminates in quadrant III since $\frac{-3.8}{8.7}$ = -.44 partial revolution and -.50C < -.44C < -.25C .

If an arc in standard position exceeds one revolution, the quadrant in which it terminates can be determined from its excess partial revolution.

EXAMPLE: An arc in standard position measures 78.4π units on a circle with radius 5 . Find the quadrant in which it terminates.

SOLUTION: C = 2π(5) = 10π units. $\frac{78.4\pi}{10\pi}$ = 7.84 revolutions. Subtracting the 7 complete revolutions leaves .84 partial revolution. The arc terminates in quadrant IV since .75C < .84C < 1.00C .

EXAMPLE: An arc in standard position is generated by -14.76 revolutions on a circle with radius 2 . Find the measure of the arc and the quadrant in which it terminates.

SOLUTION: C = 2πr = 2π(2) = 4π . s = (-14.76)C = (-14.76)(4π) = -185.48 . The arc has an excess partial revolution of -.76 . Therefore the arc terminates in quadrant I since -1.00C < -.76C < -.75C .

THE UNIT CIRCLE AND ITS ARC MEASURES

The most convenient circle to use as a reference in trigonometry has a radius of 1 unit. Therefore, it is called the <u>unit circle</u>. Since the radius is 1 , the circumference is 2π . $C = 2\pi r = 2\pi(1) = 2\pi$.

If the circumference of the unit circle is divided into quadrants each arc has a length of $\frac{\pi}{2}$ units. The positive and negative division points are shown below:

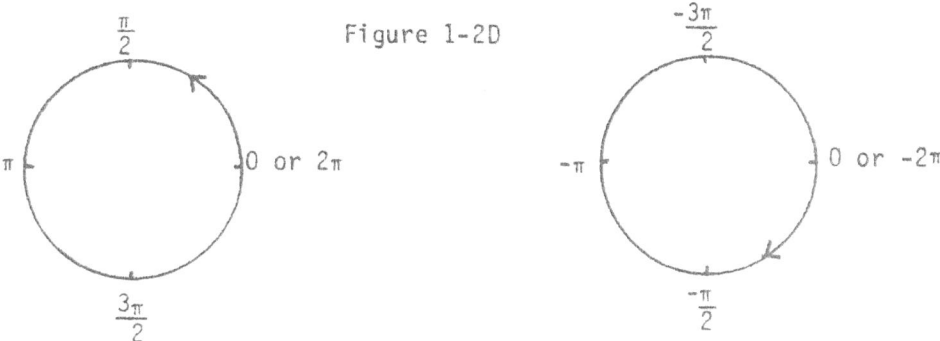

Figure 1-2D

For reasons that will be examined later, divisions of the circumference into 8 and 12 parts are frequently employed to illustrate trigonometric ideas. These divisions produce arcs of $\frac{\pi}{4}$ and $\frac{\pi}{6}$ units respectively:

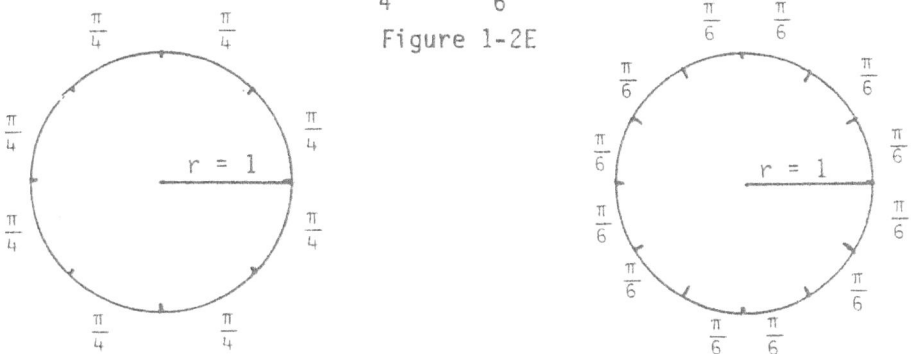

Figure 1-2E

The 8 and 12 part divisions of the unit circle are combined in positive (or negative) increments as shown in Figure 1-2F.

Figure 1-2F

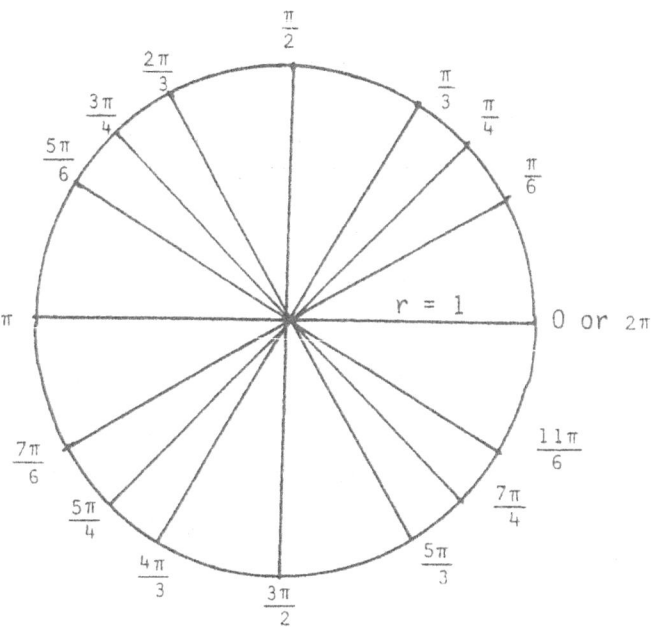

THE REAL NUMBER LINE AND THE UNIT CIRCLE

If a real number line is placed tangent to a unit circle so that their zero points coincide, as in Figure 1-2G, then it is possible to establish a correspondence between the points on the line and the points on the circle.

Figure 1-2G

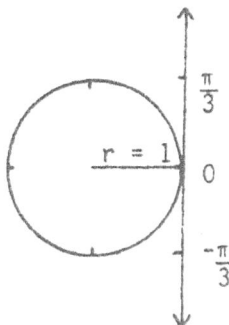

Imagining the real number line as a flexible string and the circle as a spool, the line may be wrapped around the circle as in Figure 1-2H, generating any length of arc desired. This wrapping results in a many-to-one correspondence since points on the number line that lie 2π units apart will end up in the same position on the circle. The understanding of this correspondence will assume greater importance later when the trigonometric functions are defined in Chapter 4 . For the present this model will reinforce understanding of revolutions and coterminal arcs.

Figure 1-2H

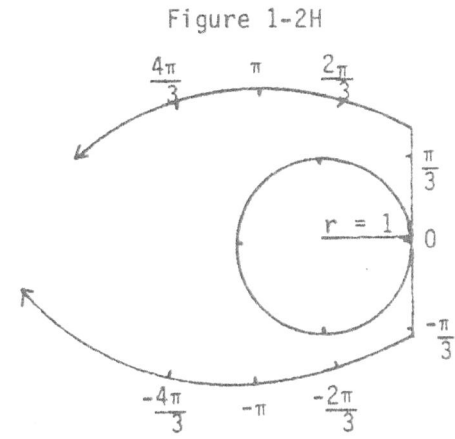

EXERCISE SET 1-2.

Identify the quadrant on the unit circle in which the following arcs terminate if they are in standard position:

1) $\frac{5\pi}{3}$ 2) $\frac{7\pi}{5}$ 3) $\frac{4\pi}{5}$ 4) $\frac{8\pi}{3}$ 5) $\frac{-3\pi}{4}$ 6) $\frac{-7\pi}{4}$ 7) $\frac{-13\pi}{3}$

8) $\frac{-19\pi}{7}$ 9) 4.26   10) 3.87   11) -2.65   12) -14.32   13) 15.78

14) 104.68π   15) -37.8π   16) 2π - 4   17) $\frac{2\pi}{3} + \frac{5\pi}{6}$ 18) $\frac{2\pi}{5} - \frac{3\pi}{4}$

19) 3π - 8   20) $\frac{4\pi}{7} - 17$

In exercises 21 - 24 a circle with radius 6.2 is given. Find the measurements of the arcs given by their indicated revolutions:

21) 3.78   22) 13.78π   23) -14.73π   24) -84.96

25) Locate the arcs measuring 1, 2, and 3 between the positive division points shown in Figure 1-2F.

26) Locate the arcs measuring 4, 5, and 6 between the positive division points shown in Figure 1-2F.

1-3. CIRCLES AND THEIR CENTRAL ANGLES

A central angle of a circle is one that has its vertex at the center.

Figure 1-3A

In the figure two central angles are determined by the line segments $\overline{OP}$ and $\overline{OQ}$. These are the angles subtended (cut off) by the minor arc PnQ and the major arc QwP. To avoid ambiguity, the central angle from $\overline{OP}$ to $\overline{OQ}$ will be described as "the central angle of the minor arc from P to Q" and the central angle from $\overline{OQ}$ to $\overline{OP}$ as "the central angle of the major arc from Q to P" if both angles are positively directed. If the central angles are directed clockwise they have negative measures. In this case the central angle from $\overline{OP}$ to $\overline{OQ}$ will be described as "the central angle of the major arc from P to Q" and the central angle from $\overline{OQ}$ to $\overline{OP}$ as the "central angle of the minor arc from Q to P. Corresponding to arcs, central angles in standard position have their initial sides formed by a radius to the zero point.

Consider now a central angle θ (read "theta") in the unit circle formed by radii to 0 and $\frac{\pi}{6}$:

Figure 1-3B

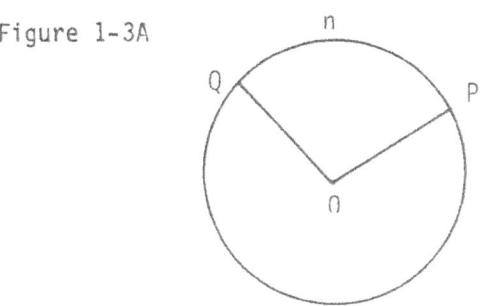

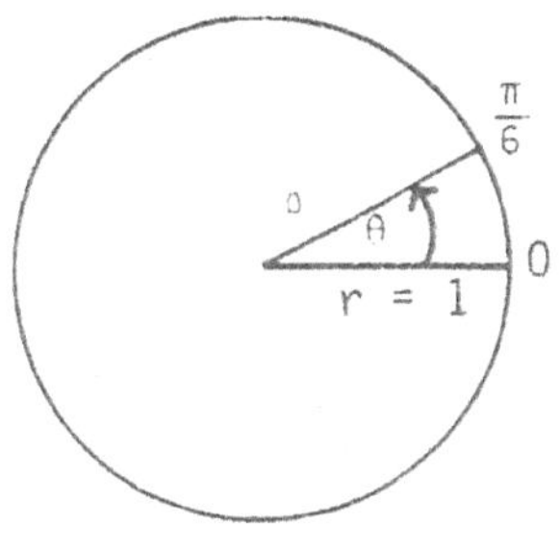

The angle is in standard position and is measured as a positive quantity from the radius at 0 to the radius terminating at the division point $\frac{\pi}{6}$.

DEGREE MEASURE OF A CENTRAL ANGLE

Any circle may be divided into 360 equal arcs. The central angles obtained by drawing radii to the 360 division points each have a measure of 1^O (1 degree). We can obtain the degree measure of θ in Figure 1-3B by the following proportion: $\frac{\theta}{360^O} = \frac{s}{C}$ where s is the arc subtended and C is the circumference. For the angle above,

$$\frac{\theta}{360^O} = \frac{\frac{\pi}{6}}{2\pi} \quad \text{(remember } 2\pi \text{ is the circumference of the unit circle)}$$

$$\theta = 360^O \cdot \frac{\frac{\pi}{6}}{2\pi} = 360^O \cdot \frac{\pi}{6} \cdot \frac{1}{2\pi} = \frac{360^O}{12} = 30^O$$

In a like manner, the central angle from 0 to $\frac{5\pi}{3}$ (Figure 1-3C) is found as follows:

$$\frac{\theta}{360^O} = \frac{\frac{5\pi}{3}}{2\pi}$$

$$\theta = 360^O \cdot \frac{5\pi}{3} \cdot \frac{1}{2\pi} = 300^O$$

Figure 1-3C

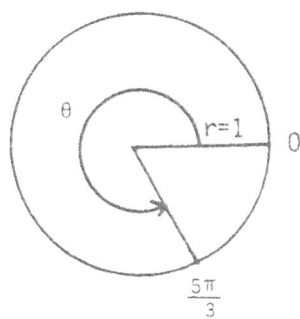

Note that in the computations above, the factors $360°$ and $\frac{1}{2\pi}$ are always present as multiples of the terminal number. Now, $360° \cdot \frac{1}{2\pi} = \frac{180°}{\pi}$. This constant factor can alway be used in finding the central angles more quickly.

EXAMPLE: Find the central angle from 0 to $\frac{7\pi}{6}$.

SOLUTION: $\Theta = \frac{7\pi}{6} \cdot \frac{180°}{\pi} = 210°$

The calculation of central angles can be extended to any positioning of the angles in the unit circle. If t_1 is an initial point of an arc subtended by a central angle Θ on the unit circle and t_2 is the terminal point of the arc, then the measure of the central angle from t_1 to t_2 (Figure 1-3D) is given by $\Theta = (t_2 - t_1) \cdot \frac{180°}{\pi}$.

FIGURE 1-3D

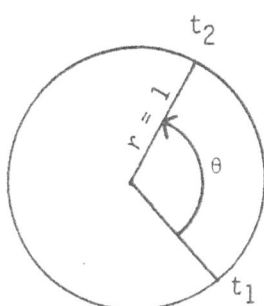

EXAMPLE: Find the measure of the central angle from $\frac{3\pi}{4}$ to $\frac{7\pi}{6}$ on the unit circle.

SOLUTION: $\Theta = (\frac{7\pi}{6} - \frac{3\pi}{4}) \cdot \frac{180°}{\pi} = (\frac{14\pi - 9\pi}{12}) \cdot \frac{180°}{\pi} = \frac{5\pi}{12} \cdot \frac{180°}{\pi} = 75°$.

CAUTION: Remember that the factor $\frac{180°}{\pi}$ can only be used to calculate central angles in the <u>unit circle</u>.

EXAMPLE: Find the measure of the central angle from $\frac{3\pi}{4}$ to $\frac{7\pi}{6}$ in a circle with radius 5.

SOLUTION: $\frac{\theta}{360^{0}} = \frac{(\frac{7\pi}{6} - \frac{3\pi}{4})}{(5)(2\pi)}$ (Circumference of a circle with radius 5).

$$\theta = \frac{(\frac{7\pi}{6} - \frac{3\pi}{4})}{(5)(2\pi)} \cdot \frac{360^{0}}{1} = \frac{\frac{5\pi}{12}}{5} \cdot \frac{180^{0}}{\pi} = \frac{180^{0}}{12} = 15^{0}$$

The example given above shows how to generalize finding the measure of any central angle in degrees. If t_1 and t_2 are the initial and terminal points respectively of an arc of a circle with radius r, then the central angle from t_1 to t_2 is given by $\theta = (t_2 - t_1) \cdot \frac{180^{0}}{\pi r}$.

NEGATIVE ANGLES AND ARCS

Angles and arcs that are generated in a <u>clockwise</u> direction are measured as negative quantities. Figure 1-3E shows a representative sample of negative angles and their arcs for the unit circle.

Figure 1-3E

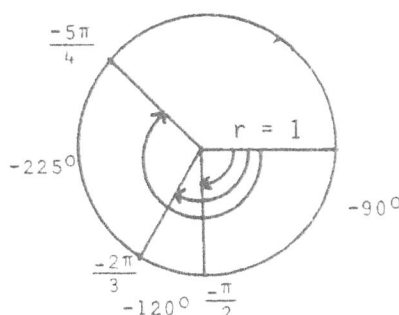

With the possibility of both positive and negative central angles we can speak of the <u>directed</u> angle from initial point t_1 to terminal point t_2 on a circle.

EXAMPLE: Find the degree measure of the directed angle from $\frac{7\pi}{4}$ to $\frac{2\pi}{3}$ on the unit circle.

SOLUTION: $\theta = (t_2 - t_1) \cdot \frac{180^0}{\pi} = (\frac{2\pi}{3} - \frac{7\pi}{4}) \cdot \frac{180^0}{\pi}$

$= (\frac{8\pi - 21\pi}{12}) \cdot \frac{180^0}{\pi} = \frac{-13\pi}{12} \cdot \frac{180^0}{\pi} = -195^0$

EXAMPLE: What is the degree measure of the directed angle from $\frac{7\pi}{4}$ to $\frac{2\pi}{3}$ in a circle with radius 3?

SOLUTION: Same as in the previous example with $\frac{180^0}{3\pi}$ as a factor instead of $\frac{180^0}{\pi}$. The result is $\frac{-195^0}{3} = -65^0$

The direction of the central angle is solely dependent upon the difference factor $t_2 - t_1$. Consider the following:

EXAMPLE: What is the degree measure of the central angle from $-\frac{\pi}{4}$ to $\frac{2\pi}{3}$ on the unit circle?

SOLUTION:

Figure 1-3F

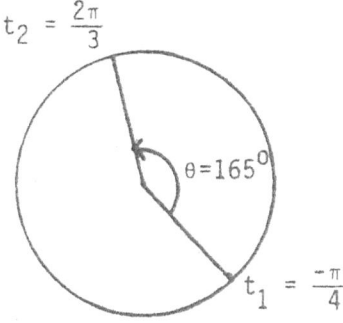

$$\theta = (t_2 - t_1) \cdot \frac{180^0}{\pi} = (\frac{2\pi}{3} - \frac{-\pi}{4}) \cdot \frac{180^0}{\pi} = (\frac{2\pi}{3} + \frac{\pi}{4}) \cdot \frac{180^0}{\pi}$$

$$= \frac{11\pi}{12} \cdot \frac{180^0}{\pi} = 165^0$$

Note that this example has division points on the circle located in the same positions as the example above where $t_1 = \frac{7\pi}{4}$ and $t_2 = \frac{2\pi}{3}$. However, in the previous example the solution yielded the negative central angle of the major arc (-195°), while in the present example the central angle of the minor arc (165°) is obtained.

If $t_2 > t_1$ the directed angle will always be positive. To obtain the negatively directed angle from t_1 to t_2, subtract 360°.

If $t_2 < t_1$, the directed angle will always be negative. To obtain the positively directed angle from t_1 to t_2, add 360°.

EXERCISE SET 1-3.

In exercises 1-10, find the degree measure of the central angle for the following arcs on the unit circle:

1) $\frac{2\pi}{3}$ 2) $\frac{-\pi}{4}$ 3) $\frac{7\pi}{5}$ 4) $\frac{11\pi}{10}$ 5) $\frac{-3\pi}{8}$

6) 2.3 7) -4.25 8) -1.38 9) $\frac{14\pi}{9}$ 10) 3.6

In exercises 11-15, two points on the unit circle are given. Find the directed central angle in degree measure from the first point to the second point:

11) $\frac{2\pi}{5}$, $\frac{3\pi}{4}$ 12) $\frac{11\pi}{6}$, $\frac{2\pi}{3}$ 13) $\frac{5\pi}{3}$, 2π 14) $\frac{7\pi}{4}$, $\frac{\pi}{2}$

15) 2.78, 6.15

In exercises 16-20, an arc and radius of a circle are given. Find the degree measure of the central angle:

16) $\frac{4\pi}{3}$, $r = 3$ 17) $\frac{7\pi}{5}$, $r = 1.6$ 18) $\frac{-7\pi}{4}$, $r = 2\pi$

19) -3.4, $r = 4.8$ 20) 6, $r = 6$

1-4. DEGREES AND RADIANS

The general formula $\theta = (t_2 - t_1) \cdot \frac{180^O}{\pi r}$ was used to determine a central angle of a circle when the radius r and the subtending arc $(t_2 - t_1)$ were known. The formula can also be used to find the arc if the angle and radius are known.

EXAMPLE: In a circle of radius 4 inches, what is the arc subtended by an angle of 72°?

SOLUTION: Substituting in the formula, $72^O = (t_2 - t_1) \cdot \frac{180^O}{\pi \cdot 4}$

Solving for the arc, $(t_2 - t_1) = \frac{(72^O)(\pi)(4)}{(180^O)} = 1.6\pi = \frac{8\pi}{5}$ inches.

The answer of course does not specify the points t_1 and t_2, only the directed magnitude of the arc between them.

Now, for simplicity, let $s = (t_2 - t_1)$. Then, since $\theta = s \cdot \frac{180^O}{\pi r}$, solving for s gives $s = \frac{\pi}{180^O} r \theta$ and this is a general formula for arc length when r and θ are given and θ is measured in degrees.

EXAMPLE: What is the arc subtended by an angle of 45° in a circle with radius 12 cm?

SOLUTION: $s = \frac{\pi}{180^O} \cdot 12$ cm $\cdot 45^O = 3\pi$ cm.

The formula for arc length can be simplified by another substitution. Let $\phi = \frac{\pi}{180^O} \cdot \theta$. Then by substitution, $s = r\phi$, where ϕ is called the radian measure of the central angle subtending the arc s.

EXAMPLE: What is the radian measure of an angle of 60°?

SOLUTION: $\phi = \frac{\pi}{180^O} \cdot 60^O = \frac{\pi}{3}$ radians.

EXAMPLE: What is the arc subtended by an angle of 4.2 radians in a circle of radius 3 cm?

SOLUTION: The formula for arc length is $s = r\phi$ where ϕ is in radians. Substituting, $s = (3$ cm$)(4.2) = 12.6$ cm.

EXAMPLE: A circle has a radius of 5 inches.   What is the central angle
subtending an arc of 12 inches?

SOLUTION: If $s = r\phi$, then $\phi = \dfrac{s}{r}$.   Substituting, s = 12 inches and r = 5 inches,
we have $\phi = \dfrac{12 \text{ inches}}{5 \text{ inches}} = \dfrac{12}{5}$ or 2.4.

NOTE: ϕ is a real number with no measurement units since the inch
units divide each other out.

The last example yields a definition of radian measure without requiring
a reference to degrees.   If r is the radius of a circle and ϕ is a central angle
subtending an arc s, then ϕ (in radians) $= \dfrac{s}{r}$ (Figure 1-4A).

Figure 1-4A

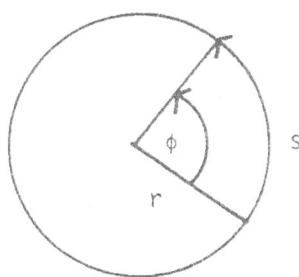

The definition for the radian measure of a central angle leads to the
following results:

1.   If ϕ is a central angle of a circle with radius r and ϕ subtends an arc
of r units, then ϕ = 1 radian.   Proof: $\phi = \dfrac{s}{r} = \dfrac{r}{r} = 1$, since $s = r$.

2.   An angle of 1 radian is equivalent to an angle of $\dfrac{180}{\pi}$ degrees.
Proof: Since $\phi = \dfrac{\pi}{180^\circ} \cdot \theta$, where θ is in degrees and ϕ is in radians,
then if $\phi = 1$, substituting $1 = \dfrac{\pi}{180^\circ} \cdot \theta$.   Solving for θ, we obtain
$\theta = \dfrac{180^\circ}{\pi}$.

The calculator approximation for $\frac{180^0}{\pi}$ is 57.3^0 rounded to tenths. For rough mental computations, the student should keep in mind that 1 radian is close to 60^0 (Figure 1-4B).

Figure 1-4B

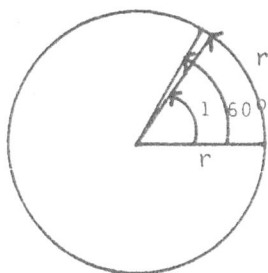

EXAMPLE: Approximate mentally the degree measure of an angle of 5 radians.

SOLUTION: An angle of 5 radians is a few degress less than (5)(60^0) = 300^0; a good approximation would be 285^0. A check with a calculator gives 286.5^0 rounded to tenths.

CONVERTING FROM RADIANS TO DEGREES

The formula $\phi = \frac{\pi}{180^0} \cdot \theta$ was used to convert degrees to radians. Solving for θ as we did in the proof above gives $\theta = \frac{180^0}{\pi} \cdot \phi$

EXAMPLE: Convert an angle of $\frac{-7\pi}{6}$ radians to degrees.

SOLUTION: $\theta = \frac{(180^0)}{\pi} \frac{(-7\pi)}{6} = -210^0$.

EXAMPLE: Convert an angle of 4.78 radians to degrees.

SOLUTION: $\theta = \frac{(180^0)}{\pi} (4.78) = 273.87^0$ by calculator rounded to hundredths.

Note that in the computations above that if the radian measure is given in multiples of π it is frequently unnecessary to resort to a calculator for the result, since the π factors divide out.

EXAMPLE: Convert an angle of 1^O to radian measure.

SOLUTION: Since $\phi = \frac{\pi}{180^O} \cdot \theta$ and $\theta = 1^O$, substituting we obtain

$$\phi = \frac{\pi}{180^O} \cdot 1^O = \frac{\pi}{180} \quad \text{radians.} \quad \text{Dividing } \pi \text{ by 180 gives } .01745$$

radians.

SETTING UP CONVERSION PROBLEMS.

Some calculators are restricted to one or the other measures for angles, either degrees or radians. Therefore, it is important for the student to know how to convert from one measure to the other rapidly. The following examples will illustrate the thought processes to guarantee correct results.

EXAMPLE: Convert 128° to radians.

SOLUTION: We want to eliminate the degree units in arriving at our answer, so the proper factor to multiply by is $\frac{\pi}{180^O}$ to divide out the degree units of measure.

$$128^O \cdot \frac{\pi}{180^O} = \frac{128\pi}{180} = \frac{32\pi}{45} \quad \text{radians (an exact answer).}$$

$\frac{32\pi}{45}$ is approximately 2.234 radians by calculator.

EXAMPLE: Convert -.378 radians to degrees.

SOLUTION: We want to introduce degree units in our answer, so the proper factor to multiply by is $\frac{180^O}{\pi}$; $(-.378)(\frac{180^O}{\pi}) = -21.66^O$ by calculator approximation.

STANDARDIZING ANSWERS TO CONVERSION PROBLEMS.

In this text, we shall follow the following rules in standardization.

1. If a radian measure answer can be left as a simple multiple of π, do so, reducing all fractions to lowest terms.

   EXAMPLE: Convert 75° to radians.

   SOLUTION: $(75^O)(\frac{\pi}{180^O}) = \frac{75\pi}{180} = \frac{5\pi}{12}$.

2. If an answer is in radians, omit the word "radians" as the measure will be understood in the context of the problem.

3. If a radian measure answer cannot be left in simple multiples of π, express answers in decimal form rounded off to hundredths.

   EXAMPLE: Convert 38.19^0 to radians.

   SOLUTION: $(38.19^0)(\frac{\pi}{180^0}) = 0.67$

4. If a degree measure answer is not an integer, round off to hundredths.

   EXAMPLE: Convert 2.683 to degrees.

   SOLUTION: $(2.683)(\frac{180^0}{\pi}) = 153.72^0$

5. All degree measures will be left in the decimal system unless the context of the problem requires subdivisions into minutes and seconds, i. e., (base 60).

   EXAMPLE: Convert 153.72^0 to degrees, minutes, and seconds.

   SOLUTION: The integer part of 153.72^0 remains unchanged in the conversion. Since each degree is divided into 60 minutes, $(.72)(1^0) = (.72)(60 \text{ minutes}) = 43.2$ minutes. Finally, each minute is divided into 60 seconds so that $(.2)(60 \text{ seconds}) = 12$ seconds. The final result is 153 degrees, 43 minutes, 12 seconds. Using standard symbols for the units, we have: $153.72^0 = 153^0 \ 43' \ 12"$.

   EXAMPLE: Convert $\frac{3\pi}{8}$ to degrees, minutes, and seconds.

   SOLUTION: First converting radians to degrees, $(\frac{3\pi}{8})(\frac{180^0}{\pi}) = 67.50^0$. Since $.50^0 = 30'$ exactly, $67.50^0 = 67^0 \ 30' \ 00"$.

   If rounding off is required, do so to the nearest second.

   EXAMPLE: Convert -4.82 to degrees, minutes, and seconds.

   SOLUTION: The steps in solving by calculator are as follows:
   $(-4.82)(\frac{180^0}{\pi}) = 276.16565^0 = -276^0 + -(0.16565)(60')$
   $= -276^0 + -9.939' = -276^0 + -9' + -(0.939)(60")$

$= -276^{0} + -9' + -56.34'' = -276^{0}$ 9' 56'' to the nearest second.

To express an angle measured in degrees, minutes, and seconds in decimal form convert the minutes to seconds and divide by 3600, the total number of seconds in a degree.

EXAMPLE: Convert 79^{0} 35' 43'' to decimal degrees.

SOLUTION: 79^{0} 35' 43'' $= 79^{0} + (\frac{35 \times 60 + 43}{3600})^{0} = 79^{0} + (\frac{2143}{3600})^{0}$

$$= 79^{0} + .59527778^{0} = 79.60^{0} \text{ rounded.}$$

CENTRAL ANGLES AND REVOLUTIONS. THE GENERAL ANGLE.

Central angles greater than 360^{0} or 2π radians may be generated in the same manner as arcs greater than the circumference. It is often convenient to describe such angles in terms of revolutions. Since the number of revolutions $= \frac{s}{C}$ where s is the arc and C is the circumference, and $\frac{s}{C} = \frac{\theta}{360^{0}}$ where θ is in degrees, it follows that $\theta = (360^{0})(\text{number of revolutions})$.

EXAMPLE: Find the degree measure of a central angle of 3.75 revolutions.

SOLUTION: $\theta = (360^{0})(3.75) = 1350^{0}$.

EXAMPLE: Find the radian measure of a central angle of -4.68 revolutions.

SOLUTION: Let ϕ = number of radians. Then $\phi = (2\pi)(-4.68) = -29.41$ rounded.

EXAMPLE: Angles in standard position of a) -1135^{0} and b) 57.86 radians are given. Determine the quadrants in which they terminate.

SOLUTION: a) $\frac{-1135^{0}}{360^{0}} = -3.15$ revolutions. The excess partial revolution is $-.15$. Since $-.25C < -.15C < 0$, -1135^{0} terminates in quadrant IV.

b) $\frac{57.86}{2\pi} = 9.21$ revolutions. The excess partial revolution is .21 . Since $0 < .21C < .25C$, 57.86 terminates in quadrant I .

ANGLES AND ARCS IN THE UNIT CIRCLE.

There is a direct correspondence between the magnitudes of any arc on the unit circle and the radian measure of the central angle subtending the arc. Figure 1-4C shows this relationship between an arc length of $\frac{\pi}{3}$ linear units and a central angle of $\frac{\pi}{3}$ radians. We have

$$\theta \text{ (radians)} = \frac{s}{r} = \frac{\frac{\pi}{3} \text{ units}}{1 \text{ unit}} = \frac{\pi}{3}$$

Figure 1-4C

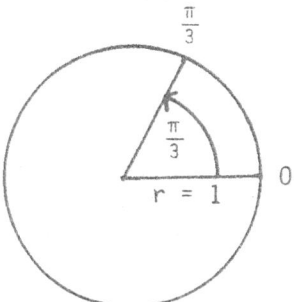

It is important to note that although the radian measures and arc measures on the unit circle are equal in magnitude, they differ in unit designation. Arc measures are given in linear units such as centimeters, meters, inches, feet, etc., whereas radian measures are real numbers without units of measure attached. By their definition, radian measures are ratio quantities. We shall return to this fact later. For the present, the following table summarizes the important arc and angle relationships for the unit circle that should be memorized for future reference.

Arc	0	$\frac{\pi}{6}$	$\frac{\pi}{4}$	$\frac{\pi}{3}$	$\frac{\pi}{2}$	$\frac{2\pi}{3}$	$\frac{3\pi}{4}$	$\frac{5\pi}{6}$	π	$\frac{7\pi}{6}$	$\frac{5\pi}{4}$	$\frac{4\pi}{3}$	$\frac{3\pi}{2}$	$\frac{5\pi}{3}$	$\frac{7\pi}{4}$	$\frac{11\pi}{6}$	2π
Angle (radians)	0	$\frac{\pi}{6}$	$\frac{\pi}{4}$	$\frac{\pi}{3}$	$\frac{\pi}{2}$	$\frac{2\pi}{3}$	$\frac{3\pi}{4}$	$\frac{5\pi}{6}$	π	$\frac{7\pi}{6}$	$\frac{5\pi}{4}$	$\frac{4\pi}{3}$	$\frac{3\pi}{2}$	$\frac{5\pi}{3}$	$\frac{7\pi}{4}$	$\frac{11\pi}{6}$	2π
Angle (degrees)	0	30	45	60	90	120	135	150	180	210	225	240	270	300	315	330	360
Quadrant		I	I	I		II	II	II		III	III	III			IV	IV	IV

The same correspondences are illustrated in Figure 1-4D:

Figure 1-4D

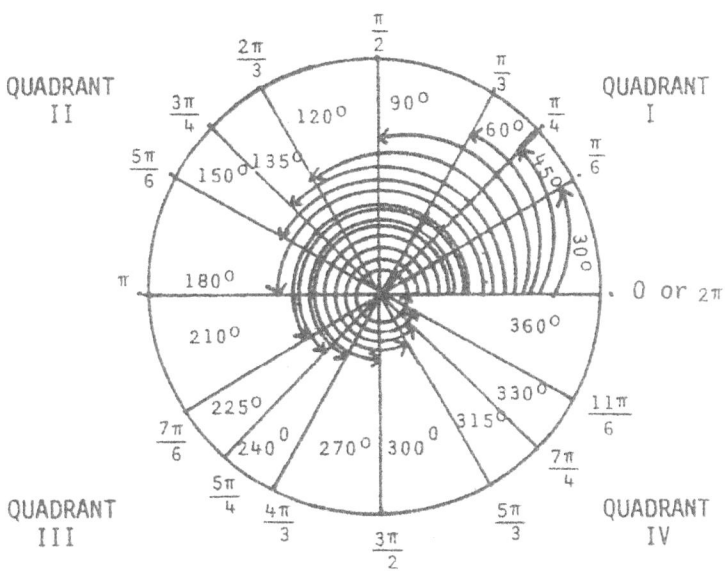

EXERCISE SET 1-4.

In exercises 1-10, convert the angles to radian measure.

1) 45^0 2) -30^0 3) 148^0 4) 180^0 5) -330^0

6) -7.8^0 7) 254^0 8) -225^0 9) 77.6^0 10) 144^0

In exercises 11-20, convert the angles to degree measure.

11) $\frac{5\pi}{6}$ 12) $\frac{7\pi}{5}$ 13) $\frac{-4\pi}{3}$ 14) $\frac{-7\pi}{6}$ 15) 2.5

16) 4.32 17) $-.07$ 18) -3.8 19) $.34$ 20) 1.3π

In exercises 21-25, find the central angle in radians for the given arc and radius.

21) $s = 4, r = 2.5$ 22) $s = \frac{-7\pi}{6}, r = 2$ 23) $s = 2\pi, r = 3$

24) $s = \frac{\pi}{3}, r = 1$ 25) $s = 1.6, r = .5$

In exercises 26-30, find the arc subtended by the given angle in a circle with the given radius.

26) $38^0, r = 1$ 27) $\frac{4\pi}{3}, r = 1$ 28) $\frac{7\pi}{4}, r = 2$

29) $-125^0, r = 2$ 30) $-240^0, r = 3$

In exercises 31-36, convert the decimal degree measures to degrees, minutes, and seconds.

31) -48.35^0 32) -117.8^0 33) 245.83^0

34) 17.47^0 35) 337.65^0 36) -145.27^0

In exercises 37-40, convert the degree, minutes, seconds measures to decimal degree measures. (Hint: convert minutes to seconds and divide by 3600 seconds).

37) 45^0 15' 22"   38) -176^0 24' 35"   39) -243^0 54' 46"

40) 354^0 46' 32"

In exercises 41 - 44 determine the quadrants in which the angles terminate:

41) 943^0 42) -1575^0 43) 15.36π 44) -24.87

1-5. CHAPTER SUMMARY

Figure 1-5A

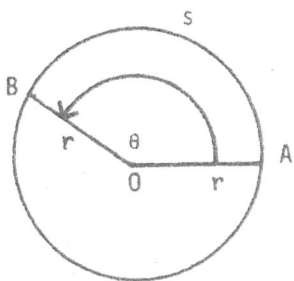

Given a circle with center at 0, radius r, minor arc s, and central angle θ subtending arc s, the following relationships hold:

1. s = rθ if θ is measured in radians.

2. s = $\frac{\pi}{180^o}$ · rθ if θ is measured in degrees.

3. $(\frac{\pi}{180^o})(\theta)$ will convert θ in degrees to radians.

4. $(\frac{180^o}{\pi})(\theta)$ will convert θ in radians to degrees.

5. Positive angles and arcs are measured in a counterclockwise direction.
    Negative angles and arcs are measured in a clockwise direction.

6. If r = s then θ is equal to 1 radian.

7. If r = 1 then we have a unit circle.

8. The circumference of any circle is given by C = 2πr.
    The circumference of a unit circle is 2π linear units.

9. The unit circle is divided into four quadrants. See Figure 1-4D.

EXERCISE SET 1-5

Find the circumference of a circle with the given radius.

1) 2.3   2) 4.56   3) 3π 4) $\pi + 4$

Find the radius of a circle with the given circumference.

5) 19.7   6) 8π 7) $2\pi - 1$ 8) 5

In what quadrant is the given point found on the unit circle?

9) $\frac{4\pi}{5}$ 10) $\frac{13\pi}{16}$ 11) $\frac{5\pi}{3}$ 12) 1.84

Locate the following points on the unit circle between multiples of $\frac{\pi}{6}$.

Example: 1.4

Solution: Enter $\frac{\pi}{6}$ into calculator memory. Multiply by successive integers until lower and upper bounds for 1.4 are found.*

$$\frac{2\pi}{6} < 1.4 < \frac{3\pi}{6}\quad \text{since}\quad \frac{2\pi}{6} = 1.0471976$$

$$\text{and}\quad \frac{3\pi}{6} = 1.5707963$$

13) 3.8   14) 5.63   15) 2.1   16) 4.75

Convert the following angles to radian measure.

17) 210^0 18) 235^0 19) 315^0 20) $-125^0\ 20'\ 35''$

21) -110.64^0 22) 75^0 23) $144^0\ 15'$ 24) -324^0

Convert the following angles to degree measure.

25) $\frac{4\pi}{5}$ 26) $\frac{7\pi}{4}$ 27) $\frac{-14\pi}{9}$ 28) -1.59 29) 2.76

30) .183   31) -4.8 32) $2\pi - 3$

An arc and radius of a circle are given. Find the radian measure of the central angle subtended by the arc.

33) $\frac{7\pi}{8}$, r = 4   34) 2.6, r = 1   35) .43, r = .84   36) $\frac{3\pi}{5}$, r = π

Find the arc subtended by the given angle in a circle with the given radius.

37) 37^0, r = 2.5   38) $145^0\ 16'$, r = 1.5   39) $\frac{4\pi}{5}$, r = 6

40) -2.48, r = 10

Write the following sequences in degree measures:

41) $\frac{\pi}{4}$, $\quad\quad \frac{3\pi}{4}$, $\quad\quad \frac{5\pi}{4}$, $\quad\quad \frac{7\pi}{4}$

42) $\frac{\pi}{6}$, $\quad\quad \frac{\pi}{3}$, $\quad\quad \frac{\pi}{2}$, $\quad\quad \frac{2\pi}{3}$, $\quad\quad \frac{5\pi}{6}$, $\quad\quad \pi$

43) $\frac{\pi}{2}$, $\quad\quad \pi$, $\quad\quad \frac{3\pi}{2}$, $\quad\quad 2\pi$

44) 2π, $\quad\quad \frac{5\pi}{3}$, $\quad\quad \frac{4\pi}{3}$, $\quad\quad \frac{2\pi}{3}$, $\quad\quad \frac{\pi}{3}$

Write the following sequences in radian measures:

45) 30°, $\quad\quad 45^{\circ}$, $\quad\quad 60^{\circ}$, $\quad\quad 90^{\circ}$

46) 180°, $\quad\quad 120^{\circ}$, $\quad\quad 150^{\circ}$, $\quad\quad 135^{\circ}$, $\quad\quad 210^{\circ}$

47) 300°, $\quad\quad 330^{\circ}$, $\quad\quad 315^{\circ}$, $\quad\quad 240^{\circ}$

48) 225°, $\quad\quad 210^{\circ}$, $\quad\quad 120^{\circ}$, $\quad\quad 45^{\circ}$

* As an alternative calculator solution to exercises 13 - 16, enter $\frac{\pi}{6}$, press X key, press K key, press 2, 3, 4, ... and press = key for multiples of $\frac{\pi}{6}$.

CHAPTER 2

THE TRIGONOMETRIC RATIOS, COSINE AND SINE

2-1. ANALYTIC GEOMETRY OF THE CIRCLE

The geometric definition of the circle can be reinterpreted by means of a coordinate system in the plane. Consider a circle with radius r positioned so that its center is at the origin of an XY-coordinate system (Figure 2-1A).

Figure 2-1A

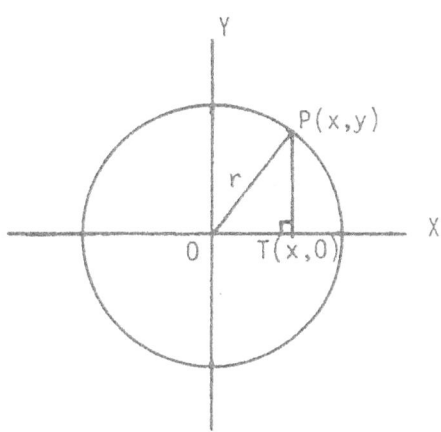

Let P(x,y) be any point on the circumference of the circle. A perpendicular line segment is drawn to the x-axis at point T(x,0). Then, by the Pythagorean theorem applied to triangle TOP, $x^2 + y^2 = r^2$. This is the defining equation of the circle. For example, the equation of a circle with its center at the origin and a radius of 5 units is $x^2 + y^2 = 25$.

The simplest way to graph $x^2 + y^2 = 25$ is to open a compass to a span of 5 units and to place the fixed point at the origin, swinging a circular arc with the drawing pencil. However, it is often important to identify specific ordered pairs (x,y) on the circumference. This can be done by solving the equation for y in terms of x:

$$x^2 + y^2 = 25$$
$$y^2 = 25 - x^2$$
$$y = \pm \sqrt{25 - x^2}$$

It can be seen that the <u>domain</u> (the set of possible real number replacements for x) lies between -5 and 5. That is, the domain = {x: -5 ≤ x ≤ 5}. Why? For each replacement of x the equation above gives two values for y, a number and its additive inverse. With the help of a calculator, a table of values rounded to tenths is given:

x	-5	-4	-3	-2	-1	0	1	2	3	4	5
y	0	±3	±4	±4.6	±4.9	±5	±4.9	±4.6	±4	±3	0

These ordered pairs are graphed in Figure 2-1B

Figure 2-1B

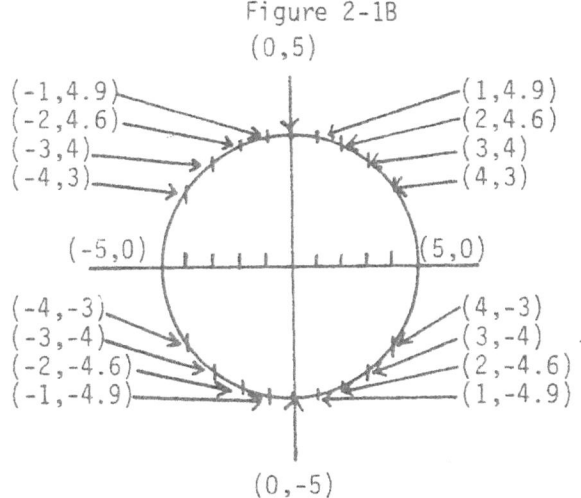

(0,5)
(-1,4.9) (1,4.9)
(-2,4.6) (2,4.6)
(-3,4) (3,4)
(-4,3) (4,3)
(-5,0) (5,0)
(-4,-3) (4,-3)
(-3,-4) (3,-4)
(-2,-4.6) (2,-4.6)
(-1,-4.9) (1,-4.9)
(0,-5)

The unit circle is defined analytically as $\{(x,y) : x^2 + y^2 = 1\}$.

As above, y can be solved in terms of x : $y = \pm\sqrt{1 - x^2}$. A table of values set up for x between -1 and 1 is given in tenths:

x	-1	-.9	-.8	-.7	-.6	-.5	-.4	-.3	-.2	-.1	0	.1	.2	.3	.4	.5	.6
y	0	±.4	±.6	±.7	±.8	±.87	±.92	±.95	±.98	±.99	±1	±.99	±.98	±.95	±.92	±.87	±.8

x	.7	.8	.9	1
y	±.7	±.6	±.4	0

Corresponding values of y are given rounded to tenths or hundredths.

The graph of the unit circle built from the table is shown in Figure 2-1C.

Figure 2-1C

The Unit Circle with X Values Graduated in Tenths.

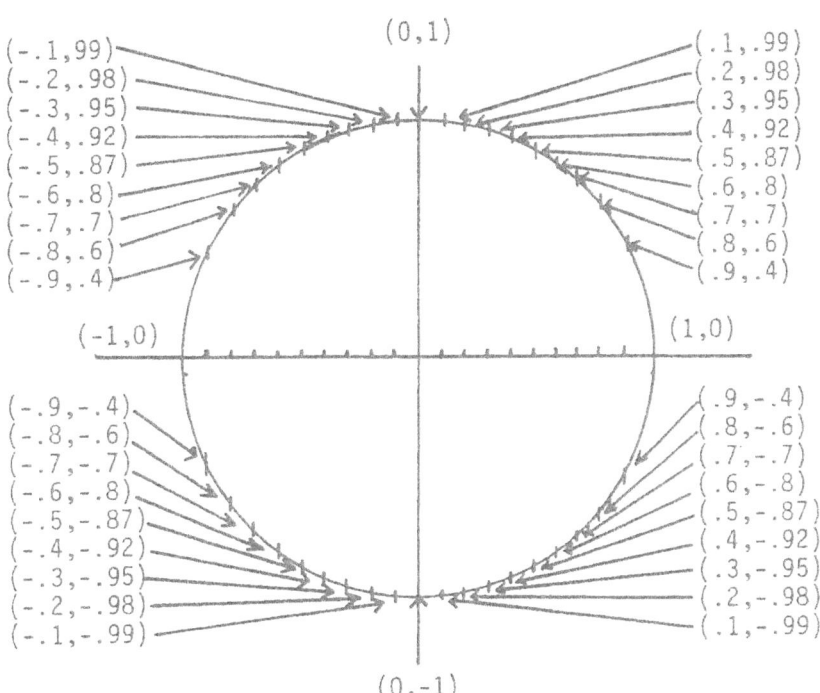

THE UNIT CIRCLE AS THE PATH OF A MOVING POINT

Another useful way of viewing the unit circle is that of a path traced out by a point moving in accordance with the relationship $x^2 + y^2 = 1$. Consider such a point located initially at (1,0) moving counterclockwise. Passage through the first quadrant moves the point from location (1,0) to (0,1). A series of positions for this movement is shown in Figure 2-1D.

Figure 2-1D

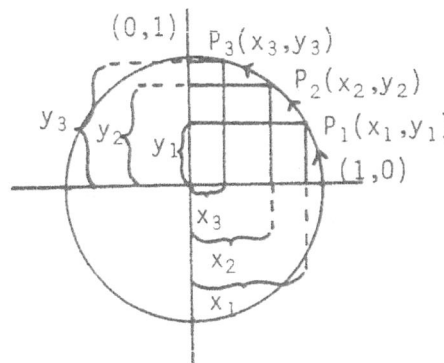

As the point moves on the circumference, the values of x decrease from 1 to 0, and the values of y increase from 0 to 1. When the point moves into the second quadrant (Figure 2-1E), the values of x continue to decrease from 0 to -1, and the values of y decrease from 1 to 0.

Figure 2-1E

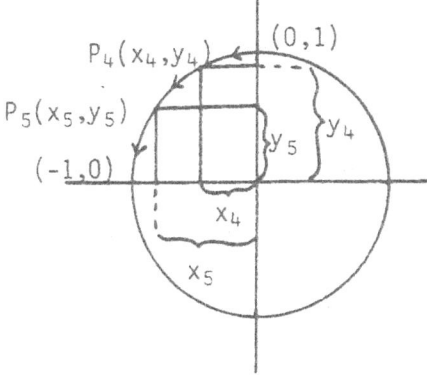

In Quadrant III, the point moves from (-1,0) to (0,-1). The values of x increase from -1 to 0 and the values of y decrease from 0 to -1. Finally, in Quadrant IV, as the point moves from (0,-1) to (1,0), the values of x increase from 0 to 1 and the values of y increase from -1 to 0 (Figure 2-1F).

Figure 2-1F

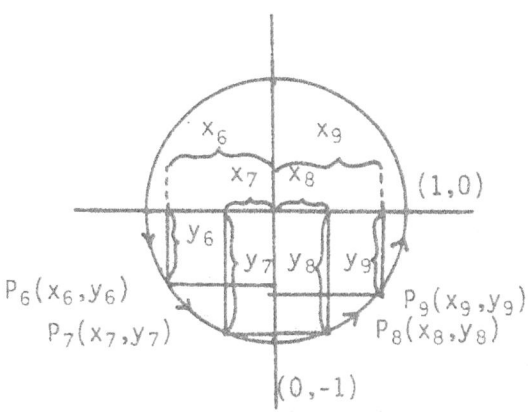

EXERCISE SET 2-1.

Find the equation of the circle with center at the origin and the given radius:

1) $r = 2$ 2) $r = \sqrt{2}$ 3) $r = \pi$ 4) $r = 4\pi$

5) $r = \sqrt{7}$ 6) $r = \frac{1}{2}$ 7) $r = .37$ 8) $r = .054$

Find the radius of the given circle with center at the origin:

9) $x^2 + y^2 = 15$ 10) $x^2 + y^2 = 225$ 11) $x^2 + y^2 = .04$

12) $x^2 + y^2 = 16\pi^2$ 13) $x^2 + y^2 = 5.8$ 14) $x^2 + y^2 = \pi$

Determine whether or not the given point is on the given circle:

15) $(-2,5)$, $x^2 + y^2 = 29$ 16) $(4,-7)$, $x^2 + y^2 = 57$

17) $(1,2)$, $x^2 + y^2 = 4$ 18) $(-3,4)$, $x^2 + y^2 = 34$

19) $(-2,3)$, $x^2 + y^2 = \sqrt{13}$ 20) $(-\pi,\pi)$, $x^2 + y^2 = 2\pi^2$

Let $P(x,y)$ be a point moving on a unit circle with center at the origin. If P moves <u>counterclockwise</u>, in what quadrant does:

21) x increase and y increase?

22) x increase and y decrease?

23) x decrease and y increase?

24) x decrease and y decrease?

Let Q(x,y) be a point moving on a unit circle with center at the origin. If Q moves <u>clockwise</u>, in what quadrant does:

25) x increase and y increase?

26) x increase and y decrease?

27) x decrease and y increase?

28) x decrease and y decrease?

2-2. THE TRIGONOMETRIC RATIOS. COSINE AND SINE.

The analytic definition of a circle with center at the origin given in Section 2-1 established a relationship between every ordered pair (x,y) on the circumference and the radius by the equation $x^2 + y^2 = r^2$. In Section 1-4 the definition of angle measure in radians established a relationship between the radius, an arc of the circle, and the angle subtended by that arc by the equation: $\theta = \frac{s}{r}$. The purpose of this section is to combine the two definitions in such a way as to yield new and more useful information. Consider any circle with center at the origin as in Figure 2-2A.

Figure 2-2A

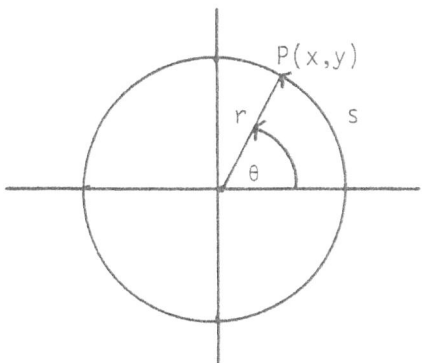

We define the ratio $\frac{x}{r}$ as the "cosine of the central angle θ" and the ratio $\frac{y}{r}$ as the "sine of the central angle θ". By abbreviating cosine as "cos" and sine as "sin", the definition may be expressed as

$$\cos \theta = \frac{x}{r}$$

$$\sin \theta = \frac{y}{r}$$

The three defining ratios, $\theta = \frac{s}{r}$, $\cos \theta = \frac{x}{r}$, and $\sin \theta = \frac{y}{r}$, involve all of the quantities connected with the point P on the circumference of the circle. In each equation, the quantities defined are pure real numbers. These numbers are independent of the units of measurement. As we shall now see, they are also independent of the size of the circle.

Consider any two circles of radii r_1 and r_2 respectively, with centers at the origin as in Figure 2-2B. Let $P_1(x_1,y_1)$ be a point on the circle with radius r_1, and $P_2(x_2,y_2)$ be a point on the circle with radius r_2. Assume , P_1 and P_2 both lie on a line drawn from the origin.

Figure 2-2B

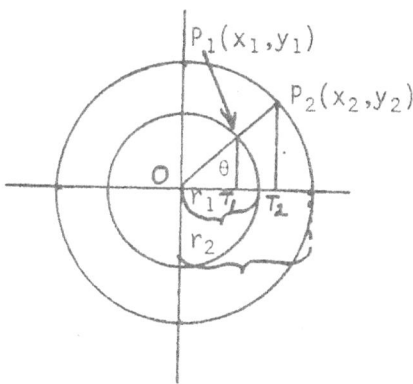

Let θ be the central angle generated from the positive part of the X-axis to the line containing P_1 and P_2. From P_1 and P_2 perpendiculars are dropped to the X-axis at T_1 and T_2 respectively, forming similar right triangles OP_1T_1 and and OP_2T_2. Since the corresponding sides of similar triangles are proportional, the following ratios are equivalent:

$$OT_1/OP_1 = OT_2/OP_2 \text{ , and } T_1P_1/OP_1 = T_2P_2/OP_2 \text{ .}$$

Since $OT_1 = x_1$, $OP_1 = r_1$, $OT_2 = x_2$, $OP_2 = r_2$, $T_1P_1 = y_1$, and $T_2P_2 = y_2$, we find by substitution that $x_1/r_1 = x_2/r_2$ and $y_1/r_1 = y_2/r_2$. These proportions lead to the conclusion that $\cos \theta$ and $\sin \theta$ are unique to the value of θ and do not vary as the dimensions of the circle are altered.

COSINES AND SINES IN THE FOUR QUADRANTS

For any circle with center at the origin, the coordinates of any point located in the first quadrant are positive. If a point is given, the radius can be determined using the equation $x^2 + y^2 = r^2$. The cosine and sine ratios can be found without knowing the central angle θ.

EXAMPLE: If P(2,5) is a point on a circle with center at the origin, find cos θ and sin θ, where θ is the central angle from the X-axis to the radius drawn to the given point (Figure 2-2C).

Figure 2-2C

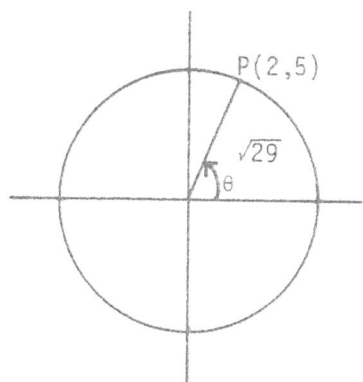

SOLUTION: $r = \sqrt{x^2 + y^2} = \sqrt{(2)^2 + (5)^2} = \sqrt{4 + 25} = \sqrt{29}$.

$$\cos \theta = \frac{x}{r} = \frac{2}{\sqrt{29}} \qquad \sin \theta = \frac{y}{r} = \frac{5}{\sqrt{29}}$$

Consider now a point on a circle in the second quadrant. The x value is negative and the y value is positive.

EXAMPLE: Find the cosine and sine of a central angle θ with initial side the positive X-axis and terminal side a line segment drawn from the origin to the point (-4,3) (Figure 2-2D).

Figure 2-2D

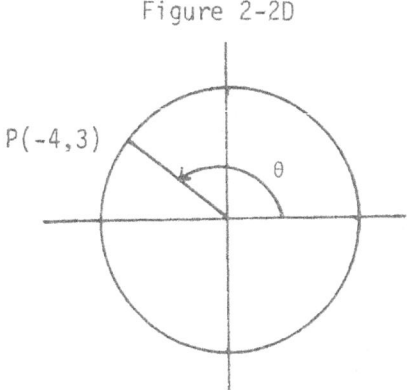

SOLUTION: $r = \sqrt{(-4)^2 + (3)^2} = \sqrt{16 + 9} = \sqrt{25} = 5.$

$\cos \theta = \frac{-4}{5}$ and $\sin \theta = \frac{3}{5}$.

Note in the above example, no mention was made of the fact that the point was located on the circumference of a circle with center at the origin. This fact can always be implied for any point in the plane. Another assumption that we shall take for granted is that all of the angles generated will have their initial sides coincident with the positive part of the X-axis, unless specifically indicated otherwise. We shall refer to this as the standard position of the central angle.

EXAMPLE: Find the cosine and sine of a central angle containing the point (-1,-1) on its terminal side.

SOLUTION: First locate the point on a graph as in Figure 2-2E.

Figure 2-2E

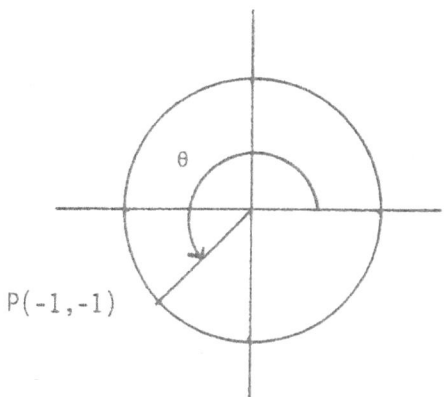

Draw a radius from the origin to the point (-1,-1) and show the angle θ terminating in the third quadrant. $r = \sqrt{(-1)^2 + (-1)^2} = \sqrt{2}$. Hence, $\cos\theta = \frac{-1}{\sqrt{2}}$, $\sin\theta = \frac{-1}{\sqrt{2}}$.

In the fourth quadrant the values of x are positive and the values of y are negative. Therefore, the cosine is positive and the sine is negative in the fourth quadrant.

EXAMPLE: Find the cosine and sine of a central angle with the point (6,-5) on its terminal side.

SOLUTION: Figure 2-2F illustrates the conditions.

Figure 2-2F

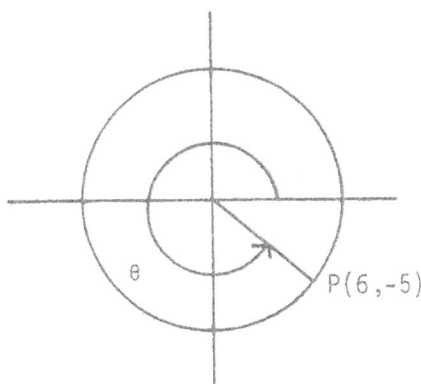

$$r = \sqrt{(6)^2 + (-5)^2} = \sqrt{36 + 25} = \sqrt{61}$$

$$\cos\theta = \frac{6}{\sqrt{61}}, \qquad \sin\theta = \frac{-5}{\sqrt{61}}$$

COSINES AND SINES OF QUADRANTAL ANGLES.

Central angles having their terminal sides coincident with the coordinate axes are called quadrantal angles. Consider any circle with radius r and center at the origin. The terminal points for radii that are coincident with the axes are shown in Figure 2-2G. Note that although r is always taken to be positive, the x and y values that are equal in magnitude to r are directed numbers.

Figure 2-2G

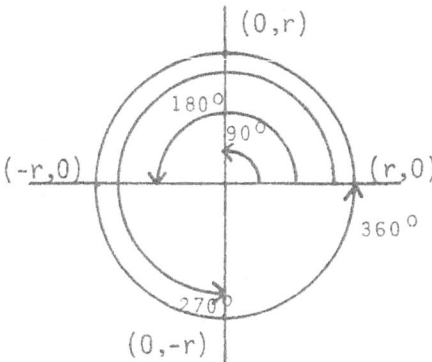

Since the size of quadrantal angles are known, we can write:

$\cos 0 = \cos 0^0 = \frac{r}{r} = 1$, $\sin 0 = \sin 0^0 = \frac{0}{r} = 0$

$\cos \frac{\pi}{2} = \cos 90^0 = \frac{0}{r} = 0$, $\sin \frac{\pi}{2} = \sin 90^0 = \frac{r}{r} = 1$

$\cos \pi = \cos 180^0 = \frac{-r}{r} = -1$, $\sin \pi = \sin 180^0 = \frac{0}{r} = 0$

$\cos \frac{3\pi}{2} = \cos 270^0 = \frac{0}{r} = 0$, $\sin \frac{3\pi}{2} = \sin 270^0 = \frac{-r}{r} = -1$

$\cos 2\pi = \cos 360^0 = \frac{r}{r} = 1$, $\sin 2\pi = \sin 360^0 = \frac{0}{r} = 0$.

Note that the cosines and sines of 0 (0^0) and 2π (360^0) are equal since their terminal sides are coincident. This is a specific example of an important generalization. For example, if the quadrantal angles are generated in a clockwise (negative) direction, we obtain the following:

$$\cos \left(-\frac{\pi}{2}\right) = \cos \frac{3\pi}{2} = 0 \text{ and } \sin \left(-\frac{\pi}{2}\right) = \sin \frac{3\pi}{2} = -1$$

EXAMPLE: Find the cosine and sine of (-180^0).

SOLUTION: Since (-180^0) is coterminal with a central angle of 180^0,

$\cos (-180^0) = \cos 180^0 = -1$ and $\sin (-180^0) = \sin 180^0 = 0$.

θ (radians)	0	$\frac{\pi}{2}$	π	$\frac{3\pi}{2}$	2π
θ (degrees)	0	90	180	270	360
cos θ	1	0	-1	0	1
sin θ	0	1	0	-1	0

UPPER AND LOWER BOUNDS FOR THE COSINE AND SINE.

The quadrantal angle values for the cosine and sine reveal another important fact. Since the x and y coordinate values of a point on a given circle can never exceed the magnitude of the radius in absolute value, the ratios $\frac{x}{r}$ and $\frac{y}{r}$ have an upper and lower bound. This leads to the fact that for any value of θ, $-1 \le \cos \theta \le 1$ and $-1 \le \sin \theta \le 1$.

EXERCISE SET 2-2.

For exercises 1 - 10, a point is given that is located on a circle with center at the origin. Find the cosine and sine of the central angle with initial side containing (r,0) and terminal side containing the given point.

1) (-3,2)   2) (2,5)   3) (-1,-3)   4) (1,1)   5) (-15,8)

6) (7,-24)   7) (-3,-4)   8) (5,-12)   9) (12,-5)   10) (-8,-15)

Identify the quadrant(s) in which the following occur:

11) $\cos \theta = \frac{-3}{7}$ 12) $\sin \theta = \frac{2}{9}$ 13) $\cos \theta = \frac{-5}{11}$, $\sin \theta = \frac{-4\sqrt{6}}{11}$

14) $\cos \theta = 2$ 15) $\sin \theta = 0$ 16) $\sin \theta = \frac{3}{5}$, $\cos \theta = \frac{-4}{5}$

17) cos θ increases, θ > 0   18) sin θ increases, cos θ decreases, θ > 0.

19) cos θ decreases, sin θ decreases, θ < 0.

20) $-1 < \cos \theta < 0$, $-1 < \sin \theta < 0$.

21) Using the definition of cosine and sine, show that for any angle θ,
$(\cos \theta)^2 + (\sin \theta)^2 = 1$.

22) For what quadrantal angles is $\frac{1}{\sin \theta}$ not defined?

23) For what quadrantal angles is $\dfrac{1}{\cos\theta}$ not defined?

24) If $-1 \le \cos\theta \le 1$, what can be said about $|\cos\theta|$?

25) If $-1 \le \sin\theta \le 1$, what can be said about $|\sin\theta|$?

Prove the following:

26) $\sin\pi = \cos\dfrac{3\pi}{2}$ 27) $\cos 270^\circ = \sin 0^\circ$ 28) $\sin(-270^\circ) = \sin 90^\circ$

29) $\cos\dfrac{\pi}{2} + \sin\dfrac{\pi}{2} = 1$ 30) $\cos\left(\dfrac{\pi}{2} + \dfrac{\pi}{2}\right) = -1$

2-3. SPECIAL ANGLES AND VALUES OF THEIR COSINE AND SINE RATIOS.

The quadrantal angles studied in the previous section are examples of special angles for which the cosine and sine are readily derived. In this section we shall examine some other simple divisions of the circle that lead to useful trigonometric ratios.

MULTIPLES OF $\pi/4$ (45^0)

To begin, Figure 2-3A shows a central angle of $\frac{\pi}{4}$ radians. From the terminal point on the radius a perpendicular is dropped to the X-axis.

Figure 2-3A

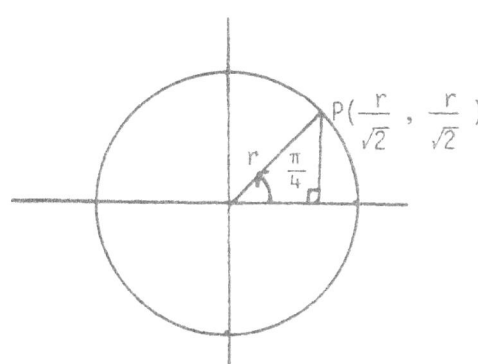

The right triangle formed is a 45^0-45^0-90^0 triangle that is isosceles. Since $x = y$ and $x^2 + y^2 = r^2$, by substitution $2x^2 = r^2$ and $\frac{x^2}{r^2} = \frac{1}{2}$. Taking square roots we obtain $\frac{x}{r} = \pm \frac{1}{\sqrt{2}}$. Since r is always positive and x is positive in the first quadrant, $\cos \frac{\pi}{4} = \frac{x}{r} = \frac{1}{\sqrt{2}}$. Since $y = x$, $\frac{y}{r} = \frac{1}{\sqrt{2}}$ also. Hence, $\cos \frac{\pi}{4} = \sin \frac{\pi}{4} = \frac{1}{\sqrt{2}}$. If θ is in degrees, $\cos 45^0 = \sin 45^0 = \frac{1}{\sqrt{2}}$.

Having derived the cosine and sine of $\frac{\pi}{4}$ we can now use arguments of symmetry to derive the ratios for $\frac{3\pi}{4}$, $\frac{5\pi}{4}$, and $\frac{7\pi}{4}$. Figure 2-3B shows the coordinate relationships for the points on the circle corresponding to these central angles.

Figure 2-3B

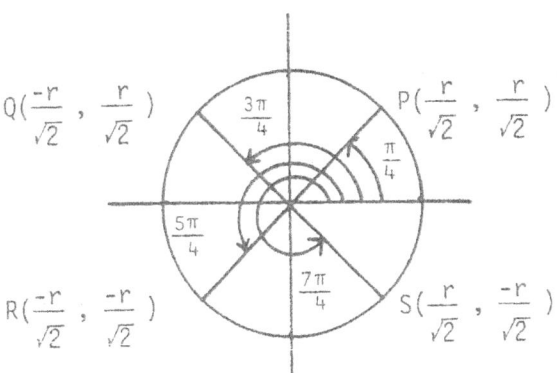

If two points on the circumference are equidistant from the Y-axis they are symmetric with respect to the Y-axis. If P(a,b) is located in the first quadrant, a point Q(-a,b) in the second quadrant will form a symmetric relationship with P and the Y-axis. The terminal points in Figure 2-3B corresponding to central angles $\frac{\pi}{4}$ and $\frac{3\pi}{4}$ exhibit this relationship. Thus, the coordinates of P are $(\frac{r}{\sqrt{2}}, \frac{r}{\sqrt{2}})$ and of Q are $(\frac{-r}{\sqrt{2}}, \frac{r}{\sqrt{2}})$.

The point S on the terminal side of the $\frac{7\pi}{4}$ angle joins the point P in being symmetric with respect to the X-axis. For these two points the X-coordinates are the same and the Y-coordinates are additive inverses. In general two points are symmetric to the X-axis if they have coordinates (c,d) and (c,-d).

From the foregoing we can see that Q and R are symmetric to the X-axis and points R and S are symmetric to the Y-axis.

Finally, it can be shown that the points P and R are symmetric with respect to the origin. This symmetry holds when both the X-coordinates and the Y-coordinates are additive inverses of each other. This same relationship holds for points Q and S. Any positive acute angle in standard position , such as $\frac{\pi}{4}$, that establishes the symmetries described above is known as a <u>reference angle.</u>

The trigonometric ratios for the odd multiples of $\frac{\pi}{4}$ are summarized in the following table:

θ	$\frac{\pi}{4}$	$\frac{3\pi}{4}$	$\frac{5\pi}{4}$	$\frac{7\pi}{4}$
$\cos \theta$	$\frac{1}{\sqrt{2}}$	$\frac{-1}{\sqrt{2}}$	$\frac{-1}{\sqrt{2}}$	$\frac{1}{\sqrt{2}}$
$\sin \theta$	$\frac{1}{\sqrt{2}}$	$\frac{1}{\sqrt{2}}$	$\frac{-1}{\sqrt{2}}$	$\frac{-1}{\sqrt{2}}$

EXAMPLE: Derive the cosine and sine of $\frac{7\pi}{4}$ from Figure 2-3B.

SOLUTION: The coordinates of the terminal point S are $x = \frac{r}{\sqrt{2}}$ and $y = \frac{-r}{\sqrt{2}}$.

$$\cos \frac{7\pi}{4} = \frac{x}{r} = \frac{\frac{r}{\sqrt{2}}}{r} = \frac{\frac{1}{\sqrt{2}}}{1} = \frac{1}{\sqrt{2}}$$

$$\sin \frac{7\pi}{4} = \frac{y}{r} = \frac{\frac{-r}{\sqrt{2}}}{r} = \frac{\frac{-1}{\sqrt{2}}}{1} = \frac{-1}{\sqrt{2}}$$

MULTIPLES OF $\pi/6$ (30°) and $\pi/3$ (60°).

In geometry we learn that if we are given an equilateral triangle measuring 1 unit on a side, an altitude will bisect a side and measure $\frac{\sqrt{3}}{2}$ units.

Figure 2-3C

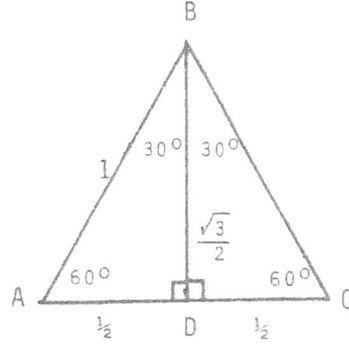

In Figure 2-3C, BD is perpendicular to AC and AB = 1, AD = $\frac{1}{2}$, BD = $\frac{\sqrt{3}}{2}$. In right triangle ADB, the side AD opposite the 30^0 ($\frac{\pi}{6}$) angle is $\frac{1}{2}$ of the hypotenuse, and the side BD opposite the 60^0 ($\frac{\pi}{3}$) angle is $\frac{\sqrt{3}}{2}$ of the hypotenuse. The Pythagorean theorem applied to triangle ABD will confirm that $(\frac{1}{2})^2 + (\frac{\sqrt{3}}{2})^2 = (1)^2$.

Transferring these relationships to our circle model (Figure 2-3D), the coordinates for a point on the terminal side of a central angle of $\frac{\pi}{6}$ are given by $(\frac{r\sqrt{3}}{2} , \frac{r}{2})$

Figure 2-3D

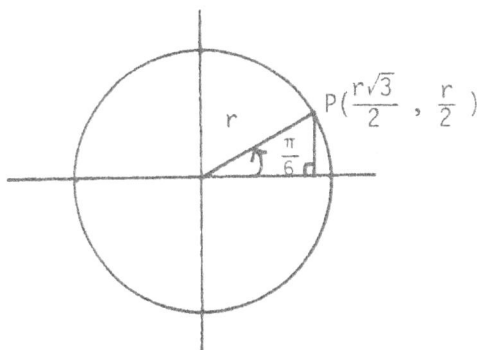

From the definitions of cosine and sine,

$$\cos \frac{\pi}{6} = \frac{x}{r} = \frac{(\frac{\sqrt{3}}{2})r}{r} = \frac{\sqrt{3}}{2} , \qquad \sin \frac{\pi}{6} = \frac{y}{r} = \frac{\frac{r}{2}}{r} = \frac{1}{2} .$$

Expressed in degrees, $\cos 30^0 = \frac{\sqrt{3}}{2}$ and $\sin 30^0 = \frac{1}{2}$.

Symmetry results in the following distribution of points on the circle: (Figure 2-3E). The reference angle, $\frac{\pi}{6}$, establishes symmetries with angles terminating in the other quadrants. In this example, they are $\frac{5\pi}{6}$, $\frac{7\pi}{6}$, and $\frac{11\pi}{6}$. The symmetries are expressed by the magnitudes of the coordinates at the end points of the terminating radii for the angles.

Figure 2-3E

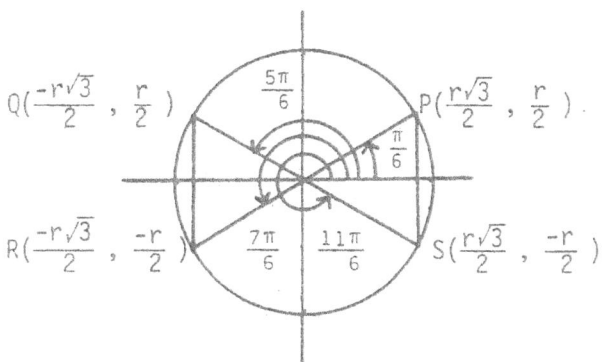

Using the same method of derivation as that for $\cos \frac{\pi}{6}$ and $\sin \frac{\pi}{6}$, we can tabulate the results as follows:

θ	$\frac{\pi}{6}$	$\frac{5\pi}{6}$	$\frac{7\pi}{6}$	$\frac{11\pi}{6}$
$\cos \theta$	$\frac{\sqrt{3}}{2}$	$\frac{-\sqrt{3}}{2}$	$\frac{-\sqrt{3}}{2}$	$\frac{\sqrt{3}}{2}$
$\sin \theta$	$\frac{1}{2}$	$\frac{1}{2}$	$\frac{-1}{2}$	$\frac{-1}{2}$

Note that in the unit circle ($r = 1$), the coordinates of the points on the circumference are equal in magnitude to the cosine and sine of the central angle. We shall use this fact for a direct reading of the cosine and sine of $\frac{\pi}{3}$ (60°). Thus, $\cos \frac{\pi}{3} = \frac{1}{2}$ and $\sin \frac{\pi}{3} = \frac{\sqrt{3}}{2}$.

Figure 2-3F

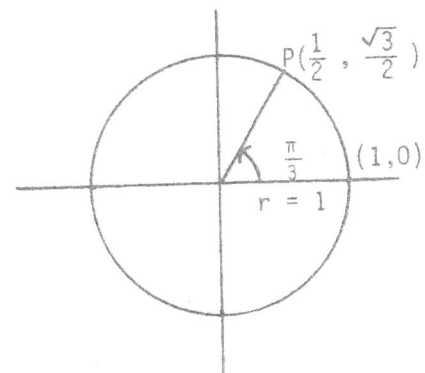

Again we may use symmetry to obtain the coordinates of terminal points for central angles of $\frac{2\pi}{3}$, $\frac{4\pi}{3}$, $\frac{5\pi}{3}$.

Figure 2-3G

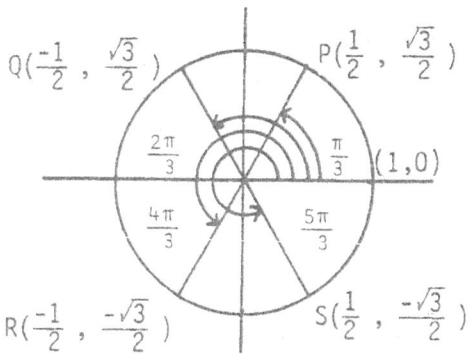

The results are summarized in the following table:

θ	$\frac{\pi}{3}$	$\frac{2\pi}{3}$	$\frac{4\pi}{3}$	$\frac{5\pi}{3}$
$\cos \theta$	$\frac{1}{2}$	$\frac{-1}{2}$	$\frac{-1}{2}$	$\frac{1}{2}$
$\sin \theta$	$\frac{\sqrt{3}}{2}$	$\frac{\sqrt{3}}{2}$	$\frac{-\sqrt{3}}{2}$	$\frac{-\sqrt{3}}{2}$

For the rest of this course we shall be using the special angles and their trigonometric ratios frequently. The student's task in learning new ideas is made much easier by familiarizing himself with the values derived in this section. The following method is helpful in case memory fails and the tables given in this section are not available.

EXAMPLE: Find $\cos \frac{5\pi}{4}$.

SOLUTION: The problem is illustrated in Figure 2-3H.

Figure 2-3H

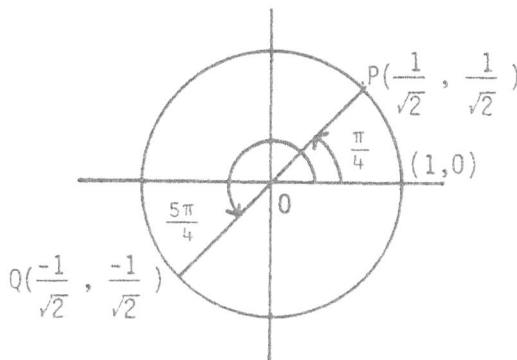

1. Sketch a unit circle.

2. Identify $\frac{5\pi}{4}$ as an odd multiple of $\frac{\pi}{4}$.

3. Sketch $\frac{\pi}{4}$ (45^0), the reference angle.

4. Put in the memorized coordinates for $\frac{\pi}{4}$: ($\frac{1}{\sqrt{2}}$, $\frac{1}{\sqrt{2}}$) .

5. Move five $\frac{\pi}{4}$ arc measures counterclockwise from (1,0). Or add π to $\frac{\pi}{4}$.

6. Draw the terminal side of the $\frac{5\pi}{4}$ angle, by extending OP backward to Q .

7. Put in the coordinates of $\frac{5\pi}{4}$ on the unit circle using symmetry: ($\frac{-1}{\sqrt{2}}$, $\frac{-1}{\sqrt{2}}$).

8. Equate $\cos \frac{5\pi}{4}$ to the x-coordinate and $\sin \frac{5\pi}{4}$ to the y-coordinate.

9. Solution is $\cos \frac{5\pi}{4} = \frac{-1}{\sqrt{2}}$.

EXAMPLE: Find $\sin \frac{4\pi}{3}$.

SOLUTION: Figure 2-3I

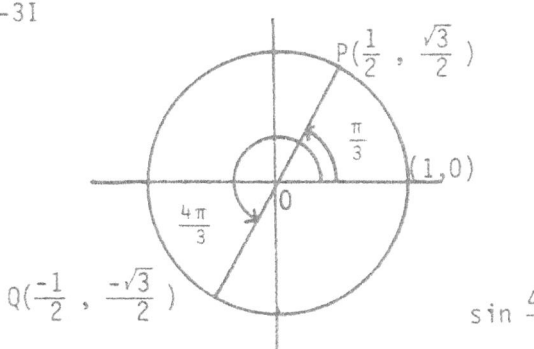

$$\sin \frac{4\pi}{3} = \frac{-\sqrt{3}}{2}$$

EXAMPLE: Find $\cos \frac{5\pi}{6}$.

SOLUTION: Figure 2-3J.

Figure 2-3J

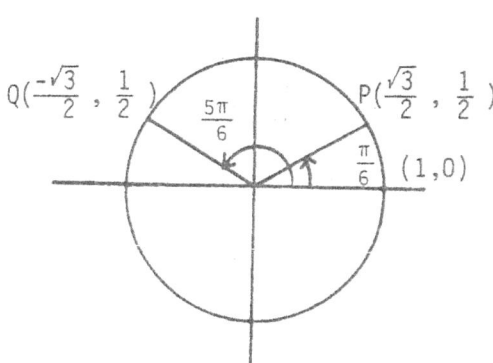

$$\cos \frac{5\pi}{6} = \frac{-\sqrt{3}}{2}$$

EXAMPLE: Find $\sin \left(\frac{-5\pi}{4}\right)$.

SOLUTION: Figure 2-3K.

Figure 2-3K

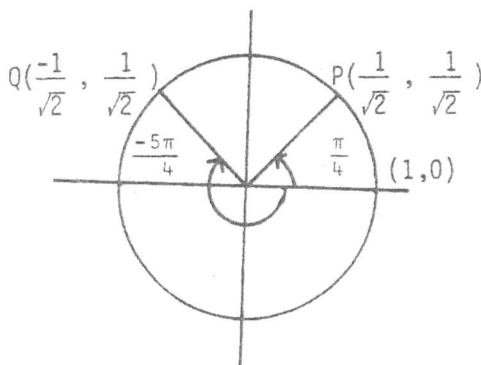

$$\sin \left(\frac{-5\pi}{4}\right) = \frac{1}{\sqrt{2}}$$

Note in the example above that $\frac{-5\pi}{4}$ is coterminal with $\frac{3\pi}{4}$. Since the cosine and sine of an angle are solely dependent upon the coordinates of the end-point on its terminal radius, it follows that $\cos \left(\frac{-5\pi}{4}\right) = \cos \frac{3\pi}{4} = \frac{-1}{\sqrt{2}}$. In the next section this principle will be extended to all coterminal angles.

TABLE OF COSINES AND SINES FOR SPECIAL ANGLES FROM 0 TO 2π.

θ (radians)	0	$\frac{\pi}{6}$	$\frac{\pi}{4}$	$\frac{\pi}{3}$	$\frac{\pi}{2}$	$\frac{2\pi}{3}$	$\frac{3\pi}{4}$	$\frac{5\pi}{6}$	π
θ (degrees)	0	30	45	60	90	120	135	150	180
$\cos\theta$	1	$\frac{\sqrt{3}}{2}$	$\frac{1}{\sqrt{2}}$	$\frac{1}{2}$	0	$\frac{-1}{2}$	$\frac{-1}{\sqrt{2}}$	$\frac{-\sqrt{3}}{2}$	-1
$\sin\theta$	0	$\frac{1}{2}$	$\frac{1}{\sqrt{2}}$	$\frac{\sqrt{3}}{2}$	1	$\frac{\sqrt{3}}{2}$	$\frac{1}{\sqrt{2}}$	$\frac{1}{2}$	0

- -

θ (radians)	$\frac{7\pi}{6}$	$\frac{5\pi}{4}$	$\frac{4\pi}{3}$	$\frac{3\pi}{2}$	$\frac{5\pi}{3}$	$\frac{7\pi}{4}$	$\frac{11\pi}{6}$	2π
θ (degrees)	210	225	240	270	300	315	330	360
$\cos\theta$	$\frac{-\sqrt{3}}{2}$	$\frac{-1}{\sqrt{2}}$	$\frac{-1}{2}$	0	$\frac{1}{2}$	$\frac{1}{\sqrt{2}}$	$\frac{\sqrt{3}}{2}$	1
$\sin\theta$	$\frac{-1}{2}$	$\frac{-1}{\sqrt{2}}$	$\frac{-\sqrt{3}}{2}$	-1	$\frac{-\sqrt{3}}{2}$	$\frac{-1}{\sqrt{2}}$	$\frac{-1}{2}$	0

EXERCISE SET 2-3.

Use the table above to find the indicated values of the following:

1) $\cos\frac{3\pi}{4}$ 2) $\sin\frac{2\pi}{3}$ 3) $\cos(2\pi - \frac{\pi}{3})$ 4) $\sin(\frac{\pi}{2} - \frac{\pi}{3})$

5) $\sin(\frac{\pi}{2} + \frac{\pi}{3})$ 6) $\cos(\frac{2\pi}{3} + \pi)$ 7) $(\cos 30^0)^2$ 8) $\sin(180^0 + 45^0)$

9) $\sin(165^0 - 45^0)$ 10) $3\sin\frac{7\pi}{4}$ 11) $\sin(-\pi)$ 12) $\cos(\frac{-3\pi}{4})$

13) $2\sin(\frac{-4\pi}{3})$ 14) $(\sin\frac{5\pi}{6})^2 + (\cos\frac{5\pi}{6})^2$ 15) $\cos(\frac{-\pi}{6}) - \sin(\frac{-\pi}{6})$

16) $(\cos 240^0)^2 - (\sin 240^0)^2$ 17) $\cos(\frac{11\pi}{6} - \frac{3\pi}{2})$ 18) $\sin(\frac{3\pi}{2} - \frac{11\pi}{6})$

19) $\dfrac{\sin\frac{4\pi}{3}}{\cos\frac{4\pi}{3}}$ 20) $\dfrac{\cos 315^0}{\sin 315^0}$

Use the table above in reverse to find values of θ such that $0 \le \theta \le 2\pi$.

EXAMPLE: $\dfrac{\sin \theta}{\cos \theta} = 1$

SOLUTION: Dividing the sine value by the cosine value for each column entry

reveals that $\theta = \dfrac{\pi}{4}$ or $\theta = \dfrac{5\pi}{4}$.

EXAMPLE: $\sin 2\theta = \dfrac{-1}{2}$

SOLUTION: The table shows that $\sin \dfrac{7\pi}{6} = \dfrac{-1}{2}$ and $\sin \dfrac{11\pi}{6} = \dfrac{-1}{2}$.

Then it must follow that $2\theta = \dfrac{7\pi}{6}$ or $2\theta = \dfrac{11\pi}{6}$. Hence $\theta = \dfrac{7\pi}{12}$ or $\dfrac{11\pi}{12}$.

However, these are not the only possible solutions for θ between

0 and 2π . In the next section it will be shown that if 2θ is al-

lowed to be greater than 2π , then $\theta = \dfrac{19\pi}{12}$ or $\theta = \dfrac{23\pi}{12}$ are also solu-

tions to the given equation.

21) $\cos \theta = \dfrac{-1}{\sqrt{2}}$ 22) $\sin \theta = \dfrac{\sqrt{3}}{2}$ 23) $\sin \theta = \dfrac{-1}{2}$ 24) $\cos \theta = \left(\dfrac{-1}{\sqrt{2}}\right)^2$

25) $\sin \theta = 1$ 26) $\cos \theta = -1$ 27) $(\sin \theta)(\cos \theta) = -1$

28) $(\sin \theta)(\cos \theta) = 0$ 29) $\dfrac{\sin \theta}{\cos \theta} = 0$ 30) $\dfrac{\cos \theta}{\sin \theta} = 1$

31) $(\sin \theta)^2 + (\cos \theta)^2 = 1$ 32) $\cos (\theta - 15^0) = \dfrac{1}{\sqrt{2}}$ 33) $\sin 2\theta = 1$

34) $\cos 2\theta = \dfrac{-1}{2}$ 35) $\cos \dfrac{\theta}{2} = \dfrac{1}{\sqrt{2}}$ 36) $\sin \dfrac{\theta}{2} = \dfrac{-1}{\sqrt{2}}$ 37) $\sin 3\theta = \dfrac{-1}{2}$

38) $\cos 3\theta = \dfrac{\sqrt{3}}{2}$ 39) $\cos (\theta + 20^0) = \dfrac{1}{2}$ 40) $(\cos \theta)^2 - (\sin \theta)^2 = 1$

2-4. <u>COSINES AND SINES OF SPECIAL ANGLES GREATER THAN 360°</u>.

In Section 2-2 we took note of the fact that cos 0° = cos 360° = 1 and sin 0° = sin 360° = 0 . Another example in Section 2-3 showed that cos $(-\frac{5\pi}{4})$ = cos $\frac{3\pi}{4}$ = $\frac{-1}{\sqrt{2}}$ and sin $(-\frac{5\pi}{4})$ = sin $\frac{3\pi}{4}$ = $\frac{1}{\sqrt{2}}$. Both of these examples are illustrations of the generalization that the cosines of coterminal angles are equal and the sines of coterminal angles are equal. In order to be coterminal, angles must differ by some positive or negative integral multiple of 360° or 2π radians. For example, $(-\frac{5\pi}{4})$ = $\frac{3\pi}{4}$ + 2π(-1) .

This principle may be extended to angles, both positive or negative, that are greater than one revolution.

<u>EXAMPLE</u>: Find an angle between 0° and 360° that is coterminal with 1370°.

<u>SOLUTION</u>: First find the number of revolutions equivalent to 1370°.

$\frac{1370^O}{360^O}$ = 3.8055556 revolutions. Next, subtract 3 to obtain .8055556 partial revolution. Multiply by 360°: (360°)(.8055556) = 290° .

Then 1370° = 290° + 3(360°), so 1370° is coterminal with 290°.

<u>EXAMPLE</u>: Find an angle between 0 and 2π that is coterminal with $(-\frac{43\pi}{6})$.

<u>SOLUTION</u>: $\frac{-43\pi}{6}$ = $-7\frac{1}{6}\pi$ = (-6π) + $(-1\frac{1}{6}\pi)$= (-6π) + $(\frac{-7\pi}{6})$. It isn't necessary to convert to revolutions in order to see that $\frac{-7\pi}{6}$ is the angle equivalent to the excess partial revolution. Next, add 2π to obtain the positive coterminal angle between 0 and 2π . $\frac{-7\pi}{6}$ + 2π = $\frac{5\pi}{6}$.

Therefore, $(\frac{-43\pi}{6})$ is coterminal with $\frac{5\pi}{6}$.

<u>TRIGONOMETRIC COORDINATES</u>

Returning to the definitions of cosine and sine, if Θ is a central angle in standard position subtending an arc from (r,0) to (x,y) on a circle with radius r, then cos Θ = $\frac{x}{r}$ and sin Θ = $\frac{y}{r}$. Solving for x and y yields the following;   x = r cos θ   and y = r sin θ .

Since two ordered pairs are equal if their first coordinates are equal and their second coordinates are equal, it follows that :

$$(x,y) \; = \; (r \cos θ \, , \, r \sin θ) \; .$$

The coordinates r cos θ and r sin θ are called <u>trigonometric coordinates</u>.

If (x,y) is a point on the unit circle, (x,y) = (cos θ , sin θ).

<u>EXAMPLE</u>: Express the coordinates of the point $(\frac{-\sqrt{3}}{2} \, , \, \frac{-1}{2})$ in trigonometric

form.

<u>SOLUTION</u>: $(\frac{-\sqrt{3}}{2} \, , \, \frac{-1}{2})$ is a point on the unit circle since $(\frac{-\sqrt{3}}{2})^2 + (\frac{-1}{2})^2$

$= \frac{3}{4} + \frac{1}{4} = 1$. Hence $\cos θ = \frac{-\sqrt{3}}{2}$ and $\sin θ = \frac{-1}{2}$. Analysis by sym-

metry with a reference angle of $\frac{π}{6}$, we find $θ = \frac{7π}{6}$ or 210° (See Fi-

gure 2-4A below). Therefore $(-\frac{\sqrt{3}}{2} \, , \, \frac{-1}{2}) = (\cos \frac{7π}{6}, \sin \frac{7π}{6})$.

Now, let θ be any angle subtending a positive arc less than or equal to the circumference of the unit circle. If θ is measured in radians:

$$(\cos θ \, , \, \sin θ) = (\cos(θ + 2πk), \sin(θ + 2πk))$$

and if θ is measured in degrees:

$$(\cos θ \, , \, \sin θ) = (\cos(θ + k·360°), \sin(θ + k·360°))$$

where k = 0 , ±1 , ±2 , ±3 , ...

The relationships above express the equality of the trigonometric coordinates for coterminal angles.

Figure 2-4A

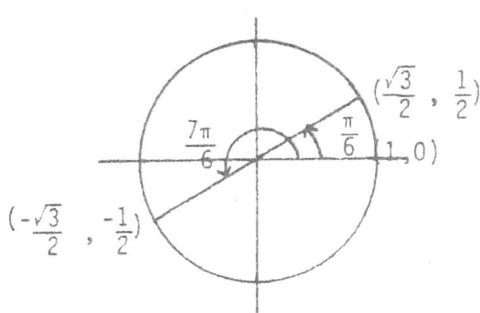

EXAMPLE: Find the cosine and sine of 480°.

SOLUTION: Let $(\cos 480^O, \sin 480^O)$ be a point on the unit circle.

Since $480^O = 120^O + 360^O$, $(\cos 480^O, \sin 480^O) = (\cos 120^O, \sin 120^O)$.

That is, 480° is coterminal with 120°. Analysis by symmetry yields

$\cos 480^O = \cos 120^O = -\cos 60^O = \dfrac{-1}{2}$, $\sin 480^O = \sin 120^O = \sin 60^O = \dfrac{\sqrt{3}}{2}$.

EXAMPLE: Find the cosine and sine of -9495°.

SOLUTION: $-9495^O \div 360^O = \qquad -26.375$ revolutions

Add 27 revolutions: $\underline{+27 \qquad \text{revolutions}}$

$\qquad\qquad$ 0.625 partial revolution

Multiply by 360°: $\underline{X \quad 360^O}$

Product $\qquad = \qquad 225^O$

Therefore -9495° is coterminal with 225°.

Analysis by symmetry yields $\cos(-9495^O) = \cos 225^O = -\cos 45^O = \dfrac{-1}{\sqrt{2}}$

and $\sin(-9495^O) = \sin 225^O = -\sin 45^O = \dfrac{-1}{\sqrt{2}}$.

EXAMPLE: Find the cosine and sine of $\dfrac{375\pi}{6}$.

SOLUTION: $\dfrac{375\pi}{6} = 62.5\pi = 62\pi + 0.5\pi$. Since 0.5π (or $\dfrac{\pi}{2}$) is the excess partial revolution angle measure, $\dfrac{375\pi}{6}$ is coterminal with $\dfrac{\pi}{2}$.

Therefore $\cos \dfrac{375\pi}{6} = \cos \dfrac{\pi}{2} = 0$ and $\sin \dfrac{375\pi}{6} = \sin \dfrac{\pi}{2} = 1$.

The following example is a slight variation of the same problem that emphasizes the symmetry relationship.

EXAMPLE: Express $\sin 17970^O$ as the sine of a positive acute angle.

SOLUTION: $17970^O \div 360^O = 49.91666...$ revolutions

Subtract 49: $\underline{-49.}$

Remainder: $\qquad 0.91666...$ partial revolution

Multiply by 360°: 360° X $0.91666... = 330^O$. Hence 17970° is coterminal with 330°. Figure 2-4B shows that the terminal radii of 330° and 30° are symmetric with respect to the X-axis. Comparing the second coordinates we find that $\sin 330^O = \frac{-1}{2} = - \sin 30^O$.

Therefore, $\sin 17970^O = - \sin 30^O$.

Figure 2-4B

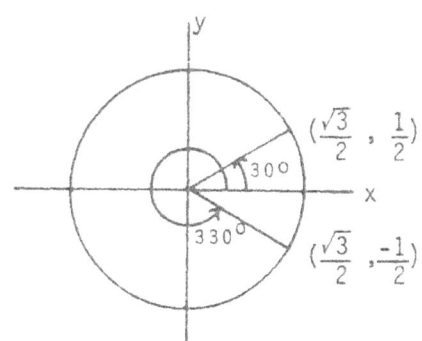

EXERCISE SET 2-4.

Find an angle θ, $0^O \le \theta < 360^O$, that is coterminal with each of the following:

1) 475° 2) 778° 3) -564° 4) -319°

5) 658° 6) 2785° 7) -436° 8) -1184°

Find an angle θ, $0 \le \theta < 2\pi$, that is coterminal with each of the following:

9) $\frac{5\pi}{2}$ 10) $\frac{10\pi}{3}$ 11) 10.8 12) $\frac{-3\pi}{4}$ 13) $\frac{-17\pi}{8}$

14) -15.8 15) 254 16) -48.6

Find the indicated values of the following:

17) $\cos 450^\circ$ 18) $\sin 1590^\circ$ 19) $\cos (-2220^\circ)$ 20) $-\cos (-765^\circ)$

21) $\sin (2295^\circ) + \cos (2295^\circ)$ 22) $(\cos 810^\circ)^2 + (\sin 810^\circ)^2$

23) $2 \sin \frac{40\pi}{3}$ 24) $\cos \frac{67\pi}{6} - \sin \frac{67\pi}{4}$ 25) $\sin (\frac{-761\pi}{3})$

26) $\cos (3\pi - \frac{37\pi}{6})$ 27) $\sin (-3510^\circ)$ 28) $\cos 7020^\circ$

29) $\sin 27.25\pi$ 30) $\cos (\frac{-550\pi}{6})$

Express the given trigonometric ratio in terms of a positive acute angle.

EXAMPLE: $\cos 150^\circ = -\cos 30^\circ$.

31) $\sin 135^\circ$ 32) $\cos 135^\circ$ 33) $\cos 120^\circ$ 34) $\sin (-120^\circ)$

35) $\cos (\frac{-3\pi}{4})$ 36) $\sin (\frac{-5\pi}{3})$ 37) $\sin 315^\circ$ 38) $\cos (-45^\circ)$

39) $\sin (-330^\circ)$ 40) $\cos (\frac{-7\pi}{4})$

Answer the following as TRUE or FALSE.

41) $\sin 315^\circ = \sin 45^\circ$ 42) $\sin (-120^\circ) = -\sin 60^\circ$.

43) $\cos 225^\circ = -\cos 45^\circ$ 44) $\cos (-225^\circ) = -\cos 45^\circ$

45) $\sin (\frac{-5\pi}{6}) = -\sin \frac{5\pi}{6}$ 46) $\cos (\frac{-5\pi}{6}) = -\cos \frac{5\pi}{6}$

47) $\sin (\frac{-5\pi}{6}) = \sin \frac{\pi}{6}$ 48) $\sin (\frac{-\pi}{3}) = \sin \frac{\pi}{3}$

49) $\cos (\frac{-\pi}{2}) = \cos \frac{\pi}{2}$ 50) $\sin \frac{3\pi}{2} = -\sin \frac{\pi}{2}$

2-5. COSINE AND SINE OF THE GENERAL ANGLE IN DEGREE MEASURES.

USE OF TABLES AND CALCULATORS. REDUCTION FORMULAS.

Our introduction to the sine and cosine ratios has been restricted so far to the special angles. This is because the geometry of the special angles leads to simple derivations of their cosine and sine values. For this reason, the special angles are a handy tool for teaching about trigonometric relationships. However, there are problems in measurement that require access to cosines and sines of any conceivable angle. This section will give an introduction to the use of tables and electronic calculators for this purpose.

EXAMPLE: Find $\cos 25^\circ$.

TABLE SOLUTION: Refer to Table II in the Appendix. Read down the left-hand column marked "Degrees" until you find 25° 00'. Read horizontally to the right until you find the entry in the cosine column (marked "Cos θ" at the top). The entry is 0.9063. Hence, $\cos 25^\circ = 0.9063$.

CALCULATOR SOLUTION: Enter 25. Press cos key. Read .90630779 in the display. Round off to four decimal places. Note: Degree mode is required.

EXAMPLE: Find $\cos 73^\circ$.

TABLE SOLUTION: Table II. Find $73^\circ 00'$ in the right-hand column marked "Degrees" at the bottom. Find the column marked "Cos θ" at the bottom. The value is read, $\cos 73^\circ = 0.2924$. Note that the way the table is structured, $\cos 73^\circ = \sin 17^\circ$. This is an example of the complementary relationship between the cosine and the sine. Since $73^\circ + 17^\circ = 90^\circ$, 73° and 17° are complementary angles. The name for the cosine derives from this complementary relationship. This will be formalized later. For the present, in using Table II, note that the angles from 0° to 45° are graduated in the left-hand column reading down and the angles from 45° to 90° are graduated in the right-hand column reading up. The trigonometric ratios for angles from 0° to 45° are listed on the top of each column and the trigonometric ratios for angles from 45° to 90° are listed on the bottom. Care must be exercised in reading the entries with this organization in mind.

CALCULATOR SOLUTION: Enter 73. Press cos key. Read display and round off to four decimals. cos 73° = 0.2924. Note: Degree mode is required.

EXAMPLE: sin 28° 40'.

TABLE SOLUTION: Read down left-hand column to 28° 40' located between 28° 00' and 29° 00'. Read entry in sin θ column, noting its abbreviated form: 797. The complete form is obtained by noting the entry opposite 28° 00'. Hence, sin 28° 40' = 0.4797.

CALCULATOR SOLUTION: Enter 40. Divide by 60. Add 28. Display 28.6666..., the decimal degree version of 28° 40'. Press sin key to display 0.4797 rounded to four decimal places.

EXAMPLE: Find sin 148°.

TABLE SOLUTION: Since Table II and some calculators only give values of the sine and cosine between 0 and 90°, a consistent method needs to be applied for working with angle values that do not terminate in the first quadrant. Figure 2-5A illustrates the situation for the second quadrant angle of 148°.

Figure 2-5A

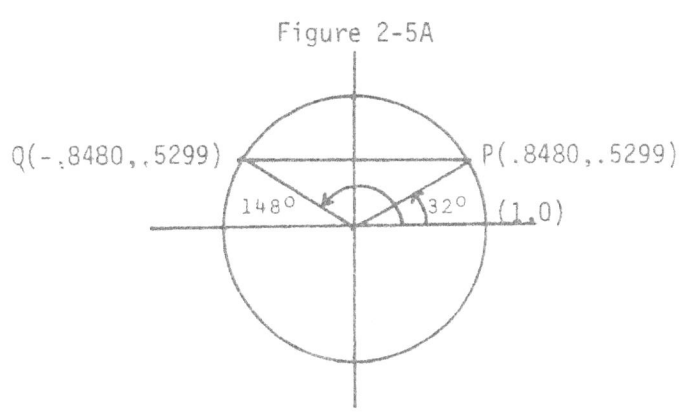

We use the same method of symmetry as applied to the special angles. On the unit circle a point P on the terminal side of a 32° angle and a point Q on the terminal side of a 148° angle are symmetric with respect to the Y-axis. This is found by noting that $32^O = 180^O - 148^O$. From Table II, sin $32^O = 0.5299$. Since the Y-coordinates of P and Q are the same, sin $148^O = 0.5299$ also. In general, if θ is a first quadrant angle, sin $(180^O - \theta) =$ sin θ. Thus, sin $148^O =$ sin $(180^O - 32^O) =$ sin 32°.

CALCULATOR SOLUTION: If you are fortunate in having a calculator that displays sines and cosines of angles greater than 90°, enter 148 and press the sin key. Otherwise, you will have to employ symmetries and/or reduction formulas in the same manner as a reader of tables. As above, subtract 148° from 180° to obtain 32°. The steps are: enter 180. Subtract 148. Press sin key. Display reads 0.5299 to four decimal places.

EXAMPLE: Find cos 239°.

TABLE SOLUTION: Figure 2-5B

Figure 2-5B

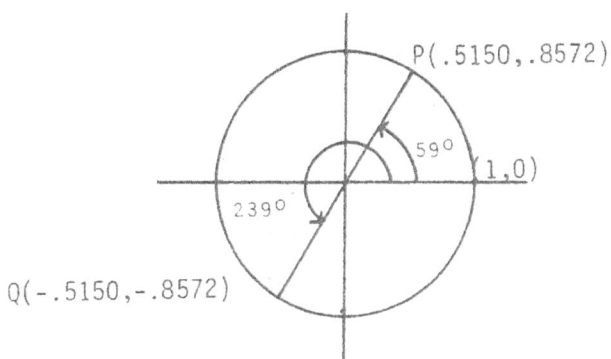

Figure 2-5B illustrates that in the third quadrant, terminal points of angles (i. e., Q) join with terminal points in the first quadrant (i. e., P) to produce symmetry with respect to the origin. To obtain the first quadrant angle, subtract 180° from 239°. $239^{\circ} - 180^{\circ} = 59^{\circ}$. The coordinates of P are $(\cos 59^{\circ}, \sin 59^{\circ}) = (0.5150, 0.8572)$ from Table II. (Remember, read values from bottom.) The coordinates of Q are the additive inverses of P. Hence, $\cos 239^{\circ} = -0.5150$. In general, if θ is a first quadrant angle,

$$\cos (180^{\circ} + \theta) = - \cos \theta.$$

Thus, $\cos 239^{\circ} = \cos (180^{\circ} + 59^{\circ}) = - \cos 59^{\circ} = -0.5150$.

We can also see from Figure 2-5B that

$$\sin 239^{\circ} = \sin (180^{\circ} + 59^{\circ}) = - \sin 59^{\circ} = -0.8572.$$

In general, if θ is a first quadrant angle, $\sin (180^{\circ} + \theta) = - \sin \theta$.

CALCULATOR SOLUTION: Enter 239. Press cos key. Or, proceed as described above.

EXAMPLE: Find $\sin 194^{\circ}$.

SOLUTION: $\sin 194^{\circ} = \sin (180^{\circ} + 14^{\circ}) = - \sin 14^{\circ} = -0.2419$.

We now proceed to a fourth quadrant angle where symmetry with respect to the X-axis is applied.

EXAMPLE: Find $\cos 295^{\circ}$ and $\sin 295^{\circ}$.

SOLUTION: Figure 2-5C.

Figure 2-5C

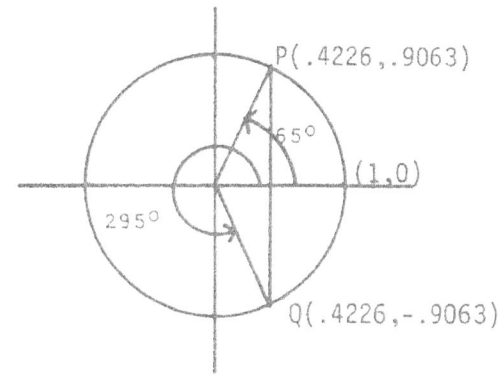

Figure 2-5C shows that cos 295° = cos 65° = 0.4226, and sin 295° = - sin 65° = -0.9063. In general, if θ is a first quadrant angle, cos (360° - θ) = cos θ and sin (360°- θ) = - sin θ. Applying these <u>reduction</u> formulas to 295°, cos 295° = cos (360° - 65°) = cos 65° = 0.4226. (To obtain 65°, simply subtract 295° from 360°.) Sin 295° = sin (360° - 65°) = - sin 65° = -0.9063.

<u>CALCULATOR SOLUTIONS</u>: Enter 295. Press cos key. Read display. Again, enter 295. Press sin key. Read display. Otherwise, use reduction formulas as above.

<u>EXAMPLE</u>: Find cos (-214°) and sin (-214°).

<u>CALCULATOR SOLUTION</u>: For calculators that take any angle entry, enter 214. Press +/- key. Press cos (sin) key. Read -0.8290 (0.5592). Round to four decimal places as shown. For other calculators or Table II, proceed as follows:

1. Locate (-214°) in the second quadrant.

Figure 2-5D

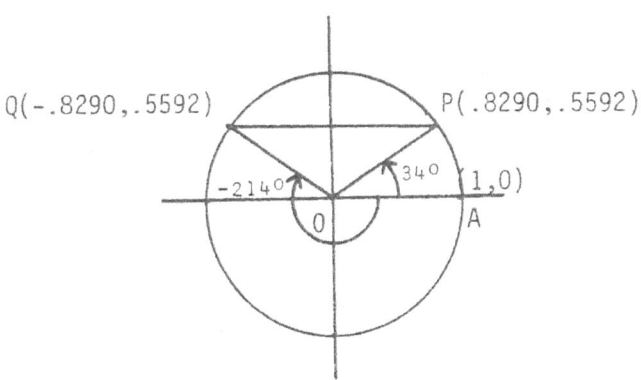

2. Label terminal point Q.

3. Find by symmetry point P in the first quadrant.

4. Find angle AOP by subtracting 180° from 214°. The measure of this first quadrant angle is 34°.

5. By table or calculator, coordinates of P are (0.8290, 0.5592).

6. Coordinates of Q by symmetry with respect to the Y-axis are
(-0.8290, 0.5592).

7. Therefore, cos (-214°) = -0.8290

$$\sin (-214°) = 0.5592$$

EXAMPLE: Find cos (-400° 20') and sin (-400° 20').

SOLUTION: 1. By calculator. Enter 20. Divide by 60. Add 400. Press = key.
Press +/- key. Press STO key for storage in memory. Press cos
key. Read display. cos (-400° 20') = 0.7623. Press RCL key
(recalls 400.3333). Press sin key. Read display.
sin (-400° 20') = -0.6472.

2. By restricted calculator. Enter 20. Divide by 60. Add 400.
Subtract 360. Read 40.333 in display. Press storage key. Press
cos key. Read 0.7623. Press memory recall key. Read 40.3333
in display. Press sin key. Read 0.6472. Interpret results as
explained below.

3. By Table II.

Figure 2-5E

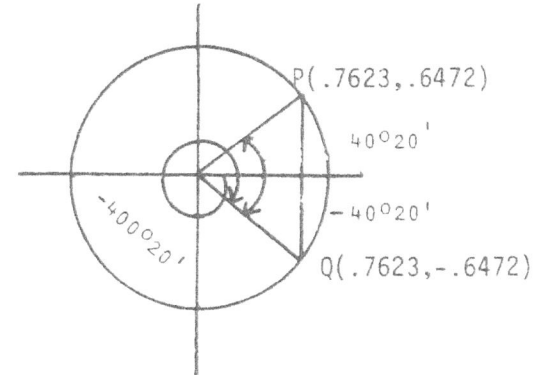

An angle of (-400° 20') is generated clockwise and terminates at the same radius as an angle of -40° 20'. This can be found by adding 360° to -400° 20'. Table II gives the cosine and sine of 40° 20'. By symmetry with respect to the X-axis, point Q has coordinates (0.7623, -0.6472). Hence, cos (-400° 20') = 0.7623, sin (-400° 20') = -0.6472.

EXAMPLE: Find sin (-7084°) and cos (-7084°).

SOLUTION: For calculators that can handle angles of large absolute value,
proceed as in the example above. For other calculators or table
users, the following is recommended:

1.   Divide (-7084) by 360. (-7084) ÷ 360 = -19.6777...
     This gives the number of clockwise (negative) revolutions.

2.   Subtract 19 complete revolutions, leaving -.6777....

3.   Multiply by 360°. (360°)(-.6777...) = (-244°).

4.   (-244°) terminates in the second quadrant. (Figure 2-5F).

Figure 2-5F

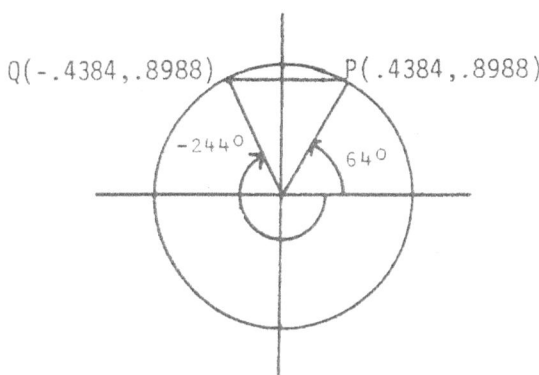

5.   Subtract 180° from 244°. 244° - 180° = 64°, the reference angle
     for symmetry in the first quadrant.

6.   From Table II, cos 64° = 0.4384, sin 64° = 0.8988.

7.   By symmetry, cos (-244°) = -0.4384, sin (-244°) = 0.8988.

8. Hence, cos (-7084^0) = -0.4384, sin (-7084^0) = 0.8988, since (-7084^0) is coterminal with (-244^0).

SYMMETRY REDUCTION FORMULAS FOR COSINES AND SINES.

All of the calculations involving angles terminating in the second, third, and fourth quadrants have required symmetry relationships with reference angles in the first quadrant in order to make use of Table II. These relationships can be formalized in terms of reduction formulas as follows:

If $0^0 < \theta < 90^0$,

cos $(180^0 - \theta)$ = -cos θ and sin$(180^0 - \theta)$ = sin θ.

These formulas are used for angles terminating in quadrant II.

EXAMPLE: Find cos 154^0 and sin 154^0.

SOLUTION: cos 154^0 = cos $(180^0 - 26^0)$ = -cos 26^0 = -.8988

sin 154^0 = sin $(180^0 - 26^0)$ = sin 26^0 = .4384

For angles terminating in quadrant III, cos $(180^0 + \theta)$ = -cos θ and sin $(180^0 + \theta)$ = -sin θ.

EXAMPLE: Find cos 247^0 and sin 247^0.

SOLUTION: cos 247^0 = cos $(180^0 + 67^0)$ = - cos 67^0 = -.3907

sin 247^0 = sin $(180^0 + 67^0)$ = - sin 67^0 = -.9205

For angles terminating in quadrant IV, cos $(360^0 - \theta)$ = cos θ and sin $(360^0 - \theta)$ = - sin θ.

EXAMPLE: Find cos 342.18^0 and sin 342.18^0.

SOLUTION: cos 342.18^0 = cos $(360^0 - 17.82^0)$ = cos 17.82^0 = .9520

sin 342.18^0 = sin $(360^0 - 17.82^0)$ = - sin 17.82^0 = -.3060

Another pair of useful formulas allow the cosines and sines of negative angles to be expressed in terms of positive angles. For any angle θ,

cos $(-\theta)$ = cos θ and sin $(-\theta)$ = -sin θ

EXAMPLE: Express as cosines or sines of positive angles,

$\cos(-185^0)$, $\sin(-35^0)$, $\cos(-137^0)$.

SOLUTION: $\cos(-185^0) = \cos 185^0$

$\sin(-35^0) = -\sin 35^0$

$\cos(-137^0) = \cos 137^0$.

The above can be combined with the previous reduction formulas to express the results in terms of positive acute angles.

EXAMPLE: Express $\sin(-243^0)$, $\cos(-194^0)$, and $\sin(-345^0)$ in terms of sines or cosines of positive acute angles.

SOLUTION: $\sin(-243^0) = -\sin 243^0 = -\sin(180^0 + 63^0) = -[-\sin 63^0]$

$= \sin 63^0$

$\cos(-194^0) = \cos 194^0 = \cos(180^0 + 14^0) = -\cos 14^0$

$\sin(-345^0) = -\sin 345^0 = -\sin(360^0 - 15^0) = -[-\sin 15^0]$

$= \sin 15^0$.

It can further be shown that the restriction of θ to the first quadrant in the earlier reduction formulas is unnecessary. That is, for any angle θ,

$\cos(180^0 - \theta) = -\cos\theta,$ $\qquad \sin(180^0 - \theta) = \sin\theta$

$\cos(180^0 + \theta) = -\cos\theta,$ $\qquad \sin(180^0 + \theta) = -\sin\theta$

$\cos(360^0 - \theta) = \cos\theta,$ $\qquad \sin(360^0 - \theta) = -\sin\theta$

$\cos(-\theta) = \cos\theta,$ $\qquad \sin(-\theta) = -\sin\theta$

EXAMPLE: Prove that for any angle θ, $\cos(180^0 - \theta) = -\cos\theta$.

SOLUTION: If $0 < \theta < 90^0$, we have already shown that symmetry with respect to the Y-axis verifies the formula. Suppose now that $90^0 < \theta < 180^0$ as in Figure 2-5G.

Figure 2-5G

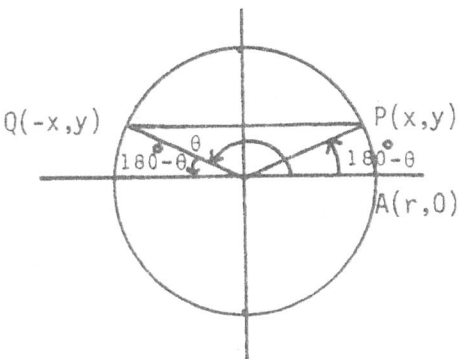

We can see again that the same symmetry works in this case. For the second quadrant angle, $\cos \theta = \frac{-x}{r}$ and $\cos (180^\circ - \theta) = \frac{x}{r}$. Therefore, $\cos (180^\circ - \theta) = - \cos \theta$.

Suppose $180^\circ < \theta < 270^\circ$.

Figure 2-5H

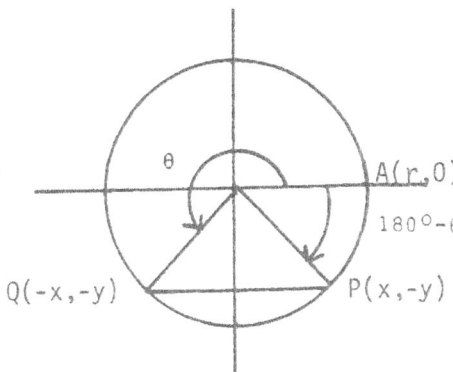

In this case, $180^\circ - \theta$ is a negative angle terminating in the fourth quadrant. For example, if $\theta = 235^\circ$, $180^\circ - 235^\circ = -55^\circ$. Again, the coordinates by symmetry give $\cos (180^\circ - \theta) = - \cos \theta$. The demonstration for $270^\circ < \theta < 360^\circ$ is left to the student.

EXERCISE SET 2-5.

In exercises 1 - 20, find the trigonometric ratio using Table II. Check your results with a calculator.

1) $\sin 12^\circ 40'$ 2) $\sin 75.5^\circ$ 3) $\cos 38^\circ 50'$ 4) $\cos (-73^\circ 10')$

5) $\sin (-85^\circ 20')$ 6) $\sin 117^\circ$ 7) $\cos (223.67^\circ)$ 8) $\cos (-312.33^\circ)$

9) $\sin (-175.5^\circ)$ 10) $\sin 346^\circ$ 11) $\cos (-378^\circ)$ 12) $\cos 538^\circ$

13) $\sin 654^\circ$ 14) $\sin (-1085^\circ)$ 15) $\cos 2483^\circ$ 16) $\cos (-857^\circ 10')$

17) $\sin 17.17^\circ$ 18) $\sin (-6.83^\circ)$ 19) $\cos (-1^\circ)$ 20) $\sin \frac{180^\circ}{\pi}$

In exercises 21 - 28, look up the given ratio in the cosine or sine column of Table II. Find θ if $0 < \theta < 90^\circ$.

21) $\cos \theta = 0.8141$ 22) $\sin \theta = 0.4695$ 23) $\sin \theta = 0.8355$

24) $\cos \theta = 0.6862$ 25) $\cos \theta = 0.6450$ 26) $\sin \theta = 0.9983$

27) $\sin \theta = 0.0116$ 28) $\cos \theta = 0.1080$

In exercises 29 - 32, $-90^\circ < \theta < 90^\circ$. Find θ given the following sine ratios. Use Table II.

29) $\sin \theta = -0.0843$ 30) $\sin \theta = 0.3173$ 31) $\sin \theta = 0.9377$

32) $\sin \theta = -0.8141$

In exercises 33 - 36, $0 < \theta < 180^\circ$. Find θ given the following cosine ratios. Use Table II.

33) $\cos \theta = -0.8704$ 34) $\cos \theta = 0.6820$ 35) $\cos \theta = -0.7547$

36) $\cos \theta = -0.9827$

37) Show that if $270^\circ < \theta < 360^\circ$, $\cos (180^\circ - \theta) = - \cos \theta$.

38) Show that if $90^\circ < \theta < 180^\circ$, $\cos (180^\circ + \theta) = -\cos \theta$.

39) Show that if $90^\circ < \theta < 180^\circ$, $\sin (180^\circ + \theta) = -\sin \theta$.

40) Show that if $270^\circ < \theta < 360^\circ$, $\sin (360^\circ - \theta) = -\sin \theta$.

2-6. <u>COSINE AND SINE OF THE GENERAL ANGLE IN RADIAN MEASURE</u>.

While working the problems in the previous section, you may have wondered how the calculator produces the results found in Table II so quickly. Before the invention of the calculus, tables of the trigonometric ratios were compiled by strenuous geometric analysis. With the advent of the calculus, expressions for the sine and cosine were developed in terms of infinite series. Two such formulas are given as follows. If θ is any real number (that is, an angle measured in radians),

$$\sin \theta = \theta - \frac{\theta^3}{1\cdot2\cdot3} + \frac{\theta^5}{1\cdot2\cdot3\cdot4\cdot5} - \frac{\theta^7}{1\ 2\ 3\ 4\ 5\ 6\ 7} + \ldots$$

$$\text{and } \cos \theta = 1 - \frac{\theta^2}{1\cdot2} + \frac{\theta^4}{1\cdot2\cdot3\cdot4} - \frac{\theta^6}{1\cdot2\cdot3\cdot4\cdot5\cdot6} + \ldots$$

The expressions on the right are called infinite power series in θ and involve only exponentiation, addition, subtraction, multiplication, and division. These operations are handled with ease by electronic calculators, but are very arduous if done by hand.

<u>EXAMPLE</u>: Find an approximation to $\sin \frac{\pi}{4}$ by summing up four terms in the power series.

<u>SOLUTION</u>: $\sin \frac{\pi}{4} = \frac{\pi}{4} - \frac{(\frac{\pi}{4})^3}{1\cdot2\cdot3} + \frac{(\frac{\pi}{4})^5}{1\cdot2\cdot3\cdot4\cdot5} - \frac{(\frac{\pi}{4})^7}{1\cdot2\cdot3\cdot4\cdot5\cdot6\cdot7}$

$\sin \frac{\pi}{4}$ = .78539816 - .08074551 + .00249039 - .00003658 = 0.70710647. Comparing this result with a direct reading from the calculator, enter π. Divide by 4. Press for radians. Press sin key. Display reads .70710678. The results agree to the sixth decimal place without rounding off!

For the remainder of this section we shall solve a variety of examples for cosines and sines of angles measured in radians using tables or the electronic calculator.

EXAMPLE: Find sin 0.35

SOLUTION: By Table I. Read down the radians column until the entry 0.35 is found. Read across to the sin column. Find sin 0.35 = .3429.

CALCULATOR SOLUTION: Enter radian mode. Enter .35 . Press sin key. Read .3429 rounded off to four decimal places.

CALCULATOR WITHOUT RADIAN MODE: Enter .35 . Multiply by 180. Divide by π . Read 20.05 degrees. Press sin key. Read .3429.

Note that Table I is limited to angles between 0 and 1.57 radians. In order to obtain cosines and sines of angles beyond the table without the calculator, the symmetry reduction formulas must be used. If $0 \leq \theta \leq \frac{\pi}{2}$ then

$\cos (\pi - \theta) = - \cos \theta$ $\qquad\qquad$ $\sin (\pi - \theta) = \sin \theta$

$\cos (\pi + \theta) = - \cos \theta$ $\qquad\qquad$ $\sin (\pi + \theta) = - \sin \theta$

$\cos (2\pi - \theta) = \cos \theta$ $\qquad\qquad$ $\sin (2\pi - \theta) = - \sin \theta$

Also, for all θ , $\cos (-\theta) = \cos \theta$ and $\sin (-\theta) = - \sin \theta$.

EXAMPLE: Find cos 2.34

SOLUTION: By Table I . Since 2.34 lies in the second quadrant, the reduction formula $\cos (\pi - \theta) = - \cos \theta$ is used. $\cos (2.34) = \cos (\pi - .80)$ $= - \cos .80 = -0.6967.$

CALCULATOR SOLUTION: Radian mode. Enter 2.34. Press cos key. Read -.6956 rounded. Note there is a discrepancy in the results obtained by calculator (-.6956) and the result obtained by Table I . This occurs because of the approximation of 3.14 used for π in the table results.

EXAMPLE: Find $\sin \left(\frac{-5\pi}{8}\right)$.

SOLUTION: By calculator. Radian mode. Enter 5 . Multiply by π . Divide by 8. Press = key. Change sign (± key). Press sin key. Read -.9239 . By Table I and symmetry. Sketch the given angle. Extend the radius backward to terminate in the first quadrant. Figure the size of the first quadrant reference angle by noting the supplementary relationship. That is, $\pi - \frac{5\pi}{8} = \frac{3\pi}{8}$. Since $\frac{3\pi}{8} = 1.18$ rounded, find this value in Table I . Read the entry in the sin column. sin 1.18 = .9246.

By symmetry with respect to the origin, sin $(\frac{-5\pi}{8}) = -$ sin 1.18 = $-$.9246.

Figure 2-6A

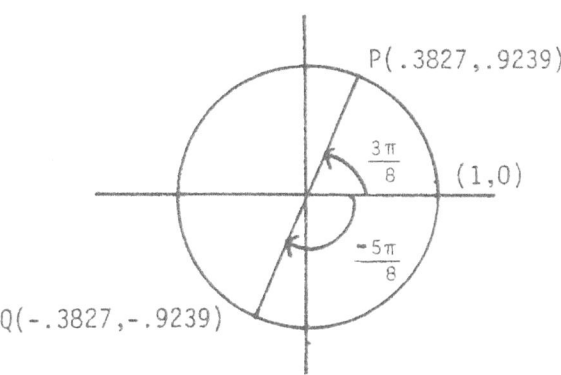

To summarize the above using the symmetry reduction formulas:

$$\sin (\frac{-5\pi}{8}) = - \sin (\frac{5\pi}{8}) = - \sin (\pi - \frac{3\pi}{8}) = - \sin \frac{3\pi}{8}$$

For large angles in radian measures, reduction to coterminal angles by the use of excess partial revolutions is a convenient method.

EXAMPLE: Find sin 42.67 .

SOLUTION: By calculator. Radian mode. Enter 42.67 . Press sin key.

Read $-$.9668. Therefore sin 42.67 = $-$.9668.

By table I . 42.67 ÷ 2π = 6.79 revolutions. Subtract 6 leaving .79 partial revolution. Multiply by 2π: (.79)(2π) = 4.96 an angle coterminal with 42.67 located in quadrant IV. By symmetry, sin 42.67 = sin 4.96 = sin (2π $-$ 1.32) = $-$ sin 1.32 = $-$.9687 by table I .

In the next example a negative angle with a large absolute value is reduced to a positive coterminal angle before applying the symmetry reduction formula.

EXAMPLE: Find cos (-435.86).

SOLUTION: By calculator. Radian mode. Enter -435.86. Press cos key.
Read -.6812 rounded.

By table I . (-435.86) ÷ 2π = -69.37 revolutions. Add 70 complete revolutions: -69.37 + 70 = .63 partial revolution. Multiply by 2π: (.63)(2π) = 3.96 a positive angle in quadrant III that is coterminal with -435.86. Subtract π: 3.96 - π = .82, the reference angle in quadrant I. From table I, cos .82 = .6822. Then by the symmetry reduction formula: cos (-435.86) = cos 3.96 = cos (π + .82)

$$= - \cos .82 = -.6822 .$$

In certain problems the cosine or sine is given and the angle or angles are to be found.

EXAMPLE: sin θ = 0.8305 and θ lies between 0 and $\frac{\pi}{2}$. Find θ .

SOLUTION: Locate 0.8305 in the sine column of table I . In the same row the entry in the radian column is 0.98. Therefore, θ = 0.98 .

If values for θ in the other quadrants are to be found also then the symmetry reduction formulas are used.

EXAMPLE: cos θ = .4976 and 0 < θ < 2π . Find θ .

SOLUTION: Locate .4976 in the cosine column of Table I . Read θ = 1.05 radians.
Now, by symmetry, cos (2π - 1.05) = cos 1.05 as shown in Figure 2-6B.

Figure 2-6B.

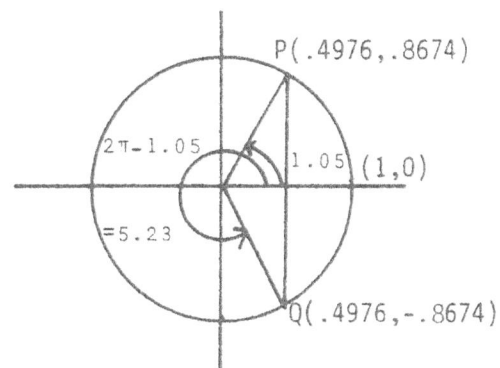

Therefore, 1.05 and 2π - 1.05 are the two angles between 0 and 2π such that their cosines are equal to .4976. (Note:2π - 1.05 = 6.28 - 1.05 = 5.23.)

EXAMPLE: $\sin \theta$ = -.1692 and $0 < \theta < 2\pi$. Find θ.

SOLUTION: Locate .1692 in the sine column of Table I   Read θ = 0.17. This is the <u>reference</u> angle. The sine is negative in the third and fourth quadrants. Therefore, $\theta = \pi + 0.17$ or $\theta = 2\pi - 0.17$. Simplifying, θ = 3.14 + 0.17 = 3.31 or θ = 6.28 - 0.17 = 6.11.

CALCULATOR SOLUTION: Radian mode. Enter .1692. Press INV key or ARC key. Press sin key. Read .17001791. Round off to hundredths. Reference angle is 0.17. Proceed as above for negative sine.

<u>EXERCISE SET 2-6</u>.

Read Table I for the following values:

1) cos 0.83   2) sin 1.25   3) cos 0.11   4) cos 1.49

5) sin 0.23

Read Table I to find θ in radians given: $(0 < \theta < \frac{\pi}{2})$

6) cos θ = .9638   7) cos θ = .5319   8) sin θ = .7643.   9) sin θ = .9425

Round θ off to the nearest hundredths of a radian given: $(0 < \theta < \frac{\pi}{2})$

10) cos θ = .2000   11) sin θ = .8435   12) sin θ = .1800

13) cos θ = .6325

Use a calculator or Table I to find the following:

14) sin (-0.35)   15) cos (-1.28)   16) cos (2.95)

17) sin (-4.86)   18) sin (-5.36)   19) cos (12.9)

20) sin 154.8   21) cos (-238.74)   22) cos 14.48

23) sin $\frac{7\pi}{8}$ 24) cos $\frac{23\pi}{5}$ 25) cos $(\frac{-\pi}{5})$

26) cos (π - 2)   27) cos (2 - π)   28) sin $(\frac{\pi}{2} + 1)$

29) sin $(\frac{\pi}{2} - 1)$ 30) cos 1   31) sin 2

32) cos 1776

Using Table I or a calculator, find θ, $0 < \theta < 2\pi$ such that *

33) sin θ = -.6131   34) cos θ = -.4357   35) cos θ = .5570

36) sin θ = -.9799   37) sin θ = .8573   38) cos θ = -.9888

39) cos θ = .7776   40) sin θ = .9284

* Note that some calculators will designate the INV or ARC key in a different manner. For example the sin key may be also lettered as the $\sin^{-1}$ key and require the user to press an f^{-1} key in order to activate this mode. Consult your manual on how to use these functions. In all cases, however, input the absolute value of the given cosine or sine in order to obtain the first quadrant reference angle before applying the symmetry reduction formulas.

2-7.   CHAPTER SUMMARY

1.   The analytic definition of the circle with center at the origin and radius r:

$\{(x,y) : x^2 + y^2 = r^2\}.$

2.   The unit circle $:\{(x,y) : x^2 + y^2 = 1.\}$

3.   If P(x,y) is a point on a circle located in the first quadrant, there are
     points Q(-x,y), R(-x,-y), and S(x,-y) located symmetrically in quadrants
     II, III, and IV, respectively, as shown in Figure 2-7A.

Figure 2-7A

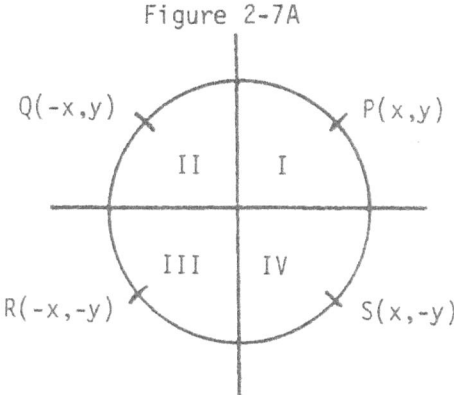

4.   If P(x,y) is any point on a circle with radius r, center at the origin:

$$\cos \theta = \frac{x}{r} \quad \text{and} \quad \sin \theta = \frac{y}{r}.$$

where θ is the central angle bounded by radii to A(r,0) and P(x,y) as in
Figure 2-7B.

Figure 2-7B

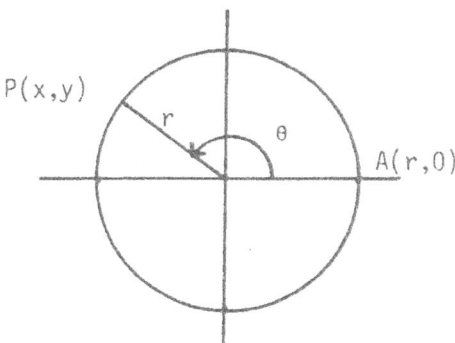

5. If P(x,y) is any point on a unit circle, r = 1 and cos θ = x and sin θ = y, where the central angle θ is in <u>standard</u> <u>position</u> as in Figure 2-7B.

6. The cosine and sine are positive or negative depending upon the quadrant in which the central angle θ terminates. This is illustrated in Figure 2-7C.

Figure 2-7C

sin θ is + cos θ is -	sin θ is + cos θ is +
sin θ is - cos θ is -	sin θ is - cos θ is +

7. $-1 \leq \cos \theta \leq 1$ and $-1 \leq \sin \theta \leq 1$.

8. The values of the cosine and sine for quadrantal angles are restricted to the set {-1,0,1}. These are summarized as follows:

θ (degrees)	0^0	90^0	180^0	270^0	360^0
θ (radians)	0	$\frac{\pi}{2}$	π	$\frac{3\pi}{2}$	2π
cos θ	1	0	-1	0	1
sin θ	0	1	0	-1	0

9. Besides the quadrantal angles, the special angles include multiples of $\frac{\pi}{4}$, $\frac{\pi}{6}$, and $\frac{\pi}{3}$. The cosines and sines of these special angles are summarized in Section 2-3.

10. The trigonometric ratios for coterminal angles are equal. For example, $\cos 380^0 = \cos 20^0$ and $\sin(-70^0) = \sin 290^0$.

11. Table II is used to find trigonometric ratios of angles measured in degrees to the nearest 10 minutes.

12. Table I is used to find trigonometric ratios of angles measured to hundredths of a radian.

13. Electronic calculators have varying capabilities in handling the trigonometric ratios. Some can give direct readings for any angle input that might conceivably be used in either degrees or radians. Others are restricted to degree inputs between 0^0 and 90^0 and require conversion of radians to degrees by the formula θ (degrees) $= \frac{180^0}{\pi} \cdot \theta$ (radians). Conversion of degrees, minutes, seconds to decimal degree equivalents is usually required. Check your owner's manual for specifics.

14. In using Table I or Table II , the first quadrant angles given by the tables are designated as <u>reference angles</u>. Reference angles and symmetries are used to find the cosines and sines of angles terminating in the other quadrants. Symmetry reduction formulas are given in Sections 2-5 and 2-6.

<u>EXERCISE SET 2-7.</u>

In exercises 1 - 4, find the equation of the circle containing the given point.

1) (4,3)   2) (-1,-5)   3) (4.5, 7.8)   4) (-8,9.7)

In exercises 5 - 8, find the cosines of the central angles in standard position terminating at the points given in exercises 1 - 4.

In exercises 9 - 12, cos θ is given and the quadrant in which θ terminates. Find sin θ.

9) cos θ = .9397, IV   10) cos θ = -.5807, II

11) cos θ = .9832, I   12) cos θ = -.7529, III

In exercises 13 - 16, find θ if cos θ is given as in exercises 9 - 12. ($0 \leq \theta < 360^0$)

Find the following:

17) $\sin 7\pi$ 18) $\sin (-\frac{5\pi}{4})$ 19) $\cos \frac{17\pi}{4}$

20) $\cos (\frac{-11\pi}{3})$ 21) $\sin \frac{15\pi}{8}$ 22) $\sin (\frac{-97\pi}{6})$

23) $\cos \frac{22\pi}{6}$ 24) $\cos 378\pi$ 25) $\sin 412^0$

26) $\sin (-755^0)$ 27) $\cos 515^0 \; 20'$ 28) $\cos (-611^0 \; 50')$

29) $\sin 4.67$ 30) $\sin (-57.3)$ 31) $\cos 12.35$

32) $\cos (-84.67)$

33) If $90^0 < \theta < 180^0$ and $s < \cos \theta < t$, find s and t, where s is the greatest lower bound and t is the least upper bound.

34) If $\frac{3\pi}{2} < \theta < 2\pi$ and $s < \sin \theta < t$, find s and t, where s and t are defined as in exercise 33.

TRUE OR FALSE:

35) $\cos (-\pi) = \cos 14\pi$ 36) $\sin (-11\pi) = \sin \pi$

37) $\cos \frac{14\pi}{3} = \cos \frac{\pi}{3}$ 38) $\cos \frac{\pi}{6} = \sin \frac{2\pi}{3}$

39) $2 \cos \frac{4\pi}{3} = \sin \frac{3\pi}{2}$ 40) $\sin \frac{3\pi}{2} = 2(\cos \frac{3\pi}{4})(\sin \frac{3\pi}{4})$

41) $\cos (45^0 - 15^0) = \cos 45^0 - \cos 15^0$.

42) $\sin (90^0 + 30^0) = \sin 90^0 + \sin 30^0$.

43) $(\cos 30^0)^2 - (\sin 30^0)^2 = \cos 60^0$.

44) $(\sin 60^0)(\cos 60^0) = \frac{1}{2} \sin 120^0$.

Using Table I or calculator, find θ, $0 < \theta < 2\pi$ such that:

45) $\sin \theta = 0.6889$ 46) $\sin \theta = 0.9128$

47) $\cos \theta = 0.9428$ 48) $\cos \theta = 0.2867$

49) $\sin \theta = -0.9959$ 50) $\sin \theta = -0.0799$

51) $\cos \theta = -0.9838$ 52) $\cos \theta = -0.5653$

CHAPTER 3

THE SIX TRIGONOMETRIC RATIOS
THE FUNDAMENTAL IDENTITIES

3-1. TANGENT AND COTANGENT

Given a circle with center at the origin, radius r, and a point P (x,y) on the circumference, we have by definition: $\cos \theta = \frac{x}{r}$. $\sin \theta = \frac{y}{r}$.

Figure 3-1A

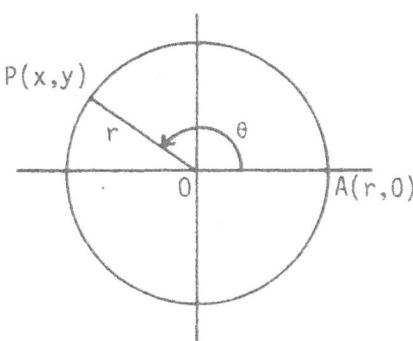

We shall now define two more trigonometric ratios. The tangent of the central angle θ (abbreviated "tan θ") from OA to OP is given by: $\tan \theta = \frac{y}{x}$. The cotangent of the central angle θ (abbreviated "cot θ") from OA to OP is given by: $\cot \theta = \frac{x}{y}$.

It follows from these definitions that if the coordinates of a terminal point P are given, tan θ and cot θ can be determined directly.

EXAMPLE: A point P(3,-2) is given on the terminal side of a central angle θ with initial side the positive X-axis. Find tan θ and cot θ.

SOLUTION: x = 3, y = -2. $\tan \theta = \frac{y}{x} = \frac{-2}{3}$, $\cot \theta = \frac{x}{y} = \frac{3}{-2} = \frac{-3}{2}$. Note that it isn't necessary to determine r since it does not enter into the definitions of the tangent and cotangent ratios.

If θ is given, there are several ways to determine tan θ and cot θ depending on the measure of θ.

EXAMPLE: Find tan 30^0 and cot 30^0.

SPECIAL ANGLE SOLUTION (PREFERRED):

1. Draw a unit circle and insert the memorized coordinates of
    terminal point $P(\frac{\sqrt{3}}{2}, \frac{1}{2})$ (Figure 3-1B).

Figure 3-1B

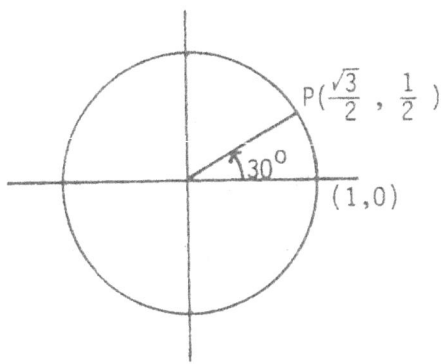

2. $\tan 30^0 = \frac{y}{x} = \dfrac{\frac{1}{2}}{\frac{\sqrt{3}}{2}} = \frac{1}{2} \cdot \frac{2}{\sqrt{3}} = \frac{1}{\sqrt{3}}$

3. $\cot 30^0 = \frac{x}{y} = \dfrac{\frac{\sqrt{3}}{2}}{\frac{1}{2}} = \frac{\sqrt{3}}{2} \cdot \frac{2}{1} = \sqrt{3}$

This is considered a preferred solution because it yields an exact
answer. It also provides us with a specific example of a relationship that is _almost_ always true for the tangent and cotangent:

FUNDAMENTAL IDENTITY: For any angle $\theta \neq k \cdot 90^0$, $k = 0, \pm1, \pm2,$.
$$\tan \theta = \frac{1}{\cot \theta} \quad \text{and} \quad \cot \theta = \frac{1}{\tan \theta} .$$

PROOF: By definition $\tan \theta = \frac{y}{x}$ and $\cot \theta = \frac{x}{y}$.

Hence, $\tan \theta = \frac{y}{x} = \dfrac{\frac{y}{y}}{\frac{x}{y}} = \frac{1}{\frac{x}{y}} = \frac{1}{\cot \theta}$.

Similarly, $\cot \theta = \dfrac{x}{y} = \dfrac{\frac{x}{x}}{\frac{y}{x}} = \dfrac{1}{\frac{y}{x}} = \dfrac{1}{\tan \theta}$.

IMPORTANT NOTE: If θ is a quadrantal angle, then either x or y is zero. This means that either $\tan \theta$ or $\cot \theta$ is equal to zero. Since the reciprocal of zero is undefined, the fundamental identity does not hold for quadrantal angles. With this exception noted, we may now apply it freely.

EXAMPLE: Given $\cot \theta = \dfrac{-5}{3}$, find $\tan \theta$.

SOLUTION: $\tan \theta = \dfrac{1}{\cot \theta} = \dfrac{1}{\frac{-5}{3}} = \dfrac{-3}{5}$.

EXAMPLE: Find the tangent and cotangent of π.

SOLUTION: The coordinates of π on the unit circle are $(-1,0)$.

$\tan \pi = \dfrac{y}{x} = \dfrac{0}{-1} = 0$, $\cot \pi = \dfrac{x}{y} = \dfrac{-1}{0}$ undefined.

Returning to an earlier example we found $\tan 30^{\circ} = \dfrac{1}{\sqrt{3}}$ and $\cot 30^{\circ} = \sqrt{3}$.

Let us examine how the calculator and table are used to find approximate values.

CALCULATOR SOLUTION: Degree mode. Enter 30°. Press tan key. Read .5774 rounded to four decimal places. Hence, $\tan 30^{\circ}$ = .5774. To find the $\cot 30^{\circ}$, press $\dfrac{1}{x}$ key. Read 1.732 rounded to four significant figures. Note that by pressing the $\dfrac{1}{x}$ key, we are using the fundamental identity $\cot 30^{\circ} = \dfrac{1}{\tan 30^{\circ}}$.

TABLE II SOLUTION: Locate 30° in the left column. Read horizontally to the tangent and cotangent column entries. Read $\tan 30^{\circ}$ = .5774 and $\cot 30^{\circ}$ = 1.732.

EXAMPLE: Find $\tan 60^{\circ}$ and $\cot 60^{\circ}$.

TABLE II SOLUTION: This example follows closely on the last example because the table is structured to give complementary angle readings. We note that 60° appears in the right hand column of Table II in the same row as the left hand complementary angle 30^0. For the tangent and cotangent of 60^0 we read these columns from the bottom. The entries are tan 60^0 = 1.732 and cot 60^0 = .5774.

The complementary angle relationship can be generalized as follows:

If $0^0 < \theta < 90^0$, tan $(90^0 - \theta)$ = cot θ and cot $(90^0 - \theta)$ = tan θ.

EXAMPLE: Express tan 72^0 as the cotangent of an angle less than 45^0.

SOLUTION: tan 72^0 = cot $(90^0 - 72^0)$ = cot 18^0.

Check: By Table II, tan 72^0 = 3.078 = cot 18^0.

We shall now consider angles in quadrants other than the first.

EXAMPLE: Find tan $\frac{3\pi}{4}$ and cot $\frac{3\pi}{4}$.

SPECIAL ANGLE SOLUTION: The coordinates of $\frac{3\pi}{4}$ in the unit circle are $(\frac{-1}{\sqrt{2}} , \frac{1}{\sqrt{2}})$.

Hence, tan $\frac{3\pi}{4} = \frac{y}{x} = \frac{\frac{1}{\sqrt{2}}}{\frac{-1}{\sqrt{2}}} = \frac{1}{-1} = -1$

$$\cot \frac{3\pi}{4} = \frac{1}{\tan \frac{3\pi}{4}} = \frac{1}{-1} = -1$$

EXAMPLE: Find tangent and cotangent of 214^0 20'.

CALCULATOR SOLUTION: Degree mode. Enter 20. Divide by 60. Add 214. Press =. Press tan key. Read .6830. tan 214^0 20' = .6830. Press $\frac{1}{x}$ key. Read 1.464. cot 214^0 20' = 1.464.

TABLE II SOLUTION: Since $180^0 < 214^0$ 20' $< 270^0$, the angle is in the third quadrant. Draw a figure for the symmetry relationship (Figure 3-1C).

Figure 3-1C.

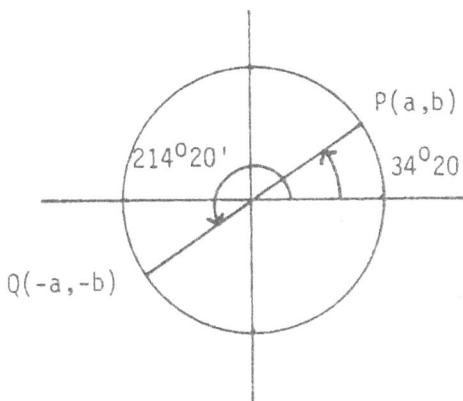

We note that if the coordinates of P are (a,b), the coordinates of Q are (-a,-b). Since the $\tan 214^{\circ} 20' = \frac{-b}{-a} = \frac{b}{a} = \tan 34^{\circ} 20'$, Table II gives the following results: $\tan 214^{\circ} 20' = \tan 34^{\circ} 20' = .6830$, and

$$\cot 214^{\circ} 20' = \cot 34^{\circ} 20' = 1.464.$$

For further simplicity, we shall summarize the signs (+ or -) of the tangent and cotangent ratios in each of the four quadrants.

Figure 3-1D.

Positive and Negative Values of the Tangent and Cotangent

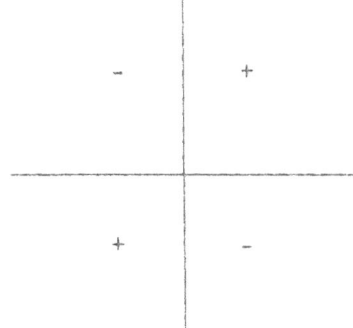

If the angle is given in radians, proceed as follows:

EXAMPLE: Find the tangent and cotangent of 5.47 .

SOLUTION BY CALCULATOR: Radian mode. Enter 5.47 . Press tan . Read -1.057.
Hence, tan 5.47 = -1.057. For the cotangent, enter 5.47. Press tan
key. Press 1/x key. Read -.9459. Therefore, cot 5.47 = -.9459.

TABLE I SOLUTION: 5.47 terminates in the fourth quadrant. By symmetry, the
first quadrant reference angle is $2\pi - 5.47 = 6.28 - 5.47 = 0.81$.
Reading Table I, tan 0.81 = 1.050 and cot 0.81 = .9520. Since the
tangent and cotangent in quadrant IV are negative, tan 5.47 = -1.050
and cot 5.47 = -.9520 .

Note again that calculator and table readings do not agree because
of the rounding off of 2π to 6.28 in the table solution.

Tangent and Cotangent of Special Angles

Given angles of 30° $(\frac{\pi}{6})$, 45° $(\frac{\pi}{4})$, and 60° $(\frac{\pi}{3})$ in standard position
on the unit circle, the coordinates of the points on the terminal sides of
these angles are $(\frac{\sqrt{3}}{2}, \frac{1}{2})$, $(\frac{1}{\sqrt{2}}, \frac{1}{\sqrt{2}})$, and $(\frac{1}{2}, \frac{\sqrt{3}}{2})$, respectively. By the definitions of the tangent and cotangent ratios,

$\tan 30^{\circ} = \frac{1}{2} / \frac{\sqrt{3}}{2} = \frac{1}{\sqrt{3}}$ $\cot 30^{\circ} = \frac{\sqrt{3}}{2} / \frac{1}{2} = \sqrt{3}$

$\tan 45^{\circ} = \frac{1}{\sqrt{2}} / \frac{1}{\sqrt{2}} = 1$ $\cot 45^{\circ} = \frac{1}{\sqrt{2}} / \frac{1}{\sqrt{2}} = 1$

$\tan 60^{\circ} = \frac{\sqrt{3}}{2} / \frac{1}{2} = \sqrt{3}$ $\cot 60^{\circ} = \frac{1}{2} / \frac{\sqrt{3}}{2} = \frac{1}{\sqrt{3}}$

Symmetry Reduction Formulas for Tangent and Cotangent

Taking into consideration the variation of signs for the x and y coordinates of the different quadrants, the following symmetry reduction formulas
apply:

If θ is given in degrees,

$\tan(180^\circ - \theta) = -\tan\theta$ $\cot(180^\circ - \theta) = -\cot\theta$

$\tan(180^\circ + \theta) = \tan\theta$ $\cot(180^\circ + \theta) = \cot\theta$

$\tan(360^\circ - \theta) = -\tan\theta$ $\cot(360^\circ - \theta) = -\cot\theta$

If θ is given in radians,

$\tan(\pi - \theta) = -\tan\theta$ $\cot(\pi - \theta) = -\cot\theta$

$\tan(\pi + \theta) = \tan\theta$ $\cot(\pi + \theta) = \cot\theta$

$\tan(2\pi - \theta) = -\tan\theta$ $\cot(2\pi - \theta) = -\cot\theta$

EXAMPLE: Find a) $\tan 225^\circ$ b) $\cot 300^\circ$ c) $\tan \frac{2\pi}{3}$ d) $\cot(-\frac{11\pi}{6})$

SOLUTION: a) 225° terminates in quadrant III.

$\tan 225^\circ = \tan(180^\circ + 45^\circ) = \tan 45^\circ = 1$

b) 300° terminates in quadrant IV.

$\cot 300^\circ = \cot(360^\circ - 60^\circ) = -\cot 60^\circ = -\frac{1}{\sqrt{3}}$

c) $\frac{2\pi}{3}$ terminates in quadrant II.

$\tan \frac{2\pi}{3} = \tan(\pi - \frac{\pi}{3}) = -\tan\frac{\pi}{3} = -\sqrt{3}$

d) $(-\frac{11\pi}{6})$ terminates in quadrant I, and is coterminal with $\frac{\pi}{6}$.

Therefore, $\cot(-\frac{11\pi}{6}) = \cot\frac{\pi}{6} = \sqrt{3}$

As an alternative solution to d) above, note that for all θ,

$\tan(-\theta) = -\tan\theta$ and $\cot(-\theta) = -\cot\theta$

Then $\cot(-\frac{11\pi}{6}) = -\cot\frac{11\pi}{6} = -\cot(2\pi - \frac{\pi}{6}) = -(-\cot\frac{\pi}{6}) = -(-\sqrt{3}) = \sqrt{3}$.

Finding the Angle When Tangent and Cotangent Values are Given

Tables I and II may be used directly to find an angle θ that corresponds to a given tangent or cotangent value.

EXAMPLE: If $0^\circ < \theta < 90^\circ$ and tan θ = 1.4725 find θ to the nearest ten minutes.

SOLUTION: Since 1.4725 > 1 = tan 45°, the location of the given number must be found by reading the TAN column upwards from the bottom of the page. The table entry nearest the given value is 1.4733 and this corresponds to the angle given in the right-hand column (reading upwards) of 55° 50' . Therefore to the nearest ten minutes $\theta = 55^\circ$ 50' .

CALCULATOR SOLUTION: Input 1.4725. Press INV key (or equivalent for your calculator such as ARC, f^{-1} or 2nd f followed by $\tan^{-1}$, etc.).* Press tan key. Read 55.81890159... . Convert decimal fraction to minutes by subtracting 55 then multiplying by 60. Round to the nearest ten minutes to obtain $\theta = 55^\circ$ 50' .

EXAMPLE: Find θ if cot θ = 1.6683 and $\pi < \theta < \frac{3\pi}{2}$.

SOLUTION: Table I . Locate 1.6683 in the Cot column reading down from the top. The radian entry on the left in the same row is 0.54. This is the reference angle. To find θ , add π . $\theta = \pi + 0.54 = 3.68$.

CALCULATOR SOLUTION: The reciprocal relationship between tan and cot must be used as well as the inverse functional keys used in the example above. Beginning with the radian mode, enter 1.6683. Press $\frac{1}{x}$ key. Read 0.5994 rounded. Press INV key or equivalent. Press tan key. Read 0.54 rounded. Add π . Read $\theta = 3.68$ as above.

---

* The significance of these calculator designations is explained in Chapter 5 . It is sufficient at this point to understand that these key functions answer the question: "What is the angle whose trigonometric ratio is given?" The process is the same as reading the tables in reverse.

EXERCISE SET 3-1.

In exercises 1 - 4 , the coordinates of a point are given on the terminal side of a central angle in standard position. Find the tangent and cotangent of the angle.

1) (5, -2)   2) (3.78,4.95)   3) $(-\frac{3}{5}, \frac{1}{4})$ 4) (-154,-33)

5) Find cot θ if tan $\theta = \frac{-5}{8}$ 6) Find tan θ if cot $\theta = \frac{4}{7}$

7) Find tan θ if cot θ = 2.647   8) Find cot θ if tan θ = 0.1376

9) Complete the table:

θ	$\frac{\pi}{2}$	π	$\frac{3\pi}{2}$	2π
tan θ	undefined			
cot θ	0			

10) Complete the table:

θ	$\frac{-\pi}{6}$	$\frac{-3\pi}{4}$	$\frac{-2\pi}{3}$	$\frac{-7\pi}{6}$
tan θ				
cot θ				

Express the following as the tangent of a positive acute angle in degrees:

Example: $\cot 473^0 = \cot (473^0 - 360^0) = \cot 113^0 = - \cot 67^0 = - \tan 23^0$

11) tan 225^0 12) tan 330^0 13) tan (-135^0) 14) tan 154^0

15) tan 475^0 16) cot 38^0 17) cot 76^0 18) tan (-14^0)

19) cot 150^0 20) cot 320^0

Express the following as the cotangent of a positive acute angle in radians:

Example: $\cot \frac{6\pi}{5} = \cot (\pi + \frac{\pi}{5}) = \cot \frac{\pi}{5}$.

21) $\cot \frac{5\pi}{4}$ 22) $\cot (\frac{-7\pi}{4})$ 23) $\cot (\frac{-3\pi}{8})$ 24) $\cot 1.85$

25) $\tan \frac{\pi}{5}$ 26) $\cot \frac{19\pi}{3}$ 27) $\tan (\frac{-5\pi}{6})$ 28) $\tan \frac{6\pi}{7}$

29) $\tan 34$ 30) $\cot 5.68$

Find the following using calculator or Table II:

31) $\tan 54^\circ \ 30'$ 32) $\cot 115^\circ \ 10'$ 33) $\tan (-415^\circ)$

34) $\tan (-142.54^\circ)$ 35) $\cot 578^\circ$ 36) $\tan 365^\circ \ 20'$

37) $\tan (-768^\circ \ 10')$ 38) $\cot (-285.63^\circ)$ 39) $\cot 1511^\circ \ 50'$

40) $\tan 92^\circ \ 50'$

Find the following using calculator or Table I .

41) $\cot 3.25$ 42) $\tan 7.56$ 43) $\tan \frac{5\pi}{8}$ 44) $\cot (\frac{-3\pi}{7})$

45) $\cot 13.8$ 46) $\tan (-3.47)$ 47) $\tan 95.6$ 48) $\cot 47.8\pi$

49) $\tan (-36.3\pi)$ 50) $\tan 1$

Use Table II to find θ, $0^\circ < \theta < 90^\circ$.

51) $\tan \theta = .1139$ 52) $\tan \theta = 4.511$ 53) $\cot \theta = 2.651$

54) $\cot \theta = .6088$ 55) $\tan \theta = 2.066$ · 56) $\cot \theta = 3.305$

Use Table I to find θ, $0 < \theta < \frac{\pi}{2}$.

57) $\cot \theta = 1.237$ 58) $\tan \theta = .2236$ 59) $\tan \theta = 10.98$

60) $\cot \theta = .3212$

Use Table II or calculator to find θ , where $90^\circ < \theta < 270^\circ$.

61) $\tan \theta = -.4592$ 62) $\tan \theta = -1.6107$ 63) $\tan \theta = 2.5605$

64) $\tan \theta = .1198$ 65) $\cot \theta = 8.5555$ 66) $\cot \theta = -3.5261$

67) $\cot \theta = -.2217$ 68) $\cot \theta = 0.6371$

3-2. <u>RELATIONSHIPS BETWEEN SINE, COSINE, TANGENT, AND COTANGENT</u>

Fundamental Identities:

$$\tan \theta = \frac{\sin \theta}{\cos \theta} , \quad \cot \theta = \frac{\cos \theta}{\sin \theta}$$

Proof: By definition, $\sin \theta = \frac{y}{r}$, $\cos \theta = \frac{x}{r}$

$$\frac{\sin \theta}{\cos \theta} = \frac{\frac{y}{r}}{\frac{x}{r}} = \frac{y}{r} \cdot \frac{r}{x} = \frac{y}{x} = \tan \theta$$

$$\frac{\cos \theta}{\sin \theta} = \frac{\frac{x}{r}}{\frac{y}{r}} = \frac{x}{r} \cdot \frac{r}{y} = \frac{x}{y} = \cot \theta$$

Another important identity can be derived from the analytic definition of the circle: $x^2 + y^2 = r^2$.

$$\text{If } x^2 + y^2 = r^2,$$

$$\frac{x^2}{r^2} + \frac{y^2}{r^2} = 1$$

$$\left(\frac{x}{r}\right)^2 + \left(\frac{y}{r}\right)^2 = 1$$

$$(\cos \theta)^2 + (\sin \theta)^2 = 1$$

We will refer to this identity as Pythagorean Identity I and use the following traditional notation for the squares of the cosine and sine:

$$\cos^2 \theta + \sin^2 \theta = 1.$$

We can now use Pythagorean Identity I and the quotient identities above in the following examples.

<u>EXAMPLE</u>: If $\tan \theta = \frac{3}{4}$ and $0 < \theta < 90^0$, find $\cot \theta$, $\sin \theta$, and $\cos \theta$.

<u>SOLUTION</u>: If $\tan \theta = \frac{3}{4}$, $\cot \theta = \frac{4}{3}$ by the reciprocal identity. Since

$\tan \theta = \frac{y}{x} = \frac{3}{4}$, let $y = 3$ and $x = 4$. Substituting in $x^2 + y^2 = r^2$,

$(4)^2 + (3)^2 = 16 + 9 = 25 = r^2$. Hence, $r = 5$.

$$\cos \theta = \frac{x}{r} = \frac{4}{5} \ , \ \sin \theta = \frac{y}{r} = \frac{3}{5} \ .$$

EXAMPLE: If $\sin \theta = \frac{2}{7}$, $0 < \theta < \frac{\pi}{2}$, find $\cos \theta$, $\tan \theta$, $\cot \theta$.

SOLUTION: By Pythagorean Identity I, $\sin^2 \theta + \cos^2 \theta = 1$, or

$$\cos^2 \theta = 1 - \sin^2 \theta = 1 - \left(\frac{2}{7}\right)^2 = 1 - \frac{4}{49} = \frac{49-4}{49} = \frac{45}{49} \ .$$

Therefore, $\cos \theta = \sqrt{\frac{45}{49}} = \frac{\sqrt{45}}{7} = \frac{3\sqrt{5}}{7} \ .$

$$\tan \theta = \frac{\sin \theta}{\cos \theta} = \frac{\frac{2}{7}}{\frac{3\sqrt{5}}{7}} = \frac{2}{3\sqrt{5}}$$

$$\cot \theta = \frac{\cos \theta}{\sin \theta} = \frac{\frac{3\sqrt{5}}{7}}{\frac{2}{7}} = \frac{3\sqrt{5}}{2}$$

EXAMPLE: If $\cot \theta = -5$ and $90^0 < \theta < 180^0$, find $\sin \theta$, $\cos \theta$, $\tan \theta$.

SOLUTION: $\tan \theta = \frac{1}{\cot \theta} = \frac{-1}{5}$

In the second quadrant, x is negative and y is positive. Since
$\tan \theta = \frac{y}{x}$, let $y = 1$ and $x = -5$. Then substituting in $x^2 + y^2 = r^2$,

$(-5)^2 + (1)^2 = r^2$, $25 + 1 = 26 = r^2$. $r = \sqrt{26}$.

Hence, $\cos \theta = \frac{x}{r} = \frac{-5}{\sqrt{26}}$, $\sin \theta = \frac{y}{r} = \frac{1}{\sqrt{26}}$.

EXAMPLE: If $\cos \theta = \frac{-5}{13}$ and $\pi < \theta < \frac{3\pi}{2}$, find $\sin \theta$, $\tan \theta$, $\cot \theta$.

SOLUTION: Since θ is in the third quadrant, $\sin \theta < 0$, $\tan \theta > 0$, and $\cot \theta > 0$.

By the Pythagorean Identity I, $\sin^2 \theta = 1 - \cos^2 \theta = 1 - \left(\frac{-5}{13}\right)^2$

$= 1 - \frac{25}{169} = \frac{169-25}{169} = \frac{144}{169}$. Taking square roots,

$\sin \theta = \pm \sqrt{\frac{144}{169}} = \pm \frac{12}{13}$. We select $\sin \theta = \frac{-12}{13}$ for the third quadrant.

By the quotient identity, $\tan \theta = \dfrac{\sin \theta}{\cos \theta} = \dfrac{\frac{-12}{13}}{\frac{-5}{13}} = \dfrac{12}{5}$

By the reciprocal identity, $\cot \theta = \dfrac{1}{\tan \theta} = \dfrac{5}{12}$.

EXAMPLE: Find θ in radians for the previous example.

SOLUTION: Since $\cos \theta = \dfrac{-5}{13}$, we can look up the first quadrant angle in

Table I that has a cosine nearest the decimal equivalent of $\dfrac{5}{13}$,

then use symmetry with respect to the origin to find θ.

$\dfrac{5}{13} = .3846$ to four significant figures. We search the cosine column

for this value and find the closest entry .3809. The angle corres-

ponding to this cosine value is 1.18 radians. To find θ in the third

quadrant, add this value to π or 3.14. We have $\theta = 1.18 + 3.14 = 4.32$.

CALCULATOR SOLUTION: Radian mode. Divide 5 by 13. Read .3846. Press INV or

ARC key. Press cos key. Read 1.18 rounded. Add 3.14. Read 4.32.

The identities we have studied will often appear in alternate forms. Some
of these are as follows:

From $\tan \theta = \dfrac{1}{\cot \theta}$, $\tan \theta \cot \theta = 1$

From $\tan \theta = \dfrac{\sin \theta}{\cos \theta}$, $\tan \theta \cos \theta = \sin \theta$

From $\cot \theta = \dfrac{\cos \theta}{\sin \theta}$, $\cot \theta \sin \theta = \cos \theta$

From $\cos^2 \theta + \sin^2 \theta = 1$, $\cos^2 \theta = 1 - \sin^2 \theta$ and

$$\sin^2 \theta = 1 - \cos^2 \theta.$$

In later applications of trigonometry to advanced mathematics it is
important to have facility in working with the fundamental identities. This
facility is gained by an exercise described as proving identities. It employs
the student's knowledge of the fundamental identities to derive other relationships and in addition, draws upon skills developed in algebra. In this

section we shall begin to develop the technique and carry it through the remainder of this text.

EXAMPLE: Prove $\cos^2 \theta \sin^2 \theta + \cos^4 \theta = \cos^2 \theta$.

The approach we shall use is to simplify the more complex left member of the identity to reduce it to be identical with the right member, $\cos^2 \theta$.

PROOF:
$$\cos^2 \theta \sin^2 \theta + \cos^4 \theta = \cos^2 \theta \, (\sin^2 \theta + \cos^2 \theta) \quad \text{factoring}$$
$$= \cos^2 \theta \, (1) \quad \text{Pythagorean Identity I}$$
$$= \cos^2 \theta \quad \text{axiom for real numbers.}$$

EXAMPLE: Prove $\sin^2 \theta \, (1 + \cot^2 \theta) = 1$

PROOF:
$$\sin^2 \theta \, (1 + \cot^2 \theta) = \sin^2 \theta \, (1 + \frac{\cos^2 \theta}{\sin^2 \theta}) \quad \text{quotient identity}$$
$$= \sin^2 \theta \, (\frac{\sin^2 \theta}{\sin^2 \theta} + \frac{\cos^2 \theta}{\sin^2 \theta}) \quad \text{common denominator}$$
$$= \sin^2 \theta \, (\frac{\sin^2 \theta + \cos^2 \theta}{\sin^2 \theta}) \quad \text{common denominator}$$
$$= \sin^2 \theta \, (\frac{1}{\sin^2 \theta}) \quad \text{Pythagorean Identity I}$$
$$= 1 \quad \text{axiom for real numbers.}$$

EXAMPLE: Prove $\cot \theta + \tan \theta = \dfrac{1}{\cos \theta \sin \theta}$

PROOF:
$$\cot \theta + \tan \theta = \frac{\cos \theta}{\sin \theta} + \frac{\sin \theta}{\cos \theta} \quad \text{quotient identities}$$
$$= \frac{\cos^2 \theta + \sin^2 \theta}{\sin \theta \cos \theta} \quad \text{common denominator and single fraction}$$
$$= \frac{1}{\sin \theta \cos \theta} \quad \text{Pythagorean Identity I}$$

EXAMPLE: Prove $\dfrac{1 + \sin \theta - \cos^2 \theta}{\cos \theta \, (1 + \sin \theta)} = \tan \theta$

PROOF:
$$\frac{1 + \sin \theta - \cos^2 \theta}{\cos \theta \, (1 + \sin \theta)} = \frac{1 - \cos^2 \theta + \sin \theta}{\cos \theta \, (1 + \sin \theta)} \quad \text{axioms of algebra}$$
$$= \frac{\sin^2 \theta + \sin \theta}{\cos \theta \, (1 + \sin \theta)} \quad \text{Pythagorean Identity I}$$

$$= \frac{\sin \theta (\sin \theta + 1)}{\cos \theta (1 + \sin \theta)} \qquad \text{factoring}$$

$$= \frac{\sin \theta}{\cos \theta} \cdot 1 \qquad \text{dividing common factors}$$

$$= \frac{\sin \theta}{\cos \theta} \qquad \text{axiom}$$

$$= \tan \theta \qquad \text{quotient identity}$$

In working on these proofs the student should keep a summary sheet of the fundamental identities for handy reference. This is one area where practice does wonders.

EXERCISE SET 3-2.

For exercises 1 - 8, a trigonometric ratio and a quadrant is given. Find the remaining unknown ratios for the sine, cosine, tangent, and cotangent of the angle. Exact values.

1) $\cot \theta = \frac{2}{3}$, Quadrant I

2) $\sin \theta = \frac{-5}{9}$, Quadrant IV

3) $\tan \theta = \frac{7}{24}$, Quadrant III

4) $\cos \theta = \frac{-3}{5}$, Quadrant II

5) $\sin \theta = \frac{8}{17}$, Quadrant II

6) $\tan \theta = \frac{-15}{17}$, Quadrant IV

7) $\cos \theta = \frac{2}{\sqrt{5}}$, Quadrant I

8) $\cot \theta = \frac{3}{4}$, Quadrant III

In exercises 9 - 16, find the value of θ, $0 < \theta < 360°$, to the nearest degree for the angles whose trigonometric ratios are given in exercises 1 - 8.

Prove the following identities:

17) $\tan \theta \cot \theta + \cos \theta \tan \theta = 1 + \sin \theta$

18) $\cos^2 \theta (1 + \tan^2 \theta) = 1$

19) $\frac{1 - \sin^2 \theta}{\cos^2 \theta} = 1$

20) $\frac{1 - \sin \theta}{\cos \theta} = \frac{\cos \theta}{1 + \sin \theta}$

21) $\frac{\cos \theta}{1 - \sin \theta} - \tan \theta = \frac{1}{\cos \theta}$

22) $\frac{1}{\sin \theta} - \cot \theta = \frac{\sin \theta}{1 + \cos \theta}$

23) $\frac{1 + \cot^2 \theta}{1 + \tan^2 \theta} = \cot^2 \theta$

24) $(\sin^2 \theta - 1)(\cot^2 \theta + 1) = -\cot^2 \theta$.

25) As θ increases from 0 to $\frac{\pi}{2}$, what happens to tan θ?

26) As θ increases from 0 to $\frac{\pi}{2}$, what happens to cot θ?

TRUE OR FALSE:

27) $\tan 120^0 < \tan 150^0$

28) $\cot 140^0 < \tan 180^0$

29) $\tan 210^0 < \cot 210^0$

30) $\tan 240^0 < \tan 330^0$

31) $\cot 300^0 < \cot 360^0$

32) $\tan 45^0 < \tan 225^0$

33) If $0 < \theta < 90^0$, $\sin \theta < \tan \theta$.

34) As θ increases from 0 to π, cot θ is always decreasing.

35) $\frac{\sin (90^0 - \theta)}{\cos (90^0 - \theta)} = \cot \theta, \quad 0 < \theta < 90^0$

36) $\frac{\cos (90^0 - \theta)}{\sin (90^0 - \theta)} = \cot \theta, \quad 0 < \theta < 90^0$

3-3. COMPLEMENTARY RELATIONSHIPS BETWEEN SINE AND COSINE, TANGENT AND COTANGENT.

We have already noted in using Table II that $\cos \theta = \sin (90^O - \theta)$ and $\tan \theta = \cot (90^O - \theta)$. For example, $\cos 57^O = \sin 33^O = .5446$ and $\tan 74^O = \cot 16^O = 3.487$. These are examples of complementary relationships. In this section we shall see how these relationships can be extended to the other quadrants. We shall prove the following geometric theorem (Figure 3-3A):

Figure 3-3A

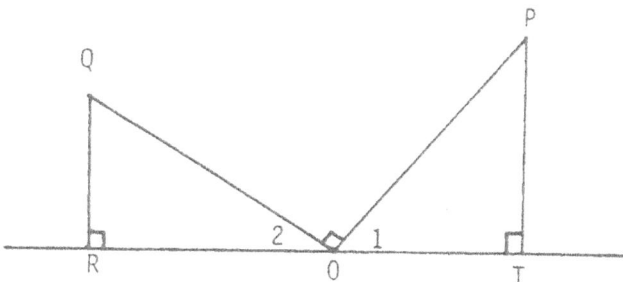

Given right triangles OTP and ORQ with OP = OQ and right angle POQ as in Figure 3-3A, then RQ = OT and RO = TP.

PROOF	REASONS
1) $\angle$ POQ is a right angle.	1) Given
2) $\angle$ 1 + $\angle$ 2 = right angle	2) $\angle$ 1 + $\angle$ 2 forms a linear pair with $\angle$ POQ
3) $\angle$ 1 + $\angle$ OPT = right angle	3) The acute angles of a right triangle are complementary
4) $\angle$ OPT = $\angle$ 2, $\angle$ OQR = $\angle$ 1	4) Complements of the same angle are equal.
5) OP = OQ	5) Given
6) $\triangle$OTP $\cong$ $\triangle$ORQ	6) ASA congruence theorem
7) RQ = OT and RO = TP	7) Corresponding sides of congruent triangles are congruent.

Transferring this result to the xy-coordinate system (Figure 3-3B), if the coordinates of P are (x,y), it must follow that the coordinates of Q are (-y,x). Hence, the following formulas must be true:

$$\cos (90^{O} + \theta) = - \sin \theta$$

$$\sin (90^{O} + \theta) = \cos \theta$$

$$\tan (90^{O} + \theta) = - \cot \theta$$

$$\cot (90^{O} + \theta) = - \tan \theta$$

Figure 3-3B

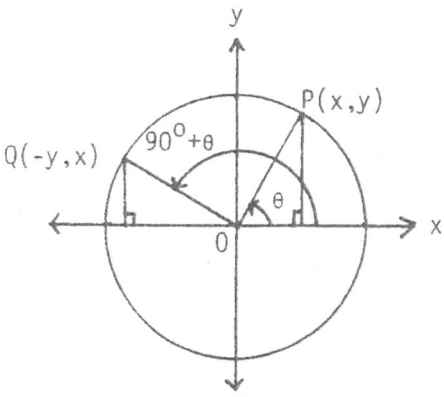

PROOF: $\cos (90^{O} + \theta) = \frac{-y}{r} = - (\frac{y}{r}) = - \sin \theta$

$\sin (90^{O} + \theta) = \frac{x}{r} = \cos \theta$

$\tan (90^{O} + \theta) = \frac{x}{-y} = - (\frac{x}{y}) = - \cot \theta$

$\cot (90^{O} + \theta) = \frac{-y}{x} = - (\frac{y}{x}) = - \tan \theta$

EXAMPLE: Express $\sin 105^{O}$ as the cosine of an angle less than 45°.

SOLUTION: $\sin 105^{O} = \sin (90^{O} + 15^{O}) = \cos 15^{O}$.

EXAMPLE: Express $\cot 155^{O}$ as the tangent of an acute angle.

SOLUTION: $\cot 155^{O} = \cot (90^{O} + 65^{O}) = - \tan 65^{O}$.

In the third and fourth quadrants, the complementary relationship is expressed in terms of the 270° quadrantal angle.

Figure 3-3C

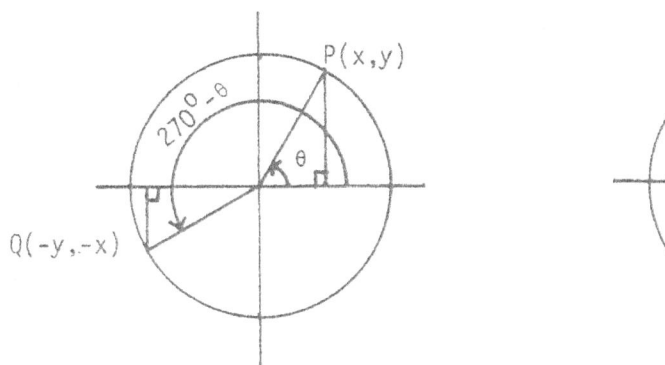 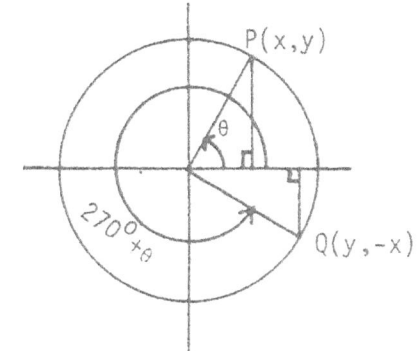

The complementary relationships are as follows:

$$\cos (270^\circ - \theta) = - \sin \theta \qquad \cos (270^\circ + \theta) = \sin \theta$$

$$\sin (270^\circ - \theta) = - \cos \theta \qquad \sin (270^\circ + \theta) = - \cos \theta$$

$$\tan (270^\circ - \theta) = \cot \theta \qquad \tan (270^\circ + \theta) = - \cot \theta$$

$$\cot (270^\circ - \theta) = \tan \theta \qquad \cot (270^\circ + \theta) = - \tan \theta$$

EXAMPLE: Prove $\cot (270^\circ - \theta) = \tan \theta$

SOLUTION: $\cot (270^\circ - \theta) = \dfrac{-y}{-x} = \dfrac{y}{x} = \tan \theta$.

EXAMPLE: Express $\cos 325^\circ$ as the sine of an acute angle.

SOLUTION: $\cos 325^\circ = \cos (270^\circ + 55^\circ) = \sin 55^\circ$.

The complementary relationships can now be joined with the reduction formulas derived earlier to express any trigonometric ratio in terms of angles between 0° and 45°:

EXAMPLE: Express $\cot 175^\circ$, $\sin (-250^\circ)$, $\tan 410^\circ$, and $\cos 645^\circ$ as trigonometric ratios of angles between 0° and 45°.

SOLUTION: $\cot 175^\circ = \cot (180^\circ - 5^\circ) = - \cot 5^\circ$

$\sin (-250^\circ) = - \sin 250^\circ = - \sin (270^\circ - 20^\circ) = - (- \cos 20^\circ) = \cos 20^\circ$

$\tan 410^\circ = \tan (360^\circ + 50^\circ) = \tan 50^\circ = \cot 40^\circ$

$\cos 645^\circ = \cos (360^\circ + 285^\circ) = \cos 285^\circ = \cos (270 + 15^\circ) = \sin 15^\circ$.

The formulas may also be applied to radian measurement.

EXAMPLE: Express cot $(\frac{-8\pi}{5})$ as a trigonometric ratio between 0 and $\frac{\pi}{4}$.

SOLUTION:

STEP	REASON
1) cot $(\frac{-8\pi}{5})$ = - cot $\frac{8\pi}{5}$.	1) cot $(-\theta)$ = - cot θ
2) $\frac{8\pi}{5}$ = 5.03. So $\frac{8\pi}{5}$ is in Quadrant IV	2) $\frac{3\pi}{2}$ < 5.03 < 2π
3) $\frac{8\pi}{5}$ is closer to $\frac{3\pi}{2}$ than to 2π	3) $\frac{8\pi}{5}$ - $\frac{3\pi}{2}$ = $\frac{\pi}{10}$
	2π - $\frac{8\pi}{5}$ = $\frac{2\pi}{5}$ = $\frac{4\pi}{10}$
4) cot $(\frac{8\pi}{5})$ = cot $(\frac{3\pi}{2} + \frac{\pi}{10})$ = - tan $\frac{\pi}{10}$	4) cot $(\frac{3\pi}{2} + \theta)$ = - tan θ
5) cot $(\frac{-8\pi}{5})$ = - cot $\frac{8\pi}{5}$ = - (- tan $\frac{\pi}{10}$)	5) Steps 1, 4
= tan $\frac{\pi}{10}$	

EXAMPLE: Express cos 71.34 as a trigonometric ratio between 0 and $\frac{\pi}{4}$.

SOLUTION:

STEP	REASON
1) 71.34 ÷ 2π = 11.35	1) Number of revolutions contained in 71.34
2) 11.35 - 11 = .35	2) Subtract complete revolutions.
3) .35 x 2π = 2.20	3) Angle between 0 and 2π that is coterminal with 71.34.
4) 2.20 is in Quadrant II	4) $\frac{\pi}{2}$ < 2.20 < π or 1.57 < 2.20 < 3.14
5) 2.20 is closer to 1.57 or $\frac{\pi}{2}$	5) 2.20 - 1.57 = 0.63 3.14 - 2.20 = 0.94
6) cos 71.34 = cos 2.20	6) coterminal angles
7) cos 2.20 = cos (1.57 + 0.63) = - sin 0.63	7) cos $(\frac{\pi}{2} + \theta)$ = - sin θ

ALTERNATIVE SOLUTION:

Follow steps 1 - 4 above to find cos 71.34 = cos 2.20 .

5. cos 2.20 = cos (π - .94) = - cos .94   Symmetry reduction formula

6. - cos .94 = - (cos $\frac{\pi}{2}$ - .63)

    = - sin .63   Complementary reduction formula

The application of step 6 requires the student to recognize that

.94 > $\frac{\pi}{4}$ = .79 rounded. However, the advantage lies in only having to

remember the complementary relationships for the first quadrant.

Another example will reinforce this alternative method.

EXAMPLE: Express tan 3849.8 as a trigonometric ratio between 0 and $\frac{\pi}{4}$.

SOLUTION: 1. 3849.8 $\div$ 2π = 612.71 revolutions.

2. .71 X 2π = 4.49 , the angle coterminal with 3849.8 .

3. tan 3849.8 = tan 4.49 = tan (π + 1.35) = tan 1.35 by symmetry.

4. 1.35 > $\frac{\pi}{4}$. Therefore the complementary relationship must be

used.

5. tan 1.35 = tan ($\frac{\pi}{2}$ - .22) = cot .22 .

6. tan 3849.8 = cot .22 as required.

The following table summarizes the complementary relationships.

Angle (degrees)	$90^0 - \theta$	$90^0 + \theta$	$270^0 - \theta$	$270^0 + \theta$
Angle (radians)	$\frac{\pi}{2} - \theta$	$\frac{\pi}{2} + \theta$	$\frac{3\pi}{2} - \theta$	$\frac{3\pi}{2} + \theta$
sin	$\cos \theta$	$\cos \theta$	$-\cos \theta$	$-\cos \theta$
cos	$\sin \theta$	$-\sin \theta$	$-\sin \theta$	$\sin \theta$
tan	$\cot \theta$	$-\cot \theta$	$\cot \theta$	$-\cot \theta$
cot	$\tan \theta$	$-\tan \theta$	$\tan \theta$	$-\tan \theta$

EXERCISE SET 3-3.

Answer the following as True or False:

1) $\sin 35^0 = \cos 55^0$ 2) $\sin (-75^0) = \cos 15^0$ 3) $\cos 112^0 = - \sin 68^0$

4) $\tan (-58^0) = - \cot 42^0$ 5) $\cot (-148^0) = \tan 58^0$

6) $\sin 213^0 = - \cos 57^0$ 7) $\cos (-332^0) = \sin 62^0$ 8) $\tan 154^0 = \cot 26^0$

9) $\sin 114^0 = \sin 24^0$ 10) $\cot (-427^0) = \tan 23^0$ 11) $\cot \frac{\pi}{3} = \tan \frac{\pi}{6}$

12) $\cos \frac{2\pi}{3} = - \sin \frac{\pi}{6}$ 13) $\sin \frac{3\pi}{4} = - \cos \frac{\pi}{4}$ 14) $\tan \frac{5\pi}{3} = \cot \frac{2\pi}{3}$

15) $\tan \frac{5\pi}{6} = - \cot \frac{\pi}{3}$ 16) $\sin \frac{\pi}{2} = - \cot \pi$ 17) $\tan (\frac{-3\pi}{4}) = \cot (\frac{-\pi}{4})$

18) $\cos (\frac{-5\pi}{6}) = - \sin \frac{2\pi}{3}$ 19) $\tan (\frac{-11\pi}{6}) = - \cot \frac{\pi}{6}$ 20) $\sin \frac{\pi}{2} = \cot \frac{5\pi}{4}$

Express the following as trigonometric ratios of angles between 0^0 and 45^0.

21) $\cos 54^0$ 22) $\sin 158^0$ 23) $\cos (-11^0)$ 24) $\cot (-113^0)$

25) $\sin (-248^0)$ 26) $\tan 583^0$ 27) $\tan (-85^0)$ 28) $\cos 302^0$

29) $\sin 219^0$ 30) $\cot 785^0$

Express the following as trigonometric ratios of angles between 0 and $\frac{\pi}{4}$:

31) $\tan \frac{5\pi}{8}$ 32) $\cot \frac{7\pi}{9}$ 33) $\cos \frac{17\pi}{7}$ 34) $\sin \frac{15\pi}{4}$

35) $\cot (\frac{-17\pi}{5})$ 36) $\tan (\frac{-5\pi}{2})$ 37) $\cos 2$ 38) $\sin (-4.37)$

39) $\cot 9.83$ 40) $\tan (-7.64)$

3-4. SECANT AND COSECANT. MORE IDENTITIES.

Given a point P(x,y) on a circle with center at the origin, radius r, and central angle θ in standard position, there are two more trigonometric ratios that can be defined:

$$\text{secant } \theta \text{ (abbreviated "sec } \theta\text{")} = \frac{r}{x}$$

$$\text{cosecant } \theta \text{ (abbreviated "csc } \theta\text{")} = \frac{r}{y}$$

From this definition we see that the secant is the reciprocal of the cosine and the cosecant is the reciprocal of the sine.

$$\sec \theta = \frac{1}{\cos \theta} \; , \; \csc \theta = \frac{1}{\sin \theta}$$

In addition to these two fundamental reciprocal identities, we can derive two more additional Pythagorean identities:

$$\tan^2 \theta + 1 = \sec^2 \theta$$

$$\text{and } \cot^2 \theta + 1 = \csc^2 \theta$$

<u>PROOF</u>: Given the equation for the circle $x^2 + y^2 = r^2$, we divide by x^2 first.

$$\frac{x^2}{x^2} + \frac{y^2}{x^2} = \frac{r^2}{x^2} \quad , \text{ if } x \neq 0$$

$$1 + \left(\frac{y}{x}\right)^2 = \left(\frac{r}{x}\right)^2$$

$$1 + \tan^2 \theta = \sec^2 \theta$$

Next, we divide $x^2 + y^2 = r^2$ by y^2 to obtain:

$$\frac{x^2}{y^2} + \frac{y^2}{y^2} = \frac{r^2}{y^2} \quad , \text{ if } y \neq 0$$

$$\left(\frac{x}{y}\right)^2 + 1 = \left(\frac{r}{y}\right)^2$$

$$\cot^2 \theta + 1 = \csc^2 \theta$$

We note that sec θ is undefined if x = 0 and csc θ is undefined if y = 0. For the secant this occurs for θ at odd multiples of $\frac{\pi}{2}$ and for the cosecant it occurs for θ at multiples of π. Thus sec $(\pm\frac{\pi}{2})$, sec $(\pm\frac{3\pi}{2})$, etc., are undefined and csc 0, csc $(\pm\pi)$, csc $(\pm2\pi)$, etc., are undefined.

Readings for values of the secant and cosecant using Table I or Table II are found as before for the other trigonometric ratios. To use the calculator for these ratios, the reciprocal identities are applied.

EXAMPLE: Find sec 54^0.

CALCULATOR SOLUTION: Input 54^0. Press cos key. Press $\frac{1}{x}$ key.
    Read sec 54^0 = 1.701.

EXAMPLE: Find csc 35^0.

CALCULATOR SOLUTION: Input 35^0. Press sin key. Press $\frac{1}{x}$ key.
    Read csc 35^0 = 1.743.

To find θ when sec θ or csc θ are given using the calculator, we proceed as follows:

EXAMPLE: Find θ, 0^0 < θ < 90^0, if sec θ = 2.854.

CALCULATOR SOLUTION: Input 2.854. Press $\frac{1}{x}$ key. Press INV key or ARC key.
    Press cos key. Read 69.49^0.

EXAMPLE: Find θ, $0 < \theta < \frac{\pi}{2}$, if csc θ = 3.712.

CALCULATOR SOLUTION: Radian mode. Input 3.712. Press $\frac{1}{x}$ key. Press INV key
    or ARC key. Press sin key. Read .27. Hence θ = .27 radians.

We note that since $|r| \geq |x|$ and $|r| \geq |y|$, then $|sec\ \theta| \geq 1$ and $|csc\ \theta| \geq 1$. Since reciprocals have the same polarity (i. e., both are positive or both are negative), the secant, like the cosine, is positive in the first and fourth quadrants and negative in the second and third quadrants. Similarly, the cosecant, like the sine, is positive in the first and second quadrants and negative in the third and fourth quadrants. The signs of all of the reciprocal pairs are illustrated in Figure 3-4A.

Figure 3-4A.

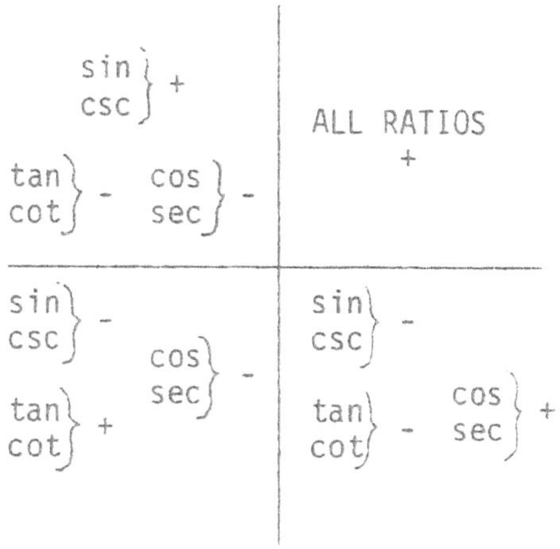

$$
\begin{array}{c|c}
\left.\begin{array}{l} \sin \\ \csc \end{array}\right\} + & \\
\left.\begin{array}{l} \tan \\ \cot \end{array}\right\} - \quad \left.\begin{array}{l} \cos \\ \sec \end{array}\right\} - & \begin{array}{c} \text{ALL RATIOS} \\ + \end{array} \\
\hline
\left.\begin{array}{l} \sin \\ \csc \end{array}\right\} - & \left.\begin{array}{l} \sin \\ \csc \end{array}\right\} - \\
\left.\begin{array}{l} \tan \\ \cot \end{array}\right\} + \quad \left.\begin{array}{l} \cos \\ \sec \end{array}\right\} - & \left.\begin{array}{l} \tan \\ \cot \end{array}\right\} - \quad \left.\begin{array}{l} \cos \\ \sec \end{array}\right\} +
\end{array}
$$

The sequence of positive ratios in the four quadrants can be memorized with the help of a mnemonic device. The sentence, "All Students Take Calculus" helps us to remember that in quadrant I all the ratios are positive. In quadrant II, the sine and its reciprocal are positive. In quadrant III, the tangent and its reciprocal are positive. Finally, in quadrant IV, the cosine and its reciprocal are positive.

Like their reciprocals, the secant and cosecant have a complementary relationship. That is,

$$\sec (90^0 - \theta) = \csc \theta, \qquad \csc (90^0 - \theta) = \sec \theta$$
$$\sec (90^0 + \theta) = - \csc \theta, \qquad \csc (90^0 + \theta) = \sec \theta$$
$$\sec (270^0 - \theta) = - \csc \theta, \qquad \csc (270^0 - \theta) = - \sec \theta$$
$$\sec (270^0 + \theta) = \csc \theta, \qquad \csc (270^0 + \theta) = - \sec \theta$$

The other reduction formulas hold as well:

$$\sec (180^0 - \theta) = - \sec \theta, \qquad \csc (180^0 - \theta) = \csc \theta$$
$$\sec (180^0 + \theta) = - \sec \theta, \qquad \csc (180^0 + \theta) = - \csc \theta$$
$$\sec (360^0 - \theta) = \sec \theta, \qquad \csc (360^0 - \theta) = - \csc \theta$$
$$\sec (- \theta) = \sec \theta, \qquad \csc (-\theta) = - \csc \theta$$

EXAMPLE: Express the following as a trigonometric ratio of an angle θ between 0^0 and 45^0 : sec 115^0, sec 328^0, csc (-58^0), csc 623^0.

SOLUTION: sec 115^0 = sec $(90^0 + 25^0)$ = $-$ csc 25^0

sec 328^0 = sec $(360^0 - 32^0)$ = sec 32^0

csc (-58^0) = $-$ csc 58^0 = $-$ csc $(90^0 - 32^0)$ = $-$ sec 32^0

csc 623^0 = csc $(360^0 + 263^0)$ = csc 263^0 = csc $(270^0 - 7^0)$ = $-$ sec 7^0.

The special angle values for the secant and cosecant are best remembered as reciprocals. These are tabulated as follows:

θ (radians)	0	$\frac{\pi}{2}$	π	$\frac{3\pi}{2}$	2π	$\frac{\pi}{6}$	$\frac{5\pi}{6}$	$\frac{7\pi}{6}$	$\frac{11\pi}{6}$
θ (degrees)	0	90^0	180^0	270^0	360^0	30^0	150^0	210^0	330^0
cosine	1	0	-1	0	1	$\frac{\sqrt{3}}{2}$	$\frac{-\sqrt{3}}{2}$	$\frac{-\sqrt{3}}{2}$	$\frac{\sqrt{3}}{2}$
secant	1	–	-1	–	1	$\frac{2}{\sqrt{3}}$	$\frac{-2}{\sqrt{3}}$	$\frac{-2}{\sqrt{3}}$	$\frac{2}{\sqrt{3}}$
sine	0	1	0	-1	0	$\frac{1}{2}$	$\frac{1}{2}$	$\frac{-1}{2}$	$\frac{-1}{2}$
cosecant	–	1	–	-1	–	2	2	-2	-2

θ (radians)	$\frac{\pi}{4}$	$\frac{3\pi}{4}$	$\frac{5\pi}{4}$	$\frac{7\pi}{4}$	$\frac{\pi}{3}$	$\frac{2\pi}{3}$	$\frac{4\pi}{3}$	$\frac{5\pi}{3}$
θ (degrees)	45^0	135^0	225^0	315^0	60^0	120^0	240^0	300^0
cosine	$\frac{1}{\sqrt{2}}$	$\frac{-1}{\sqrt{2}}$	$\frac{-1}{\sqrt{2}}$	$\frac{1}{\sqrt{2}}$	$\frac{1}{2}$	$\frac{-1}{2}$	$\frac{-1}{2}$	$\frac{1}{2}$
secant	$\sqrt{2}$	$-\sqrt{2}$	$-\sqrt{2}$	$\sqrt{2}$	2	-2	-2	2
sine	$\frac{1}{\sqrt{2}}$	$\frac{1}{\sqrt{2}}$	$\frac{-1}{\sqrt{2}}$	$\frac{-1}{\sqrt{2}}$	$\frac{\sqrt{3}}{2}$	$\frac{\sqrt{3}}{2}$	$\frac{-\sqrt{3}}{2}$	$\frac{-\sqrt{3}}{2}$
cosecant	$\sqrt{2}$	$\sqrt{2}$	$-\sqrt{2}$	$-\sqrt{2}$	$\frac{2}{\sqrt{3}}$	$\frac{2}{\sqrt{3}}$	$\frac{-2}{\sqrt{3}}$	$\frac{-2}{\sqrt{3}}$

EXAMPLE: Find <u>exact</u> values for the following:

$$\sec \left(\frac{\pi}{4} - \pi\right), \; \csc \frac{\pi}{3} + \csc \frac{\pi}{6} \; , \; \sec^2 \frac{\pi}{6} + \csc^2 \frac{\pi}{6}$$

SOLUTION:

$$\sec \left(\frac{\pi}{4} - \pi\right) = \sec \left(\frac{-3\pi}{4}\right) = \sec \frac{3\pi}{4} = -\sqrt{2}$$

$$\csc \frac{\pi}{3} + \csc \frac{\pi}{6} = \frac{2}{\sqrt{3}} + 2 = \frac{2\sqrt{3}}{3} + \frac{6}{3} = \frac{2\sqrt{3} + 6}{3}$$

$$\sec^2 \frac{\pi}{6} + \csc^2 \frac{\pi}{6} = \left(\frac{2}{\sqrt{3}}\right)^2 + (2)^2 = \frac{4}{3} + 4 = \frac{16}{3}$$

EXAMPLE: A point P (-4,7) is given on the circumference of a circle. Find the six trigonometric ratios of the central angle in standard position determined by the radius to the given point.

SOLUTION: The radius of the circle is found by $(-4)^2 + (7)^2 = r^2 = 16 + 49 = 65$ Therefore, $r = \sqrt{65}$.

$$\cos \theta = \frac{x}{r} = \frac{-4}{\sqrt{65}} \; , \; \sin \theta = \frac{y}{r} = \frac{7}{\sqrt{65}} \; , \; \tan \theta = \frac{y}{x} = \frac{7}{-4} = \frac{-7}{4}$$

$$\cot \theta = \frac{x}{y} = \frac{-4}{7} \; , \; \sec \theta = \frac{r}{x} = \frac{\sqrt{65}}{-4} = \frac{-\sqrt{65}}{4} \; , \; \csc \theta = \frac{r}{y} = \frac{\sqrt{65}}{7}$$

THE EIGHT FUNDAMENTAL IDENTITIES.

We can now summarize the fundamental identities that link together the six trigonometric ratios:

A. The Reciprocal Identities

$$\cot \theta = \frac{1}{\tan \theta} \; , \; \sec \theta = \frac{1}{\cos \theta} \; , \; \csc \theta = \frac{1}{\sin \theta}$$

B. The Quotient Identities

$$\tan \theta = \frac{\sin \theta}{\cos \theta} \; , \; \cot \theta = \frac{\cos \theta}{\sin \theta}$$

C. The Pythagorean Identities

$$\sin^2 \theta + \cos^2 \theta = 1, \; \tan^2 \theta + 1 = \sec^2 \theta, \; \cot^2 \theta + 1 = \csc^2 \theta.$$

The fundamental identities are used to prove other identities.

EXAMPLE: Prove the identity $\dfrac{\cos\theta}{\sec\theta} + \dfrac{\sin\theta}{\csc\theta} = 1$.

PROOF: $\dfrac{\cos\theta}{\sec\theta} + \dfrac{\sin\theta}{\csc\theta} \overset{(1)}{=} \dfrac{\cos\theta}{\frac{1}{\cos\theta}} + \dfrac{\sin\theta}{\frac{1}{\sin\theta}} \overset{(2)}{=} \cos^2\theta + \sin^2\theta \overset{(3)}{=} 1$

 Reasons: (1) Reciprocal identity

 (2) Simplifying

 (3) Pythagorean identity

EXAMPLE: Prove the identity $\dfrac{\sec^2\theta - 1}{\csc^2\theta - 1} = \tan^4\theta$

PROOF: $\dfrac{\sec^2\theta - 1}{\csc^2\theta - 1} \overset{(1)}{=} \dfrac{\tan^2\theta}{\cot^2\theta} \overset{(2)}{=} \dfrac{\tan^2\theta}{\frac{1}{\tan^2\theta}} \overset{(3)}{=} \tan^4\theta$

 Reasons: (1) Pythagorean identities

 (2) Reciprocal identity

 (3) Simplifying

Note that in the above example, it is useful to recognize equivalent forms of the Pythagorean identities. For example, if $\tan^2\theta + 1 = \sec^2\theta$, then $\tan^2\theta = \sec^2\theta - 1$.

EXAMPLE: Express $\tan\theta$ in terms of $\cos\theta$.

SOLUTION: $\tan\theta \overset{(1)}{=} \dfrac{\sin\theta}{\cos\theta} \overset{(2)}{=} \dfrac{\pm\sqrt{1 - \cos^2\theta}}{\cos\theta}$

 Reasons: (1) Quotient identity

 (2) Since $\sin^2\theta + \cos^2\theta = 1$, $\sin^2\theta = 1 - \cos^2\theta$ and
 $\sin\theta = \pm\sqrt{1 - \cos^2\theta}$.

Note that the sign (+ or −) of the square root will depend upon the quadrant in which θ terminates.

EXAMPLE: Show that $\tan\dfrac{3\pi}{4} = \dfrac{\sqrt{1 - \cos^2\frac{3\pi}{4}}}{\cos\frac{3\pi}{4}}$

SOLUTION: $\tan \dfrac{3\pi}{4} = -1$, $\quad \cos \dfrac{3\pi}{4} = -\dfrac{1}{\sqrt{2}}$, $\quad \cos^2 \dfrac{3\pi}{4} = \left(\dfrac{-1}{\sqrt{2}}\right)^2 = \dfrac{1}{2}$

Since $\tan \dfrac{3\pi}{4} = -1$ and $\dfrac{\sqrt{1 - \cos^2 \dfrac{3\pi}{4}}}{\cos \dfrac{3\pi}{4}} = \dfrac{\sqrt{1 - \dfrac{1}{2}}}{\dfrac{-1}{\sqrt{2}}} = \dfrac{\sqrt{\dfrac{1}{2}}}{\dfrac{-1}{\sqrt{2}}} = \dfrac{\dfrac{1}{\sqrt{2}}}{\dfrac{-1}{\sqrt{2}}} = -1$, it follows

that $\tan \dfrac{3\pi}{4} = \dfrac{\sqrt{1 - \cos^2 \dfrac{3\pi}{4}}}{\cos \dfrac{3\pi}{4}}$

EXERCISE SET 3-4.

Find the indicated values:

1) $\sec 49^\circ\ 10'$ 2) $\sec 158^\circ\ 20'$ 3) $\csc 74^\circ 50'$ 4) $\csc 242^\circ\ 30'$

5) $\sec 3.47$ 6) $\sec 5.23$ 7) $\csc 2.86$ 8) $\csc 4.52$

9) $\sec 17.63$ 10) $\csc 23.17$ 11) $\sec \left(-\dfrac{5\pi}{4}\right)$ 12) $\sec \dfrac{13\pi}{5}$

13) $\csc \left(-\dfrac{5\pi}{6}\right)$ 14) $\csc \dfrac{75\pi}{4}$ 15) $\sec (-6.77)$ 16) $\sec \left(-\dfrac{23\pi}{6}\right)$

17) $\csc 93\pi$ 18) $\csc 100$ 19) $\sec 7845.3^\circ$ 20) $\csc 4785.7^\circ$

In exercises 21 - 28, a trigonometric ratio is given. Find the remaining five trigonometric ratios if θ terminates in the given quadrant.

21) $\sin \theta = \dfrac{4}{5}$, II   22) $\tan \theta = \dfrac{-7}{24}$, IV   23) $\csc \theta = \dfrac{\sqrt{5}}{2}$, I

24) $\sec \theta = \dfrac{-8}{3}$, III   25) $\cos \theta = \dfrac{1}{4}$, IV   26) $\sin \theta = \dfrac{-8}{17}$, III

27) $\csc \theta = 2\sqrt{2}$, II   28) $\cot \theta = \dfrac{5}{12}$, I

Find the smallest positive value of θ for the following: (answer in radians)

29) $\csc \theta = 4.837$ 30) $\sec \theta = 2.543$ 31) $\sec \theta = -1.634$

32) $\csc \theta = -8.375$

Prove the following identities:

33) $\cos^2 \theta - \sin^2 \theta = 2 \cos^2 \theta - 1$ 34) $\sin^2 \theta - \cos^2 \theta = 2 \sin^2 \theta - 1$

35) $(1 + \sin \theta)(1 - \sin \theta) = \cos^2 \theta$ 36) $(\csc \theta + 1)(\csc \theta - 1) = \dfrac{1}{\tan^2 \theta}$

37) $\dfrac{\sin\theta}{1-\cos\theta} + \dfrac{\sin\theta}{1+\cos\theta} = 2\csc\theta$

38) $\tan\theta + \cot\theta = \csc\theta\sec\theta$

39) $\cos^2\theta\cot^2\theta = \cot^2\theta - \cos^2\theta$

40) $(\sin^2\theta - 1)(\cot^2\theta + 1) = 1 - \csc^2\theta$

Show that each of the following can be expressed in terms of one of the trigonometric ratios:

41) $\dfrac{1+\cos\theta}{1+\sec\theta}$ 42) $\dfrac{1+\tan^2\theta}{\tan^2\theta}$ 43) $\dfrac{\sec^3\theta}{\tan^2\theta}$ 44) $1 + \dfrac{\tan^2\theta}{\sec\theta + 1}$

Express the following as a trigonometric ratio of an angle θ between 0 and 45° (or between 0 and $\dfrac{\pi}{4}$ radians.)

45) $\sec 237^O$ 46) $\csc 194^O$ 47) $\csc 294^O$ 48) $\sec 95^O 20'$

49) $\sec(-265^O\ 10')$ 50) $\csc 3.84$ 51) $\sec\dfrac{5\pi}{8}$ 52) $\csc(-175^O\ 14')$

Find <u>exact</u> values for the following:

53) $\sec\dfrac{7\pi}{4} - \cos\dfrac{3\pi}{4}$ 54) $\csc^2\dfrac{5\pi}{6} - \cot^2\dfrac{5\pi}{6}$

55) $2\sec^2\dfrac{4\pi}{3} - 1$ 56) $\tan\dfrac{5\pi}{3} - \sec\dfrac{5\pi}{3}$

3-5. CHAPTER SUMMARY.

1) The six trigonometric ratios are defined as follows:

$$\cos \theta = \frac{x}{r} , \quad \sin \theta = \frac{y}{r} , \quad \tan \theta = \frac{y}{x} , \quad \cot \theta = \frac{x}{y} , \quad \sec \theta = \frac{r}{x} ,$$

$$\csc \theta = \frac{r}{y}$$

where P (x,y) is a point on a circle with center at the origin O(0,0), and θ is the central angle with initial side the positive x axis and terminal side a radius OP.

2) The Eight Fundamental Identities

Reciprocal Identities: $\csc \theta = \frac{1}{\sin \theta}$, $\sec \theta = \frac{1}{\cos \theta}$, $\tan \theta = \frac{1}{\cot \theta}$

Quotient Identities: $\tan \theta = \frac{\sin \theta}{\cos \theta}$, $\cot \theta = \frac{\cos \theta}{\sin \theta}$

Pythagorean Identities: $\sin^2 \theta + \cos^2 \theta = 1$

$$1 + \tan^2 \theta = \sec^2 \theta$$

$$1 + \cot^2 \theta = \csc^2 \theta$$

3) The following table gives the signs of the ratios in the four quadrants:

	Quadrant I	Quadrant II	Quadrant III	Quadrant IV
cos θ	+	−	−	+
sin θ	+	+	−	−
tan θ	+	−	+	−
cot θ	+	−	+	−
sec θ	+	−	−	+
csc θ	+	+	−	−

4)   Reduction Formulas

There are two kinds of reduction formulas. If θ is a first quadrant angle, then for angles in the second quadrant,

$$\cos (\pi - \theta) = - \cos \theta, \qquad\qquad \cos \left(\tfrac{\pi}{2} + \theta\right) = - \sin \theta$$

$$\sin (\pi - \theta) = \sin \theta, \qquad\qquad \sin \left(\tfrac{\pi}{2} + \theta\right) = \cos \theta$$

$$\tan (\pi - \theta) = - \tan \theta, \qquad\qquad \tan \left(\tfrac{\pi}{2} + \theta\right) = - \cot \theta$$

$$\cot (\pi - \theta) = - \cot \theta, \qquad\qquad \cot \left(\tfrac{\pi}{2} + \theta\right) = - \tan \theta$$

$$\sec (\pi - \theta) = - \sec \theta, \qquad\qquad \sec \left(\tfrac{\pi}{2} + \theta\right) = - \csc \theta$$

$$\csc (\pi - \theta) = \csc \theta, \qquad\qquad \csc \left(\tfrac{\pi}{2} + \theta\right) = \sec \theta$$

For angles in the third quadrant,

$$\cos (\pi + \theta) = - \cos \theta, \qquad\qquad \cos \left(\tfrac{3\pi}{2} - \theta\right) = - \sin \theta$$

$$\sin (\pi + \theta) = - \sin \theta, \qquad\qquad \sin \left(\tfrac{3\pi}{2} - \theta\right) = - \cos \theta$$

$$\tan (\pi + \theta) = \tan \theta, \qquad\qquad \tan \left(\tfrac{3\pi}{2} - \theta\right) = \cot \theta$$

$$\cot (\pi + \theta) = \cot \theta \qquad\qquad \cot \left(\tfrac{3\pi}{2} - \theta\right) = \tan \theta$$

$$\sec (\pi + \theta) = - \sec \theta, \qquad\qquad \sec \left(\tfrac{3\pi}{2} - \theta\right) = - \csc \theta$$

$$\csc (\pi + \theta) = - \csc \theta, \qquad\qquad \csc \left(\tfrac{3\pi}{2} - \theta\right) = - \sec \theta$$

For angles in the fourth quadrant,

$$\cos (2\pi - \theta) = \cos \theta, \qquad\qquad \cos \left(\tfrac{3\pi}{2} + \theta\right) = \sin \theta$$

$$\sin (2\pi - \theta) = - \sin \theta, \qquad\qquad \sin \left(\tfrac{3\pi}{2} + \theta\right) = - \cos \theta$$

$$\tan (2\pi - \theta) = - \tan \theta, \qquad\qquad \tan \left(\tfrac{3\pi}{2} + \theta\right) = - \cot \theta$$

$$\cot (2\pi - \theta) = - \cot \theta, \qquad\qquad \cot \left(\tfrac{3\pi}{2} + \theta\right) = - \tan \theta$$

$$\sec (2\pi - \theta) = \sec \theta, \qquad\qquad \sec \left(\tfrac{3\pi}{2} + \theta\right) = \csc \theta$$

$$\csc (2\pi - \theta) = - \csc \theta, \qquad\qquad \csc \left(\tfrac{3\pi}{2} + \theta\right) = - \sec \theta$$

The complementary relationships for angles in the first quadrant are:

$$\cos\left(\frac{\pi}{2} - \theta\right) = \sin\theta$$

$$\sin\left(\frac{\pi}{2} - \theta\right) = \cos\theta$$

$$\tan\left(\frac{\pi}{2} - \theta\right) = \cot\theta$$

$$\cot\left(\frac{\pi}{2} - \theta\right) = \tan\theta$$

$$\sec\left(\frac{\pi}{2} - \theta\right) = \csc\theta$$

$$\csc\left(\frac{\pi}{2} - \theta\right) = \sec\theta$$

The formulas above can be generalized as identities for any value of θ except for $\theta = \frac{k\pi}{2}$, $k = 0, \pm1, \pm2, \ldots$.

For any angle θ,

$$\cos(-\theta) = \cos\theta, \qquad \sin(-\theta) = -\sin\theta$$

$$\tan(-\theta) = -\tan\theta, \qquad \cot(-\theta) = -\cot\theta$$

$$\sec(-\theta) = \sec\theta, \qquad \csc(-\theta) = -\csc\theta$$

5) Special Values of the Six Trigonometric Ratios in the Four Quadrants

θ (radians)	0	$\frac{\pi}{6}$	$\frac{\pi}{4}$	$\frac{\pi}{3}$	$\frac{\pi}{2}$	$\frac{2\pi}{3}$	$\frac{3\pi}{4}$	$\frac{5\pi}{6}$	π
θ (degrees	0	30	45	60	90	120	135	150	180
cosine	1	$\frac{\sqrt{3}}{2}$	$\frac{1}{\sqrt{2}}$	$\frac{1}{2}$	0	$\frac{-1}{2}$	$\frac{-1}{\sqrt{2}}$	$\frac{-\sqrt{3}}{2}$	-1
sine	0	$\frac{1}{2}$	$\frac{1}{\sqrt{2}}$	$\frac{\sqrt{3}}{2}$	1	$\frac{\sqrt{3}}{2}$	$\frac{1}{\sqrt{2}}$	$\frac{1}{2}$	0
tangent	0	$\frac{1}{\sqrt{3}}$	1	$\sqrt{3}$	$-$	$-\sqrt{3}$	-1	$\frac{-1}{\sqrt{3}}$	0
cotangent	$-$	$\sqrt{3}$	1	$\frac{1}{\sqrt{3}}$	0	$\frac{-1}{\sqrt{3}}$	-1	$-\sqrt{3}$	$-$
secant	1	$\frac{2}{\sqrt{3}}$	$\sqrt{2}$	2	$-$	-2	$-\sqrt{2}$	$\frac{-2}{\sqrt{3}}$	-1
cosecant	$-$	2	$\sqrt{2}$	$\frac{2}{\sqrt{3}}$	1	$\frac{2}{\sqrt{3}}$	$\sqrt{2}$	2	$-$

θ (radians)	$\frac{7\pi}{6}$	$\frac{5\pi}{4}$	$\frac{4\pi}{3}$	$\frac{3\pi}{2}$	$\frac{5\pi}{3}$	$\frac{7\pi}{4}$	$\frac{11\pi}{6}$	2π
θ (degrees)	210	225	240	270	300	315	330	360
cosine	$-\frac{\sqrt{3}}{2}$	$\frac{-1}{\sqrt{2}}$	$\frac{-1}{2}$	0	$\frac{1}{2}$	$\frac{1}{\sqrt{2}}$	$\frac{\sqrt{3}}{2}$	1
sine	$-\frac{1}{2}$	$\frac{-1}{\sqrt{2}}$	$\frac{-\sqrt{3}}{2}$	-1	$\frac{-\sqrt{3}}{2}$	$\frac{-1}{\sqrt{2}}$	$\frac{-1}{2}$	0
tangent	$\frac{1}{\sqrt{3}}$	1	$\sqrt{3}$	$-$	$-\sqrt{3}$	-1	$\frac{-1}{\sqrt{3}}$	0
cotangent	$\sqrt{3}$	1	$\frac{1}{\sqrt{3}}$	0	$\frac{-1}{\sqrt{3}}$	-1	$-\sqrt{3}$	$-$
secant	$\frac{-2}{\sqrt{3}}$	$-\sqrt{2}$	-2	$-$	2	$\sqrt{2}$	$\frac{2}{\sqrt{3}}$	1
cosecant	-2	$-\sqrt{2}$	$\frac{-2}{\sqrt{3}}$	-1	$\frac{-2}{\sqrt{3}}$	$-\sqrt{2}$	-2	$-$

EXERCISE SET 3-5.

In exercises 1 - 8, determine the values of the remaining trigonometric ratios given that

1) $\cos \theta = \frac{3}{5}$, $\sin \theta < 0$ 2) $\tan \theta = \frac{-12}{5}$, $\sin \theta > 0$

3) $\csc \theta = \sqrt{5}$, $\sec \theta > 1$ 4) $\cot \theta = \frac{4}{3}$, $\sec \theta < -1$

5) $\sin \theta = \frac{-\sqrt{3}}{2}$, $\tan \theta > 0$ 6) $\tan \theta = \cot \theta$, $\sin \theta < 0$, $\cos \theta > 0$

7) $\sin \theta = -1$ 8) $\sec \theta = \frac{-25}{7}$, $\cot \theta < 0$

Answer True or False to the following statements:

9) $\sin \left(\frac{\pi}{2} + \frac{\pi}{5}\right) = \cos \frac{\pi}{5}$ 10) $\tan (180^0 - 42^0) = - \tan 42^0$

11) $\sec \left(\frac{3\pi}{2} + \pi\right) = - \csc \pi$ 12) $\cot (360^0 - 14^0) = - \tan 14^0$

13) $\csc 385^\circ = \sec 25^\circ$

14) $\tan (\pi - 3) = -\tan 3$

15) $\tan (2 - \pi) = \tan 2$

16) $\sin (\frac{\pi}{4} - \frac{3\pi}{2}) = -\sin \frac{\pi}{4}$

17) $\tan 125^\circ = -\cot 35^\circ$

18) $\cos 303^\circ = \sin 33^\circ$

19) $\sec 186^\circ = -\csc 6^\circ$

20) $\sin \frac{3\pi}{10} = \cos \frac{\pi}{5}$

Prove the following identities:

21) $\cot \theta \sec \theta = \csc \theta$

22) $\frac{1 - \tan^2 \theta}{1 + \tan^2 \theta} = \cos^2 \theta - \sin^2 \theta$

23) $\sin \theta \cot \theta + \cos \theta \tan \theta = \sin \theta + \cos \theta$

24) $\frac{1 + \tan \theta}{1 + \cot \theta} = \tan \theta$

25) $\frac{1}{\tan \theta + \cot \theta} = \sin \theta \cos \theta$

26) $\sin^4 \theta - \cos^4 \theta = 1 - 2 \cos^2 \theta$

27) $\frac{1 + \sec \theta}{\sec \theta} = \frac{\sin^2 \theta}{1 - \cos \theta}$

28) $\frac{\csc \theta + \sec \theta}{\tan \theta + 1} = \csc \theta$

Find values of the following:

29) $\sec 78^\circ 40'$

30) $\cos 3.84$

31) $\tan (-\frac{7\pi}{5})$

32) $\sin 3049.83^\circ$

33) $\cot (-41.86)$

34) $\csc (-\frac{12\pi}{13})$

35) $\tan 115 \pi$

36) $\sec (-548^\circ)$

Find _exact_ values of the following:

37) $1 - \sin^2 \frac{\pi}{3}$

38) $2 \cos^2 \frac{7\pi}{4} - 1$

39) $\sqrt{\frac{1 - \cos \pi}{2}}$

40) $2 \cos \frac{11\pi}{6} \sin \frac{11\pi}{6}$

41) $\sin 135^\circ - \tan 330^\circ$

42) $\sec^2 \frac{3\pi}{4} - \tan^2 \frac{3\pi}{4}$

43) $\tan \frac{5\pi}{4} \cos \frac{5\pi}{4}$

44) $\csc^2 \frac{5\pi}{6} - 1$

Express the following ratios in terms of ratios of angles between 0° and 45°:

45) $\csc 112^\circ$

46) $\sin (-348^\circ)$

47) $\tan 289^\circ$

48) $\sec (-78^\circ)$

49) $\cot 570^\circ$

50) $\cos (-784^\circ)$

51) $\sin 312.6^\circ$

52) $\tan (-856.3^\circ)$

CHAPTER 4

THE TRIGONOMETRIC FUNCTIONS

4-1. THE FUNCTION CONCEPT. COSINE AND SINE FUNCTIONS.

In mathematics, a function is a process of matching numbers from one set with those from another set. If we start with a set of real numbers called the domain and establish a rule for matching these numbers to those in another set, called the range, the function process can be illustrated as in Figure 4-1A.

Figure 4-1A

Domain   Range

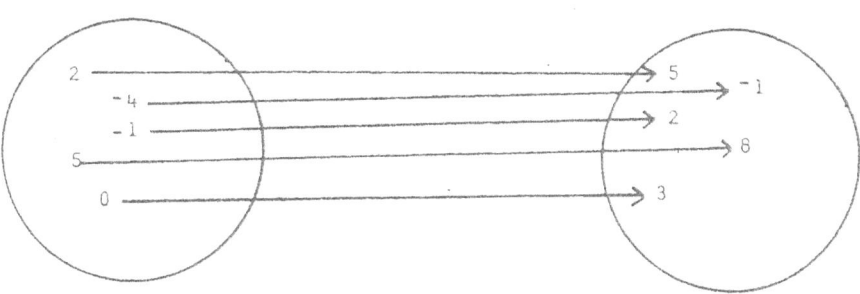

The function illustrated above is an example of a one-to-one mapping. The mapping is always from the domain to the range. In the illustrated example, the rule is established as follows: Let x be a number in the domain and y be a number in the range. Then, for each x, y = x + 3. The notation that is used is written y = f(x) = x + 3 and is read as "y is a function of x and is equal to x + 3."

Another way of looking at a function is as a set of ordered pairs with the first component of each pair coming from the domain and the second component from the range. The notation for the function in this presentation is:

$$f = \{(x,y) : y = x + 3\}$$

If the domain is expanded to the set of all real numbers, the ordered pairs contained in the function set can be matched to points in a Cartesian (rectangular) coordinate plane forming the graph of a straight line as shown in Figure 4-1B.

Figure 4-1B

Graph of the function, $f(x) = x + 3$

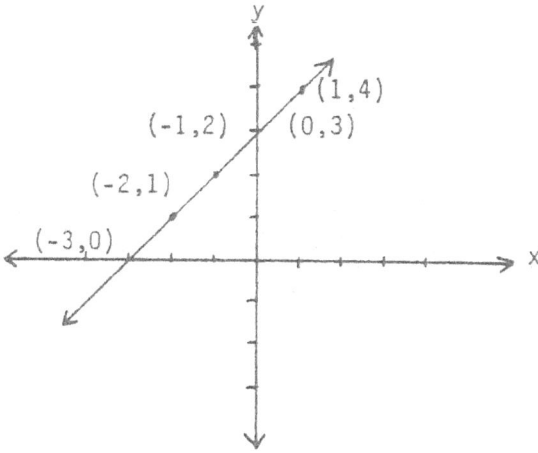

Another function studied in algebra illustrates what is called a "many-to-one" mapping (Figure 4-1C).

Figure 4-1C

Domain   Range

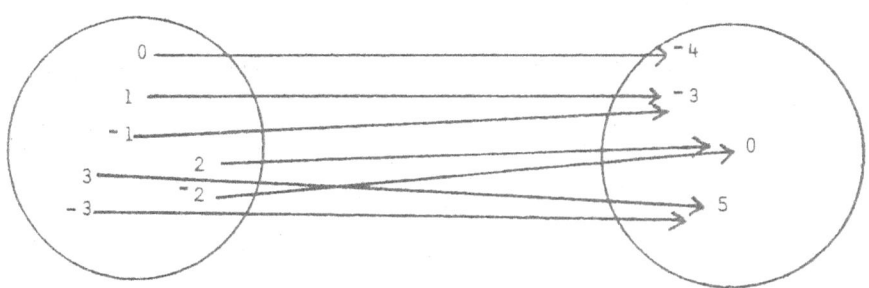

With the exception of 0 to -4, every other number in the range is matched with two numbers from the domain. The rule for the mapping is given by $f(x) = x^2 - 4$ and is an example of a quadratic function. The set of ordered pairs shown in Figure 4-1C is {(0,-4), (1,-3), (-1,-3),(2,0), (-2,0), (3,5), (-3,5)}. This is a subset of the function defined by $f = \{(x,y) : y = x^2 - 4\}$. The graph of this function is a parabola shown in Figure 4-1D.

Figure 4-1D

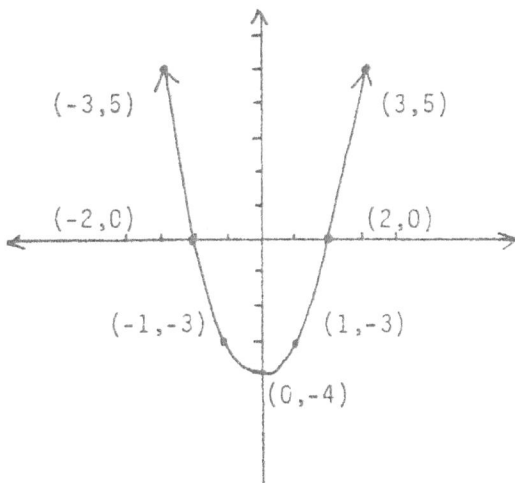

A mapping that is <u>not</u> a function is illustrated in Figure 4-1E.

Figure 4-1E

Domain   Range

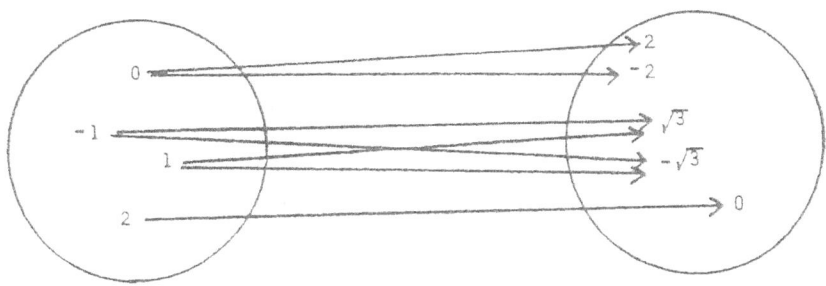

This example shows a <u>one-to-many</u> mapping. For each x in the domain, x ≠ 2, there are two values in the range. The rule for this relation is given by $y = \pm\sqrt{4 - x^2}$. If we square both sides of this equation and rearrange terms, the more familiar open sentence, $x^2 + y^2 = 4$ is obtained. The graph of this relation is shown in Figure 4-1F with ordered pairs obtained from the mapping of Figure 4-1E. The graph of any relation that contains two or more points that have the same x-coordinate and different y-coordinates such as $(1,\sqrt{3})$ and $(1,-\sqrt{3})$ is not the graph of a function.

Figure 4-1F

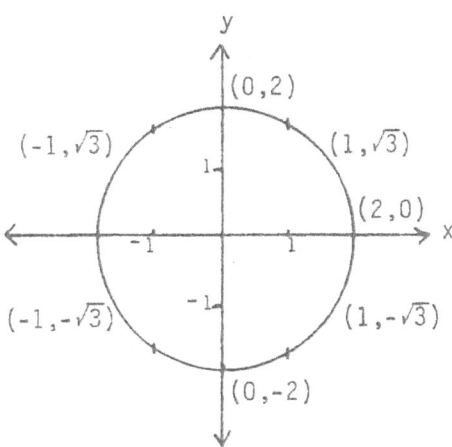

Another way to recognize the graph of a function is to pass a vertical line through it. If, at any position, the vertical line intersects the graph of the relation in more than one point, the relation is not a function. This is demonstrated in Figure 4-1G.

Figure 4-1G

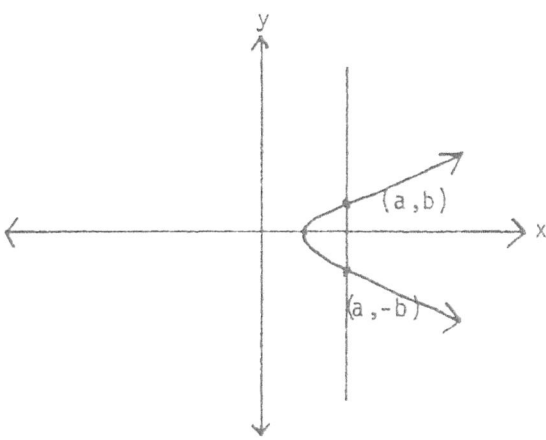

TRANSFORMING THE TRIGONOMETRIC RATIOS INTO TRIGONOMETRIC FUNCTIONS.

Consider the cosine ratio as defined earlier. Given a circle with center at the origin, radius r, and central angle θ, with P(x,y) the endpoint of the radius which terminates θ (Figure 4-1H), cos θ = $\frac{x}{r}$.

Figure 4-1H

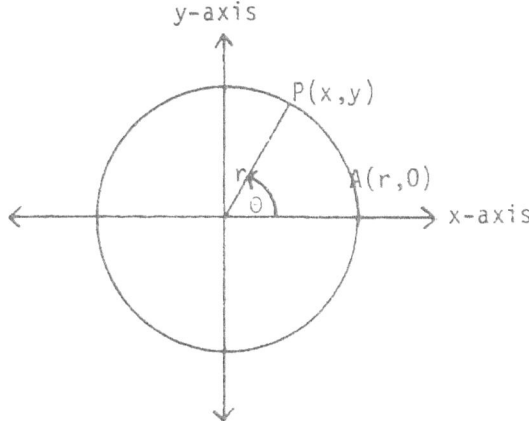

Since θ can be a real number (i.e., an angle measured in radians), and $\frac{x}{r}$ is a real number, a mapping from values of θ to values of $\frac{x}{r}$ qualifies as a function. With values of θ in the domain and values of $\frac{x}{r}$ in the range, the mapping may be pictured as follows:

Figure 4-1I

Domain   Cosine Function   Range

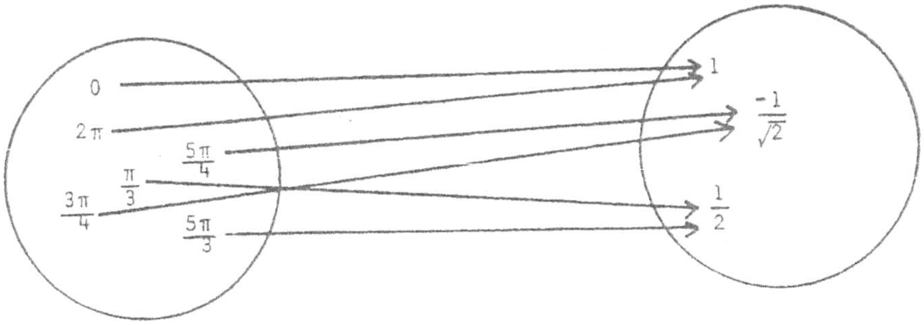

Based on experience with the cosine ratios, it can be seen that this mapping is many-to-one (Figure 4-1I). The cosine function can be thought of as a function of the central angle θ. That is,

$$f(\theta) = \cos \theta = \frac{x}{r}.$$

As a set of ordered pairs, $\cos = \{(\theta, \frac{x}{r})\}$: θ is a real number and $-1 \le \frac{x}{r} \le 1\}$. Implied in this definition is that x is the abscissa of an ordered pair (x,y) satisfying the relationship $x^2 + y^2 = r^2$.

GRAPHING THE COSINE FUNCTION.

The graph of the cosine function can be constructed by setting the horizontal axis to values of θ and the vertical axis to values of $\cos \theta = \frac{x}{r}$. With the help of a calculator and knowledge of the special angle values of the cosine, a table of values for θ from -7 to 7 can be obtained.

θ	-7	-6.5	-2π	-6	-5.5	-5π/3	-5	-3π/2	-4.5	-4π/3	-4	-3.5	-π	-3
X/r	.75	.98	1	.96	.71	.5	.28	0	-.21	-.5	-.65	-.94	-1	-.99

θ	-2.5	-2π/3	-2	-π/2	-1.5	-π/3	-1	-.5	0	.5	1	π/3	1.5	π/2	2
X/r	-.8	-.5	-.42	0	.07	.5	.54	.88	1	.88	.54	.5	.07	0	-.42

θ	2π/3	2.5	3	π	3.5	4	4π/3	4.5	3π/2	5	5π/3	5.5	6	2π	6.5	7
X/r	-.5	-.8	-.99	-1	-.94	-.65	-.5	-.21	0	.28	.5	.71	.96	1	.98	.75

The ordered pairs in the table may now be graphed in the coordinate plane and a smooth curve drawn between the points:

Figure 4-1J

Cosine Function

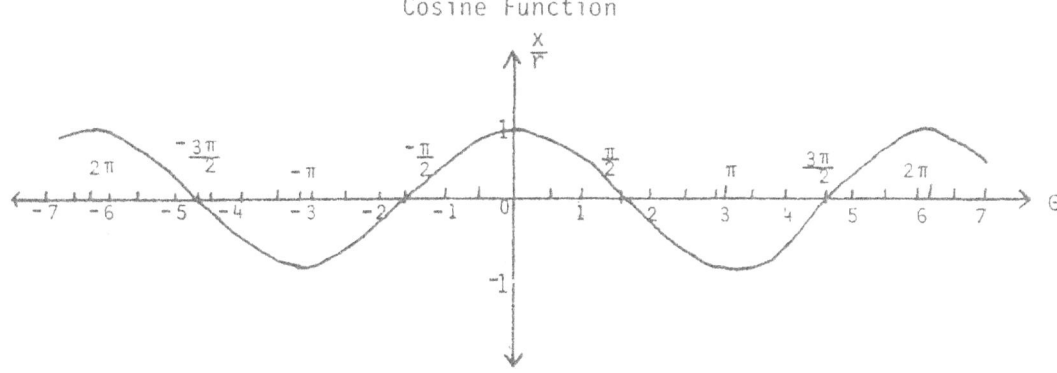

The graph of the cosine function reveals several important features:

1. Continuity. There is no break in the graph. For every real number θ, positive or negative, there is a corresponding value for $\frac{X}{r}$ between -1 and 1 inclusive.

2. Periodicity. The shape of the graph repeats itself after every interval of 2π units along the θ axis. For example, the shape of the graph between -π and π is identical to the shape between π and 3π. Likewise, the shape

between 0 and 2π is identical to the shape between -2π and 0. This is formalized by the statement cos θ = cos ($\theta \pm 2k\pi$), k = 1, 2, 3,.... This statement is equivalent to the earlier observation that the cosines of coterminal angles are equal.

3. Symmetry. The graph is symmetrical to the $\frac{x}{r}$ axis. Any function whose graph exhibits this feature is called an <u>even function</u>. Formally, if f ($-\theta$) = f (θ) then f is an even function. This statement for the cosine function is equivalent to the earlier observation that cos ($-\theta$) = cos θ.

<u>THE SINE FUNCTION</u>.

The sine function is defined as follows:

$$\sin = \{(\theta, \tfrac{y}{r}) : \theta \text{ is a real number and } -1 \le \tfrac{y}{r} \le 1\}.$$

Again, it is implied that y is the ordinate of an ordered pair (x,y) satisfying the relation $x^2 + y^2 = r^2$.

The graph of the sine function is outlined using its special angle values. The table below gives these values between $-\pi$ and π.

Figure 4-1K

Sine Function

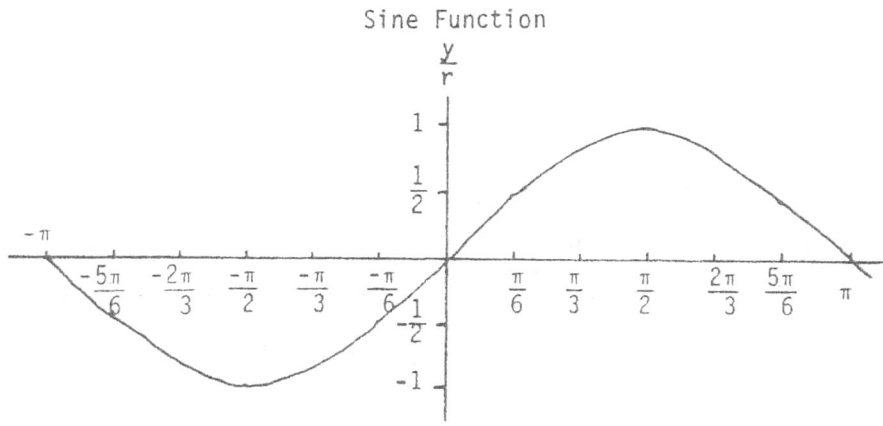

θ	$-\pi$	$\frac{-5\pi}{6}$	$\frac{-\pi}{2}$	$\frac{-\pi}{3}$	$\frac{-\pi}{6}$	0	$\frac{\pi}{6}$	$\frac{\pi}{3}$	$\frac{\pi}{2}$	$\frac{2\pi}{3}$	$\frac{5\pi}{6}$	π
$\frac{y}{r}$	0	$\frac{-1}{2}$	-1	$\frac{-\sqrt{3}}{2}$	$\frac{-1}{2}$	0	$\frac{1}{2}$	$\frac{\sqrt{3}}{2}$	1	$\frac{\sqrt{3}}{2}$	$\frac{1}{2}$	0

It may be observed that the sine graph, like the cosine, is continous and periodic with a period of 2π. That is, $\sin \theta = \sin (\theta \pm 2k\pi)$, $k = 1,2,3,...$ It is also symmetric, but unlike the cosine, it has symmetry with respect to the origin. Any function whose graph exhibits this symmetry is called an <u>odd</u> <u>function</u>. Formally, if $f(-\theta) = -f(\theta)$ then f is an odd function. This statement for the sine function is equivalent to the earlier observation that $\sin (-\theta) = - \sin \theta$. For example, the graph shows that $\sin (-\frac{\pi}{6}) = - \sin \frac{\pi}{6}$ and $\sin (\frac{-3\pi}{2}) = - \sin \frac{3\pi}{2}$.

<u>EXERCISE SET 4-1</u>.

In exercises 1 - 10, determine whether or not the given relation is a function. If it is a function, tell whether it is one-to-one or many-to-one.

1) $\{(x,y): x - 2y = 7\}$

2) $\{(x,y): y = |x|\}$

3) $\{(x,y): x^2 + y^2 = 25\}$

4) $\{(x,y): y = \sqrt{x - 3}\}$

5) $\{(x,y): x^2 - y = 10\}$

6) $\{(x,y): x = y^2 - 2\}$

7) $\{(x,y): x = 2 |y| -4\}$

8) $\{(x,y): y = x^3\}$

9) $\{(x,y): y^2 = x^2 - 1\}$

10) $\{(x,y): y = -x^2 - 3x + 2\}$

Graph the following:

11) $\cos \theta$, $-\pi \le \theta \le 3\pi$

12) $\cos \theta$, $-1 \le \theta \le 3$

13) $\sin \theta$, $-2\pi \le \theta \le \frac{\pi}{2}$

14) $\sin \theta$, $-2 \le \theta \le 2$

15) $\cos \theta$, $0 \le \theta \le 5$

16) $\sin \theta$, $2.34 \le \theta \le 7.54$

For exercises 17 - 26, determine if the functions found in exercises 1 - 10 are odd, even, or neither.

27) Find the set of all values of θ, $-2\pi \le \theta \le 2\pi$, such that $\sin \theta = 0$.

28) Find the set of all values of θ, $-2\pi \le \theta \le 2\pi$, such that $\cos \theta = 0$.

29) For how many values between $-\pi$ and 3π does $\cos \theta = \frac{1}{2}$?

3C) For how many values between -2π and π does $\sin \theta = \frac{-\sqrt{3}}{2}$?

31) Complete the table and graph the points in the coordinate plane:

θ	-5	-4	-3	-2	-1	0	1	2	3	4	5
$\cos \theta$											

32) Complete the table and graph the points in the coordinate plane:

θ	-5	-4	-3	-2	-1	0	1	2	3	4	5
$\sin \theta$											

4-2. COSINE AND SINE FUNCTIONS (CONTINUED).

Phase Shift

Another important feature of the cosine and sine functions is revealed by contrasting their graphs (Figure 4-2A).

Figure 4-2A

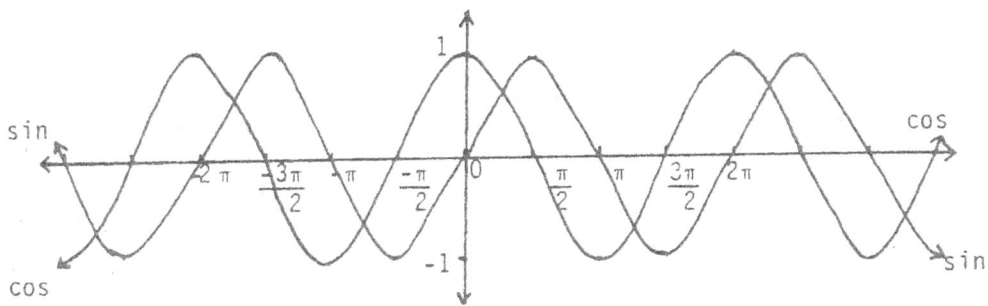

The two graphs proceed from left to right like two identical waves that closely follow each other. Indeed, the graphs are identical and only differ by what is called a phase shift. The sine and cosine are out of phase by an

interval of $\frac{\pi}{2}$ units. This feature is given formally by the relation
$\sin (\theta + \frac{\pi}{2}) = \cos \theta$ studied earlier as a complementary relationship. For
example, if $\theta = \pi$, the graph shows that $\sin (\pi + \frac{\pi}{2}) = \sin \frac{3\pi}{2} = \cos \pi$. We
say the cosine function <u>leads</u> the sine function by an interval of $\frac{\pi}{2}$. Conversely
we say that the sine function <u>lags</u> behind the cosine function by an interval of
$\frac{\pi}{2}$.

Change of Variables. A Further Step Toward Generalizing the Trigonometric Functions.

In graphing both the cosine and sine functions on one set of coordinate
axes (Figure 4-2A), the labeling of the axes was purposely omitted. The horizontal axis is obviously graduated in units of θ since the graphs were labeled
as functions of θ. But what about the vertical axis? Does it represent $\frac{x}{r}$ or
$\frac{y}{r}$? In fact, it doesn't matter as both $\frac{x}{r}$ and $\frac{y}{r}$ represent real numbers between
-1 and 1 and any appropriate letter designation would do as well for both ratios.
The ratios $\frac{x}{r}$ and $\frac{y}{r}$ were retained in the initial graphing of the cosine and sine
functions, respectively, in order to keep in mind the source of the values for
the functions. This source, the circle and the relationships established
between θ, x, y, r, and s throughout the first part of this book, will be
retained as fundamental to what shall be derived later. Indeed, it is the
reason that the trigonometric functions are sometimes referred to as the
"circular functions." However, at this point there will be a change in the
meaning of variables. This shall be done so that the graphing of the trigonometric functions can be legitimately referred to the XY-coordinate system.
To this end, the cosine and sine functions are redefined as follows:

cosine $= \{(x,y) : x$ is a real number and $y = \cos x\}$.

sine $= \{(x,y) : x$ is a real number and $y = \sin x\}$.

The range of the cosine (or sine) function now is expressed as:

$$\{y : -1 \le y \le 1\}.$$

With this new interpretation (x,y) no longer refers to a point on a circle with center at the origin. Instead the ordered pair (x,y) is now to be understood as a point on a graph of a trigonometric function.

EXAMPLE: If $x = \frac{\pi}{3}$, find cos x and sin x.

SOLUTION: $\cos x = \cos \frac{\pi}{3} = \frac{1}{2}$ and $\sin x = \sin \frac{\pi}{3} = \frac{\sqrt{3}}{2}$.

EXAMPLE: If x = .78, find cos x and sin x.

SOLUTION: Since x is a real number, .78 is a radian measure.

From Table I, cos x = cos .78 = .7109.

sin x = sin .78 = .7033.

EXAMPLE: If cos x = -.2385 and $\frac{\pi}{2} < x < \pi$, find x.

SOLUTION: From Table I, the cosine entry .2385 corresponds to a radian measure of 1.33. Since x lies between $\frac{\pi}{2}$ and π (the second quadrant), subtract 1.33 from 3.14 to obtain 1.81.

CALCULATOR SOLUTION: Radian mode. Enter -.2385. Press INV key. Press cos key. Read 1.81. Hence, x = 1.81.

The last example illustrates an important fact about the way calculators handle the trigonometric functions. Assuming a calculator is unrestricted, it can process any input from the domain and display the range value of the function without erring. For many-to-one functions, however, the calculator must be selective in processing inputs from the range to read out a corresponding domain value. For example, the inverse of the cosine function will only give values of x between 0 and π. This is illustrated as follows:

Figure 4-2B Cos

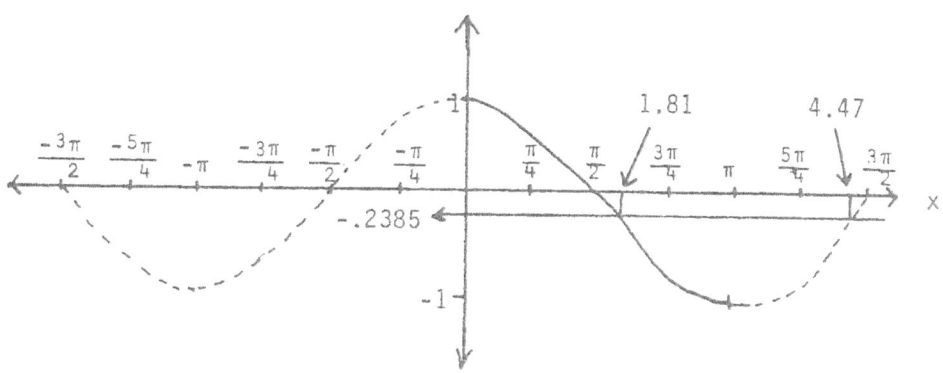

Thus, if a positive value of the cosine is input, pressing the INV (or ARC) key and the cos key will yield a value of x between 0 and $\frac{\pi}{2}$, while a negative input yields a value of x between $\frac{\pi}{2}$ and π. In finding x when cos x = -.2385, the calculator returned the domain value 1.81. Since the problem stated $\frac{\pi}{2}$ < x < π, this was the correct solution. However, if the problem had stated π < x < $\frac{3\pi}{2}$, the calculator display of 1.81 would have been incorrect. As the graph shows, the correct value of x in the interval between π and $\frac{3\pi}{2}$ is 4.47. This result can be obtained arithmetically through use of the reduction formula cos (2π - x) = cos x. If x = 1.81, then 2π - x = 6.28 - 1.81 = 4.47.

Principal Values. Cos and Sin.

The values from 0 to π form a special subset of the cosine's domain and are referred to as principal values. If the domain is restricted to these principal values a one-to-one function can be defined:

Cos   = {(x,y) : 0 ≤ x ≤ π, y = cos x}

The upper case letter, C, will be used to distinguish the restricted cosine function from the normal cosine function. We observe from the graph (Figure 4-2B) that Cos contains every value from -1 to 1 in its range. That is, if y = Cos x, then the range of Cos is defined as {y : -1 ≤ y ≤ 1}.

The principal values for the domain of the sine function is defined by {x : $-\frac{\pi}{2}$ ≤ x ≤ $\frac{\pi}{2}$}. This is illustrated in Figure 4-2C.

Figure 4-2C. Sin

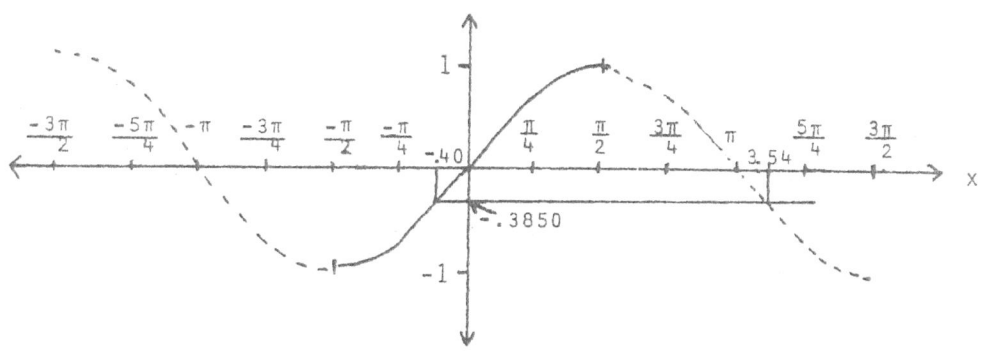

We define Sin = {(x,y): $\frac{-\pi}{2}$ ≤ x ≤ $\frac{\pi}{2}$, y = sin x}. We observe that this restricted sine function is also one-to-one and the range = {y : -1 ≤ y ≤ 1}.

The difference between Sin and sin is illustrated by the following examples.

<u>EXAMPLE</u>:   Find x, if Sin x = -.3850.

<u>CALCULATOR SOLUTION</u>: Radian mode. Input -.3850. Press INV key.

Press sin key. Read -.3952.

Therefore, x = -.40 rounded.

EXAMPLE: Find x, if sin x = -.3850 and $\pi < x < \frac{3\pi}{2}$.

CALCULATOR SOLUTION: Following the steps above x = -.40. However, if $\pi < x < \frac{3\pi}{2}$, reduction formulas are needed for the correct value of x. A glance at the graph (Figure 4-2C) will reveal that the correct value can be obtained by adding .40 to π. That is, 3.14 + .40 = 3.54. The reduction formulas being used are sin (π + x) = - sin x = sin (-x). Thus, sin 3.54 = sin (π + .40) = - sin .40 = sin (-.40) = -.3850.

The solution to this problem may be simplified by referring back to the unit circle model and using arguments of symmetry between the first and third quadrants. It is important however, that the student become familiar with reading the graphs of the functions for future applications.

Increasing and Decreasing Functions.

Our study of the restricted functions, Sin and Cos , introduces another concept of importance. Examination of Figures 4-2B and 4-2C reveals the following observation: As the eye traverses from left to right over the graph of Cos , from 0 to π, the values of the function decrease from a maximum of 1 to a minimum of -1. Similarly, traversal from left to right over the graph of Sin , from $-\frac{\pi}{2}$ to $\frac{\pi}{2}$, reveals the function values increasing from -1 to 1. For this reason, we describe Cos as a decreasing function and Sin as an increasing function.

Formally, if x_1 and x_2 are any real numbers in the domain of a function f and $x_1 < x_2$, then f is <u>increasing</u> on its domain if $f(x_1) < f(x_2)$ is always true.

Conversely, if $x_1 < x_2$ and $f(x_1) > f(x_2)$ is always true, then f is decreasing on its domain.

EXAMPLE: Graph the function $f(x) = x^3$. Tell whether the function is increasing or decreasing.

SOLUTION: A table of ordered pairs for $f(x) = x^3$ yields the following:

x	-3	-2	-1	0	1	2	3
y	-27	-8	-1	0	1	8	27

The graph of $f(x) = x^3$ shows that f is increasing throughout its domain.

Figure 4-2D

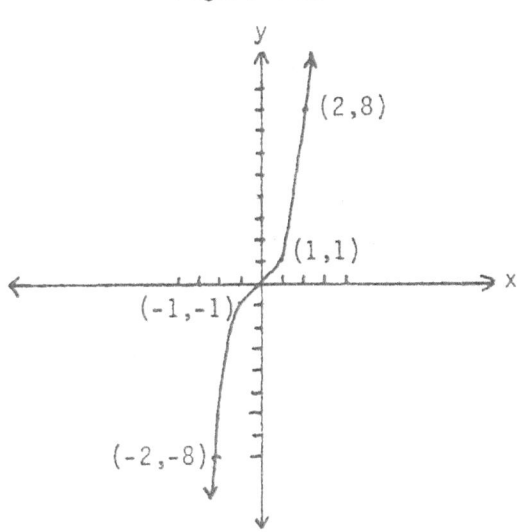

EXERCISE SET 4-2.

In exercises 1 - 10 a function and domain are given. Determine if the function is increasing, decreasing, or doing neither on the domain.

1) $f(x) = |x|$, $x < 0$

2) $g(x) = x^2$, $x > 0$

3) $h(x) = -x^3$, $x < 0$

4) $f(x) = \cos x$, $\frac{\pi}{2} < x < \pi$

5) $k(x) = \sin x$, $0 < x < \frac{\pi}{2}$

6) $t(x) = \cos x$, $\pi < x < \frac{3\pi}{2}$

7) $f(x) = \sin x$, $-\pi < x < -\frac{\pi}{2}$

8) $h(x) = -x^2 - 4$, $x > -4$

9) $g(x) = 3$, $x > 0$

10) $s(x) = \sqrt{x^2 - 1}$, $x < -1$

Graph the following:

11) $\{(x,y) : y = \text{Sin } x, \frac{-\pi}{4} \le x \le \frac{\pi}{4}\}$

12) $\{(x,y) : y = \text{Cos } x, \frac{\pi}{4} \le x \le \frac{3\pi}{4}\}$

13) $\{(x,y) : y = \cos x, -\frac{\pi}{2} \le x \le \frac{\pi}{2}\}$

14) $\{(x,y) : y = \sin x, \frac{\pi}{2} \le x \le \frac{3\pi}{2}\}$

15) For how many values of x between 0 and 3π is cos x = .3475?

16) For how many values of x between -2π and $\frac{3\pi}{2}$ is sin x = -.7?

Find the value of x in the following:

17) $\text{Sin } x = \frac{1}{2}$

18) $\text{Sin } x = -\frac{\sqrt{3}}{2}$

19) $\text{Cos } x = -\frac{1}{2}$

20) $\text{Cos } x = \frac{-1}{\sqrt{2}}$

21) $\text{Sin } x = \frac{\sqrt{2}}{2}$

22) $\text{Sin } x = 0$

23) $\text{Cos } x = -1$

24) $\text{Cos } x = 0$

25) $\text{Sin } x = -.4502$

26) $\text{Cos } x = .7805$

27) $\text{Sin } x = .48$

28) $\text{Sin } .48 = x$

29) $\text{Cos } 2.35 = x$

30) $\text{Cos } .84 = x$

31) $\text{Sin } (-1) = x$

32) $\text{Sin } (\frac{-\pi}{6}) = x$

4.3 THE TANGENT AND COTANGENT FUNCTIONS

In preparation for understanding the characteristics of the tangent and cotangent functions, the important facts learned about the tangent and cotangent ratios from Chapter 3 will be reviewed.

Consider a point $P(x,y)$ moving on a circle $x^2 + y^2 = r^2$:

Figure 4-3A

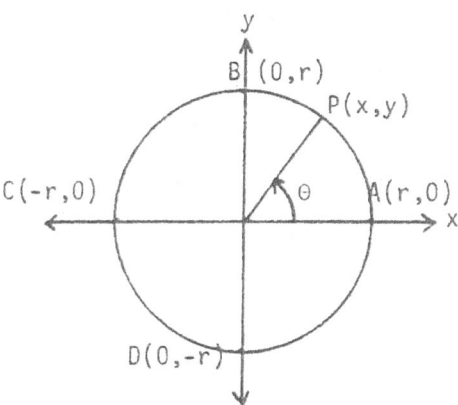

Let the central angle Θ be measured in radians. The tangent ratio is defined by $\tan \Theta = \frac{y}{x}$ and the cotangent ratio by $\cot \Theta = \frac{x}{y}$.

The quadrantal angle values are either 0 or are undefined:

$\tan 0 = \frac{0}{r} = 0$, $\cot 0 = \frac{r}{0}$ undefined

$\tan \frac{\pi}{2} = \frac{r}{0}$ undefined , $\cot \frac{\pi}{2} = \frac{0}{r} = 0$

$\tan \pi = \frac{0}{-r} = 0$, $\cot \pi = \frac{-r}{0}$ undefined

$\tan \frac{3\pi}{2} = \frac{-r}{0}$ undefined , $\cot \frac{3\pi}{2} = \frac{0}{-r} = 0$

$\tan 2\pi = \frac{0}{r} = 0$, $\cot 2\pi = \frac{r}{0}$ undefined

Since the tangent and cotangent ratios are reciprocals, their quadrant polarities are identical:

Θ	tan Θ	cot Θ
I: $0 < \Theta < \dfrac{\pi}{2}$	+	+
II: $\dfrac{\pi}{2} < \Theta < \pi$	-	-
III: $\pi < \Theta < \dfrac{3\pi}{2}$	+	+
IV: $\dfrac{3\pi}{2} < \Theta < 2\pi$	-	-

As $P(x,y)$ moves from $A(r,0)$ to $B(0,r)$, Θ increases from 0 to $\dfrac{\pi}{2}$. The ordinate y increases from 0 to r and the abscissa x decreases from r to 0. With increasing y values and decreasing x values the tangent ratio, $\dfrac{y}{x}$, must increase and the cotangent ratio, $\dfrac{x}{y}$, must decrease. This fact is confirmed by reading Table I values for tangent and cotangent.

The calculator may be used to investigate the nature of these changes. First, set the calculator in the radian mode. Complete the following table:

Θ	0	.01	.02	.03	.04	.05	.06	.07	.08	.09	.10
tan Θ	0	.01	.02		.04	.05		.07		.09	
cot Θ	+∞	99.9		33.3			16.6		12.5		9.96

Note that for small values of Θ, tan Θ and Θ agree very closely. For values of Θ very close to zero in the first quadrant cot Θ is very large. Indeed, for $0 < \Theta < .10$, cot Θ decreases from $+\infty$ to 9.96, an enormous change over such a small interval.

So far, Θ and tan Θ appear to be linearly related. At what point do they begin to deviate? Complete the following table:

Θ	.10	.2	.3	.4	.5	.6	.7	.8	.9	1.0
tan Θ		.2	.31		.55		.84		1.3	
cot Θ	9.96	4.9		2.4		1.5		.97		.64

The table indicates that tan Θ = cot Θ at some value of Θ between .7 and .8. What is that value?

The next portion of the table will reveal how tan Θ increases to $+\infty$ and cot Θ approaches 0 as Θ approaches $\frac{\pi}{2}$. Fill in the missing values:

Θ	1.0	1.1	1.2	1.3	1.4	1.5	1.55	1.56	1.57	$\frac{\pi}{2}$
tan Θ		1.96		3.6		14.1		92.6		$+\infty$
cot Θ	.64		.39		.17		.02		.0008	0

Note that the calculator will evaluate tan $\frac{\pi}{2}$ as 1.8312×10^9, or approximately 1,800,000,000.* Theoretically, tan $\frac{\pi}{2}$ is infinitely large, but the calculator must give an approximation to the irrational number $\frac{\pi}{2}$. Hence, the value it displays is finite. However, if tan 90^0 is the input, the display will indicate an undefined value by an error message.

As Θ moves into the second quadrant both ratios become negative. The tangent jumps from $+\infty$ to $-\infty$, resulting in a break in the continuity of values. The cotangent values move continuously from 0 to numbers less than 0. Fill in the missing values:

Θ	1.571	1.575	1.6	1.7	1.8	1.9	2.0	2.1	2.2	2.3	2.4	2.5
tan Θ	-4910		-34.2		-4.3		-2.18		-1.37		-.92	
cot Θ		-.004		-.13		-.34		-.58		-.89		-1.3

* Some calculators will display an error message for tan $\frac{\pi}{2}$.

We see that for some value of Θ between 2.3 and 2.4, tan Θ = cot Θ . What is that value?

In the next table, the behavior of the tangent and cotangent as Θ approaches π becomes evident:

Θ	2.6	2.7	2.8	2.9	3.0	3.1	3.14	3.141	3.1415	π
tan Θ	-.6		-.36		-.14		-.002		-.00009	0
cot Θ		-2.1		-4.1		-24		-1687		- ∞

The construction of tables need go no further as we already know that the values of the tangent and cotangent in the third quadrant are identical to their values in the first quadrant. This we have from the reduction formulas: tan (π + Θ) = tan Θ and cot (π + Θ) = cot Θ . And, if Θ is a second quadrant angle then we also know the formula is still valid. Hence, the values of the tangent and cotangent in the fourth quadrant are identical to their values in the second quadrant. It is a consequence of this fact that the tangent and cotangent functions are periodic with a period equal to π units.

We are now ready to define and graph the tangent and cotangent functions in the xy - coordinate plane.

$$\text{tangent} = \{(x,y) : x \neq \frac{k\pi}{2} , k = \pm 1, \pm 3, \pm 5, \ldots$$
$$\text{and} \quad y = \tan x\}$$

$$\text{cotangent} = \{(x,y) : x \neq k\pi, k = 0, \pm 1, \pm 2, \ldots$$
$$\text{and} \quad y = \cot x\}$$

Figure 4-38. The Tangent Function

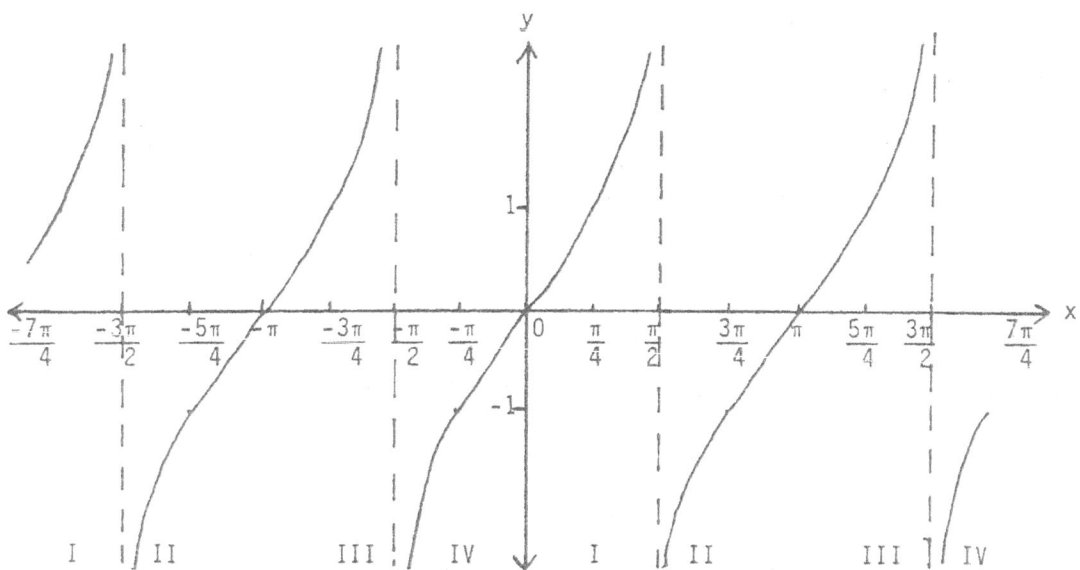

In the transformation from central angle values of θ to linear units
along the x-axis it is useful to label the quadrants for a comparative analysis.
This helps to clarify the graphing of the function for negative values of x.

Vertical asymptotes have been drawn in where the tangent function becomes
undefined. It is at these values of x that a break in the graph occurs.

Although we could have used the tables built earlier for a more detailed
graph, the simplest way to sketch the tangent is to graduate the x axis in
units of $\frac{\pi}{4}$. Since tan $\frac{k\pi}{4}$ is either 1 or -1 when k = $\pm1, \pm3, \pm5, \ldots$, and these x
values divide the quadrants into equal parts, they provide convenient points
for making a quick sketch. Also, $\frac{\pi}{4} \doteq \frac{3}{4}$, so the placement of 1 and -1 on the
y-axis can be reasonably estimated. This will yield a graph that is not too
distorted.

Graph Analysis of the Tangent Function

The following characteristics of the tangent function are illustrated by the graph:

1) Periodicity. The tangent function has a "primitive" period of π units. The primitive period is the smallest interval along the x-axis that must be covered before the function repeats itself. The formula $\tan(\pi + x) = \tan x$ incorporates this feature.

2) Discontinuity. The function is not defined for values of

$$x = \frac{k\pi}{2}, \quad k = \pm 1, \pm 3, \ldots$$

3) Odd Function. The graph is symmetric with respect to the origin so that

$$\tan(-x) = -\tan x.$$

4) Principal Values. The principal values for the tangent function are between $-\frac{\pi}{2}$ and $\frac{\pi}{2}$. If the domain is restricted to these values, a one-to-one function may be defined as follows:

$$\text{Tan} = \{(x,y) : -\frac{\pi}{2} < x < \frac{\pi}{2} \text{ and } y = \tan x\}$$

Within this domain the tangent is a continuous curve that is always increasing. The range of this Tangent function is the set of all real numbers.

EXAMPLE: Tan x = 0.4832, find x.

CALCULATOR SOLUTION: Radian mode. Input 0.4832. Press INV or ARC key.

Press Tan key. Read .45 to two decimal places.

EXAMPLE: Tan x = -1.854, find x.

CALCULATOR SOLUTION: Radian mode. Input -1.854. Press INV or ARC key.

Press Tan key. Read -1.076. Round to hundredths.

$$x = -1.08$$

EXAMPLE: Find x if tan x = 0.3247

CALCULATOR SOLUTION: Radian mode. Input 0.3247. Press INV or ARC key.

Press Tan key. Read 0.31 to hundredths.

The solution, however, is incomplete as the problem did not specify

or imply principal values. A more general solution is required.

To do this we take note of the tangent's periodicity. That is,

x = 0.31 + kπ, where k = 0,$\pm$1,$\pm$2,... .

EXAMPLE: Find x if tan x = -5.483 and $\frac{3\pi}{2} < x < 2\pi$.

CALCULATOR SOLUTION: Radian mode. Input -5.483. Press INV or ARC key.

Press Tan key. Read -1.39. Add 2π or 6.28.

Read x = 4.89 rounded to hundredths.

DISCUSSION: The solution above takes into account the fact that the calculator's

Tan key is really defined for principal values or, in other words,

the Tan function. Hence, when given a value in the range as input,

it will display only domain values between $-\frac{\pi}{2}$ and $\frac{\pi}{2}$. It was

necessary, therefore, to add 2π to -1.39 in order to obtain the

correct value of x between $\frac{3\pi}{2}$ and 2π.

Graphing the Cotangent Function

Now that we are familiar with the tangent function and its graph, we

shall use this information to graph the cotangent function. This method makes

use of the fact that these functions are reciprocals of each other.

Figure 4-3C. The Cotangent Function

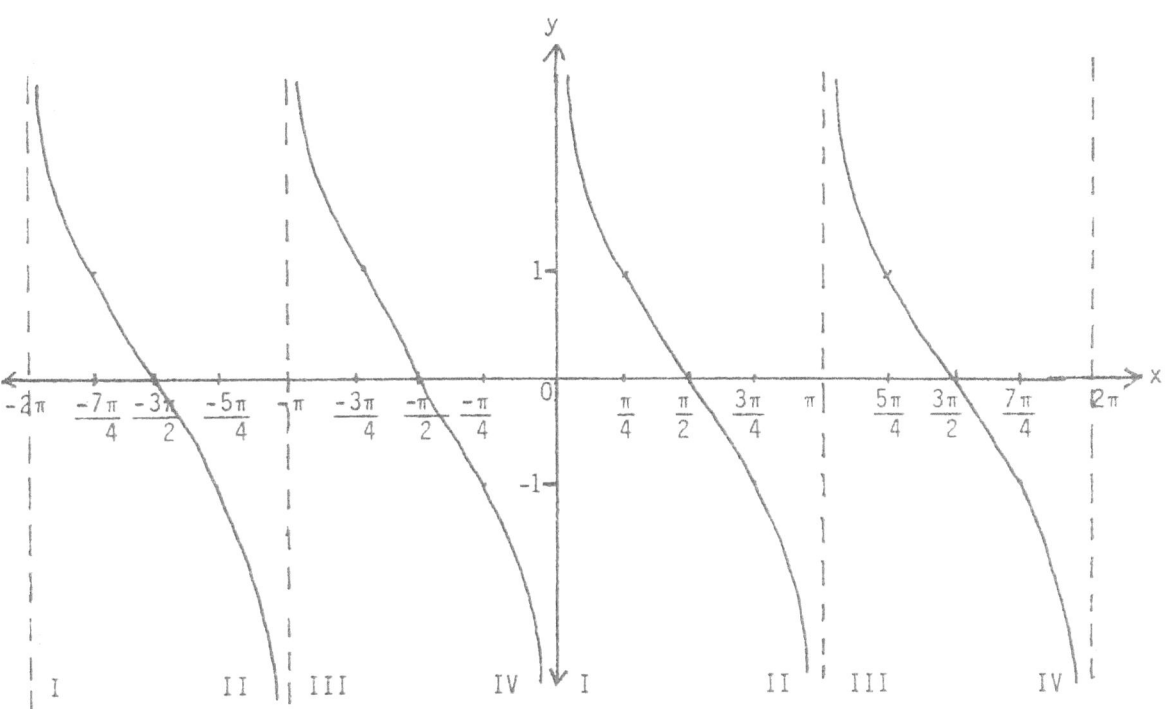

Comparison of Figures 4-3B and 4-3C will reveal that wherever tan x = 0, cot x is undefined and there is an asymptote. Conversely, if tan x is undefined, cot x = 0. For any odd multiple of $\frac{\pi}{4}$, cot x = tan x and this value is either 1 or -1. The periodicity of the cotangent function is also evident and it has the same primitive period of π units as the tangent. Observing that portion of the cotangent function which occurs between any two asymptotes indicates it is a decreasing function on that restricted domain. The cotangent is also an odd function having symmetry with respect to the origin.

In the definition of its principal values the cotangent function has more in common with the cosine function than it does with the tangent. We note that between 0 and π the cotangent is continuous and covers its entire range.

Therefore, we define

$$Cot = \{(x,y) : 0 < x < \pi \text{ and } y = \cot x\}$$

This definition creates some difficulty when applied to calculator problems. Recall that the reciprocal relationship is used to find cotangents since there is no cot key.

EXAMPLE: Find cot 1.56.

SOLUTION: Radian mode. Input 1.56. Press tan key. Press $\frac{1}{x}$ key. Read .0108 rounded. Hence, cot 1.56 = .0108.

EXAMPLE: Find Cot 2.2.

SOLUTION: Radian mode. Input 2.2. Press tan key. Press $\frac{1}{x}$ key. Read -.7279 rounded. Hence, Cot 2.2 = -.7279.

EXAMPLE: Find x if Cot x = -.7279.

SOLUTION: Input -.7279. Press $\frac{1}{x}$ key. Press INV or ARC key. Press tan key. Read -.94 rounded to hundredths. Since the domain of Cot is between 0 and π, it is necessary to add π or 3.14 to the calculator's answer. The result is 2.2 as we would expect from the previous example.

EXERCISE SET 4.3.

Answer True or False:

1) In quadrant I sin increases and tan increases.

2) In quadrant I cos decreases and cot decreases.

3) In quadrant III tan increases and cos increases.

4) In quadrant IV cot increases and sin decreases.

Use calculator and graph intersections to answer the following:

5) For what value of x , $0 < x < \frac{\pi}{2}$ does sin x = tan x?

6) For what value of x , $0 < x < \frac{\pi}{2}$ does sin x = cot x?

7) For what value of x , $0 < x < \frac{\pi}{2}$ does cos x = tan x?

8) For what values of x , $0 < x < \frac{\pi}{2}$ is cos x < cot x?

Find the following:

9) Tan -.48   10) Cot 3.87   11) Tan 74.3

12) Cot (-9.82)   13) cot (-.043)   14) tan (-1.53)

15) x , if Tan x = .7083

16) x , if Tan x = -.3021

17) x , if Cot x = 1.954

18) x , if Cot x = -.7643

19) x , if cot x = -1.594, $\frac{3\pi}{2} < x < 2\pi$

20) x , if tan x = -5.421 , $\frac{-3\pi}{2} < x < -\pi$

In how many points do the graphs intersect:

21) sin   and tan   between 0 and 2π inclusive.

22) sin   and cot   between $-\pi$ and π inclusive.

23) cos   and tan   between 0 and $\frac{3\pi}{2}$ inclusive.

24) cos   and cot   between $-\frac{\pi}{2}$ and π inclusive.

Graph the following:

25) y = tan x , $-\frac{\pi}{2} < x < \frac{3\pi}{2}$

26) y = tan x , $4 < x < 7$

27) y = cot x , $-3 < x < -1$

28) y = cot x , $-2\pi < x < \pi$

29) y = |tan x| , $0 < x < 2\pi$

30) y = |cot x| , $0 < x < 2\pi$

31) y = $\tan^2 x$, $0 < x < 2\pi$

32) y = $\cot^2 x$, $-\pi < x < \pi$

4.4 THE SECANT AND COSECANT FUNCTIONS

Graphs of the secant and cosecant functions may be obtained directly from their reciprocal relationships with the cosine and sine functions, respectively.

Figure 4-4A. The Secant Function

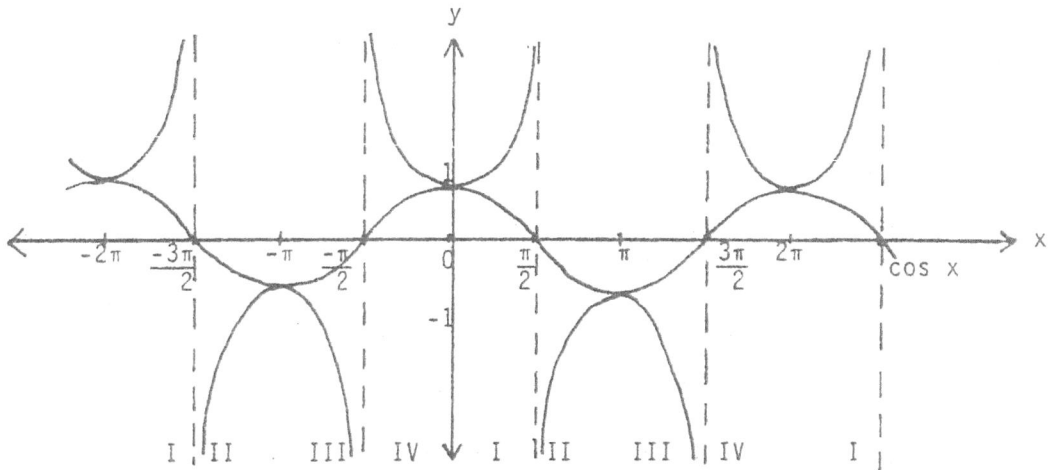

Figure 4-4B. The Cosecant Function

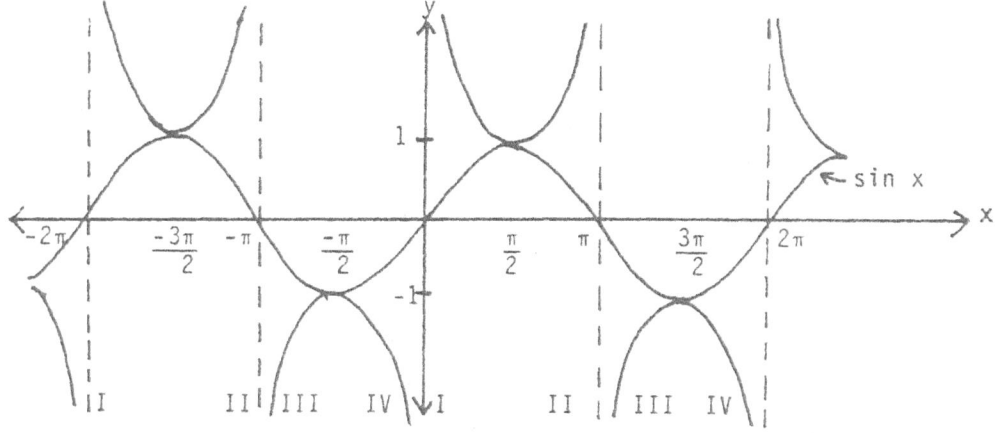

The following table summarizes the features of each graph:

Characteristic	Secant	Cosecant
Periodicity	Primitive period = 2π	Primitive period = 2π
Discontinuity	Asymptotes at $\frac{k\pi}{2}$, $k=\pm1,\pm3,\pm5,\ldots$	Asymptotes at $k\pi$, $k=0,\pm1,\pm2,\ldots$
Odd or Even	Even Symmetric with respect to y-axis sec $(-x)$ = sec x	Odd Symmetric with respect to the origin csc $(-x)$ = -csc x
Phase Relationship	Leads (with respect to csc x)	Lags. $\csc(\frac{\pi}{2} + x)$ = sec x
Principal Values*	Positive values: $0 \le x < \frac{\pi}{2}$ Negative values: $-\pi \le x < \frac{-\pi}{2}$	Positive values: $0 < x \le \frac{\pi}{2}$ Negative values: $-\pi < x \le \frac{-\pi}{2}$

*The choice of principal values for the secant and cosecant functions may appear arbitrary. The rationale for the selection of these intervals will become clear in the calculus.

Definitions, Domains, and Ranges.

secant   = $\{(x,y) : x \ne \frac{k\pi}{2},\ k = \pm1,\pm3,\pm5,\ldots$ and y = sec $x\}$

cosecant   = $\{(x,y) : x \ne k\pi,\ k = 0,\pm1,\pm2,\ldots$ and y = csc $x\}$

The domain of the secant function is the set of all real numbers with the exception of both positive and negative odd multiples of $\frac{\pi}{2}$. The exceptions are indicated by asymptotes on the graph. Similarly, for the cosecant function, the domain is the set of all real numbers with the exception of positive or negative integral multiples of π.

The graphs of both functions show that neither the secant or cosecant contains values in their ranges between -1 and 1 . That is,

Range of secant = {y : y ≤ -1 or y ≥ 1 } = Range of cosecant .

Another way of expressing this is to write:

$$|\sec x| \geq 1 \quad , \qquad |\csc x| \geq 1 \; .$$

The one-to-one principal value functions are defined as:

Sec = {(x,y) : $0 \leq x < \frac{\pi}{2}$, $-\pi \leq x < \frac{-\pi}{2}$, y = sec x }

Csc = {(x,y) : $0 < x \leq \frac{\pi}{2}$, $-\pi < x \leq \frac{-\pi}{2}$, y = csc x }

Care must be exercised while employing the calculator to find domain and range values of these functions. The examples that follow should be studied carefully.

EXAMPLE: Use a calculator to find a) sec 2.85 b) csc 4.37 c) Sec 1.43
           d) Csc 2.15 .

SOLUTION: a) Since sec 2.85 = $\dfrac{1}{\cos 2.85}$, input 2.85 in radian mode. Press

cos key. Press $\frac{1}{x}$ key. Read sec 2.85 = -1.044 rounded.

b) csc 4.37 = $\dfrac{1}{\sin 4.37}$. Input 4.37 in radian mode. Press sin key.

Press $\frac{1}{x}$ key. Read csc 4.37 = -1.062 rounded.

c) Since $0 < 1.43 < \frac{\pi}{2}$, Sec 1.43 is defined. Input 1.43 in radian mode. Press cos key. Press $\frac{1}{x}$ key. Read Sec 1.43 = 7.126.

d) Since $\frac{\pi}{2} < 2.15 < \pi$, 2.15 is not in the domain of the Csc function. Hence Csc 2.15 is undefined.

EXAMPLE: Find a) Sec (-1.612) b) Csc (-3.074) c) Sec (-5.3)

SOLUTION: a) Since $-\pi < -1.612 < \frac{-\pi}{2}$, Sec (-1.612) is defined. Radian mode. Input -1.612. Press cos key. Press $\frac{1}{x}$ key. Read Sec(-1.612)= -24.28.

b) Since $-\pi < -3.074 < \frac{-\pi}{2}$, Csc (-3.074) is defined. Input (-3.074). Press sin key. Press $\frac{1}{x}$ key. Read Csc(-3.074) = -14.81 .

c) Since $-2\pi < -5.3 < \frac{-3\pi}{2}$, -5.3 is not in the domain of the Sec function. Hence Sec (-5.3) is undefined.

In the following examples values in the ranges of these functions are given and the calculator is used fo find values that correspond in the domains.

EXAMPLE: Find x if a) sec x = 3.857   b) Sec x = 3.857 .

SOLUTION: a) Radian mode. Input 3.857 . Press $\frac{1}{x}$ key. Press INV or ARC key. Press cos key. Read 1.31 as a reference number. Since the sec function is positive in quadrants I and IV and periodic with primitive period 2π ,   x = ±1.31 + 2kπ , k = 0 , ±1 , ±2 , ...

b) Positive range values for the Sec function are given by domain numbers between 0 and $\frac{\pi}{2}$, i.e. quadrant I . Hence x = 1.31 .

EXAMPLE: Find x if a) csc x = 1.638   b) Csc x = 1.638 .

SOLUTION: a) Radian mode. Input 1.638 . Press $\frac{1}{x}$ key. Press INV key. Press sin key. Read reference number .66. The csc function is positive in quadrants I and II and periodic with primitive period 2π . Therefore, x = .66 + 2kπ or x = 2.48 + 2kπ , k = 0 , ±1 ,±2 ,... (Note that 2.48 = π - .66).

b) Positive values in the range of the Csc function are given by domain values between 0 and $\frac{\pi}{2}$. Hence x = .66 , the reference number found above.

A methodical approach to finding domain values for the sec, csc, Sec, and Csc functions when the given range values are negative draws upon the symmetry relationships discussed in Chapter 3 . The clue to this simplified approach is to proceed first in finding the reference number in quadrant I between 0 and $\frac{\pi}{2}$ and then determine the increment of π required to yield the desired value.

EXAMPLE: Find x if a) sec x = -2.594 and 0< x < π b) Sec x = -2.594 .

SOLUTION: a) Refer to Figure 4-4C . Radian mode. Input |-2.594| = 2.594. Press $\frac{1}{x}$ key. Read .3855. Press INV key. Press cos key. Read the reference number 1.175. Since sec x is negative and the given domain is between 0 and π , x must lie between $\frac{\pi}{2}$ and π . That is, x = π - 1.175 = 1.97 .

b) Follow the steps given above in part a). By the definition of Sec, x must lie between -π and -$\frac{\pi}{2}$. To obtain x, subtract from the reference number. x = 1.175 - π = -1.97 .

Figure 4-4C

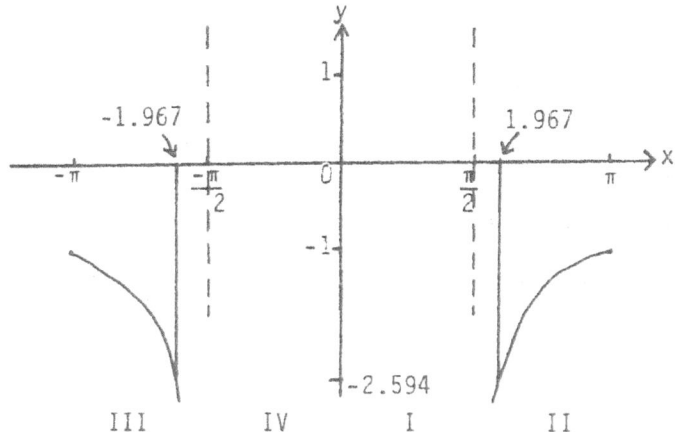

EXAMPLE: Find a) the smallest positive value of x for which csc x = -4.863

b) x if Csc x = -4.863 .

SOLUTION: a) Refer to Figure 4-4D . Radian mode. Input |-4.863| = 4.863.

Press $\frac{1}{x}$ key. Read .2056 . Press INV key . Press sin key. Read

the reference number .207 .. The smallest positive value of x for

which csc x = -4.863 will be found in quadrant III to the right of

the Y-axis. That is, x = π + .207 = 3.35 .

b) Repeat the steps above for finding the reference number, .207 .

By the definition of Csc , x must lie between -π and $-\frac{\pi}{2}$.

To obtain x , subtract π from the reference number. That is,

x = .207 - π = -2.93 .

Figure 4-4D

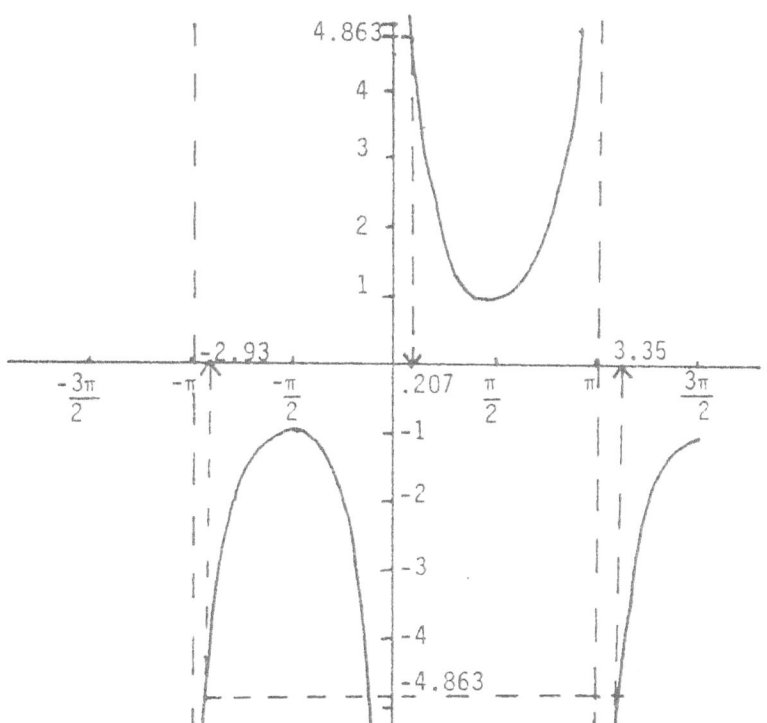

EXERCISE SET 4-4.

Refer to Figures 4-4A and 4-4B for the following:

1) In what quadrants is the secant function increasing?

2) In what quadrants is the cosecant function increasing?

3) In what quadrants is the secant function decreasing?

4) In what quadrants is the cosecant function decreasing?

5) In what quadrant do the secant and cosecant both increase?

6) In what quadrant do the secant and cosecant both decrease?

7) In what quadrant are the secant and cosecant both negative?

8) In what quadrant are the secant and cosecant both positive?

Given $-\pi < x < 2\pi$, complete the following:

9) cos and sec intersect in _____ points.

10) csc and sin intersect in _____ points.

11) sec and sin intersect in _____ points.

12) csc and cos intersect in _____ points.

13) sec and csc intersect in _____ points.

14) Find x, if Csc x = Sec x .

15) Find x, if Csc (-x) = Sec x .

16) Find x, if Csc x = 1 and sec x is undefined.

17) Find x, if csc x is undefined and sec x = -1 .

18) Between which quadrants does the secant function reach a maximum value?
    A minimum value? Remember ∞ is undefined.

19) Between which quadrants does the cosecant function reach a maximum value?
    A minimum value? Remember ∞ is undefined.

20) What is the range of the cosecant function in quadrant I ?

Use calculator or Table I to find the following:

21) sec 1.15

22) sec 2.86

23) csc .13

24) csc (-1.43)

25) Sec (-2.87)

26) Sec 4.37

27) Csc (-2.13)

28) Csc (-5.62)

Given $-\pi < x < \pi$, find x :

29) Sec x = 2.854

30) Csc x = 1.053

31) Sec x = -4.327

32) Sec x = -2.839

33) Csc x = -1.032

34) Csc x = -.3954

35) sec x = 2.541

36) sec x = 9.597

37) csc x = 1.798

38) csc x = -5.832

39) sec x = -1.123

40) sec x = -3.651

4.5 AMPLITUDE AND PERIOD. GRAPHING TECHNIQUES.

Amplitude

Figure 4-5A shows the effect of multiplying the function by a constant
factor. On the same set of axes the graphs of $y = \sin x$ and $y = 2 \sin x$ are
shown.

Figure 4-5A

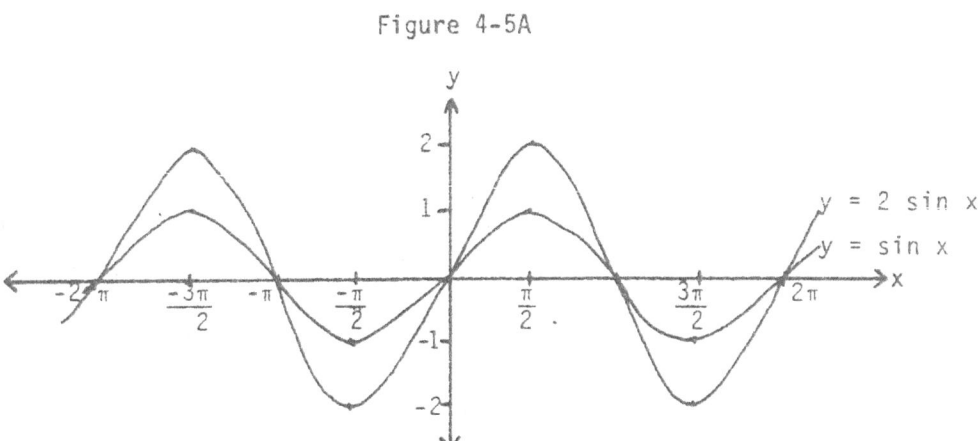

Let $f(x) = \sin x$ and $g(x) = 2 \sin x$. Then for each value of x, the
ordinate of g is twice the ordinate of f. We describe this relationship by
saying the amplitude of g is twice the amplitude of f.

DEFINITION: If A is a real number and $f(x) = A \sin x$, then the amplitude of
f is given by $|A|$.

The amplitude indicates how far above and below the x-axis the sine
(or cosine) function reaches. Essentially, it describes the upper and lower
bounds for the range of the sine or cosine functions.

EXAMPLE: What is the range of $f(x) = \frac{2}{3} \cos x$?

SOLUTION: Since $\left|\frac{2}{3}\right| = \frac{2}{3}$, the range is given by:

$\{y : -\frac{2}{3} \le y \le \frac{2}{3}\} = \left[\frac{-2}{3}, \frac{2}{3}\right]$, using interval notation for
a closed interval.

EXAMPLE: Graph the function g(x) = -2 cos x. What is the range of g?

SOLUTION: Figure 4-5B. The coefficient -2 has the effect of reflecting

the cosine function about the x axis and the amplitude,

| -2 | = 2, stretches the range to 2 units above and below the

x-axis. The standard cosine function is shown for comparison.

Figure 4-5B

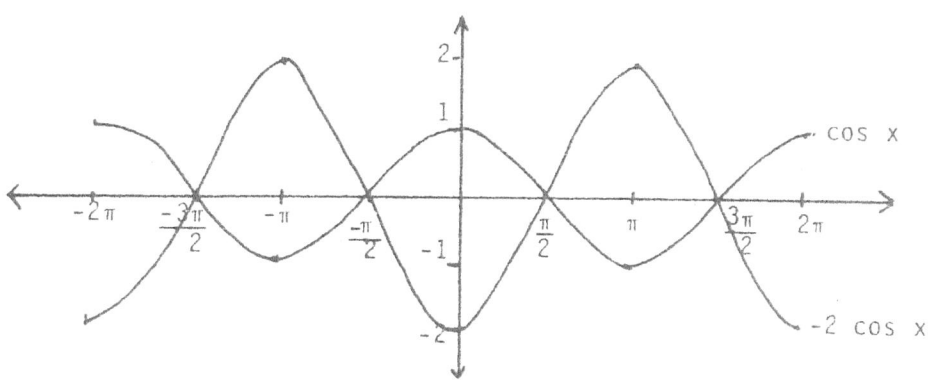

Range of g = {y : -2 ≤ y ≤ 2} = $[-2,2]$

EXAMPLE: Graph f(x) = 2 tan x. Compare its graph to that of tan x.

SOLUTION: Figure 4-5C

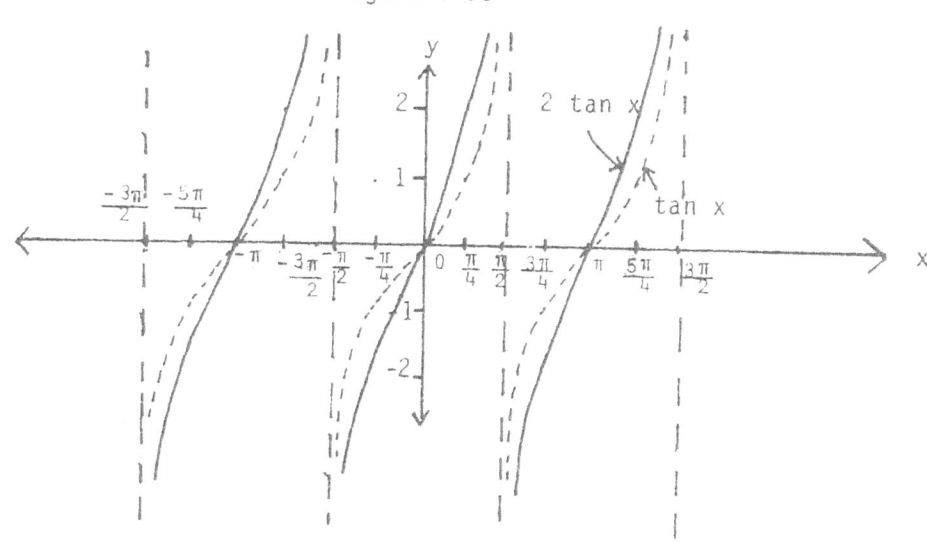

Although it isn't appropriate to describe the effect as a change in amplitude since the range of both functions are the same, the function 2 tan x * can be viewed as rising to +∞ twice as fast as tan x.

EXAMPLE: Graph g(x) = - tan x

SOLUTION: Figure 4-5D

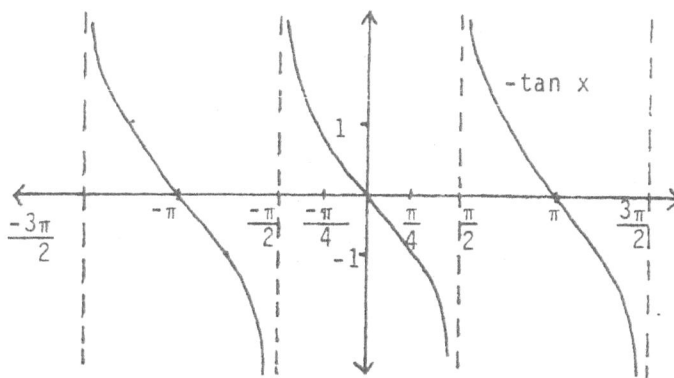

The graph above is interesting in that it reveals how the negative coefficient converts the tangent function into a cotangent function shifted $\frac{\pi}{2}$ units to the left. (Compare with Figure 4-3C.) It also confirms the complementary relationship cot $(\frac{\pi}{2} + x)$ = - tan x.

Period

In a functional expression like cos x, x is called the "argument" of the function. If the coefficient of the argument is changed, this will have a direct effect on the primitive period of the function.

* For the sake of brevity, ordinate expressions like 2 tan x are being used to represent the function itself; i.e., {(x.y): y = 2 tan x} .

<u>EXAMPLE</u>:   Compare the graphs of y = cos x and y = cos 2x.

<u>SOLUTION</u>:   Figure 4-5E

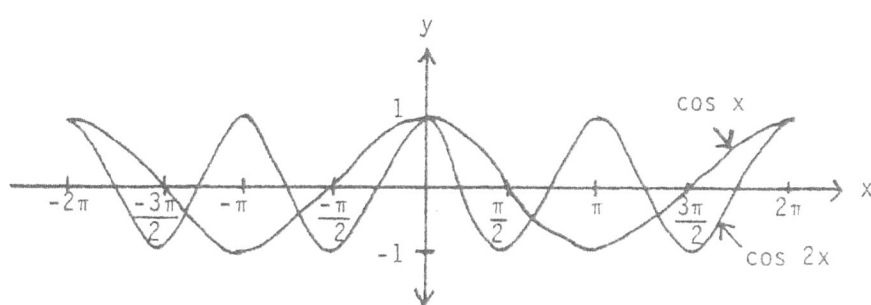

The primitive period of cos 2x is π units as compared to 2π units for cos x. In general, the primitive period of cos k x, k ≠ 0, is given by $\frac{2\pi}{|k|}$.

The same formula holds for determination of the primitive periods of sin k x, sec k x, and csc k x.

<u>EXAMPLE</u>:   Graph the function $f(x) = 2 \sec \frac{1}{2} x$.

<u>SOLUTION</u>:   Figure 4-5F

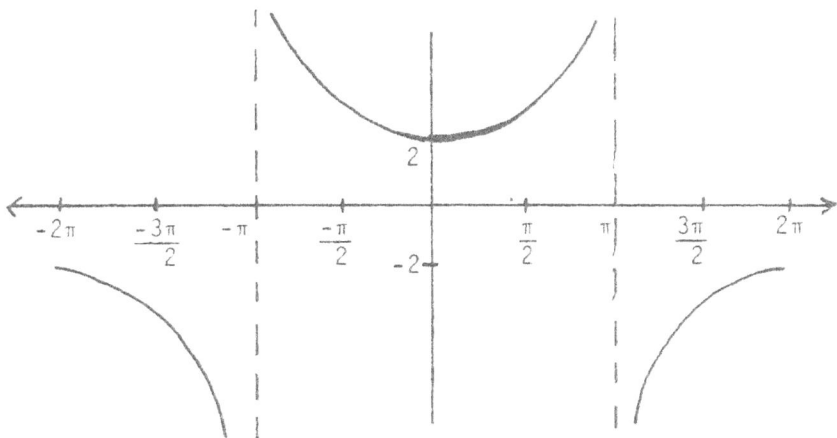

The results are obtained by reference to the standard secant function (Figure 4-4A). First we note that the "amplitude" coefficient 2 alters the range so that the minimum positive value is 2 and the maximum negative value is -2. Next, the primitive period is determined by the formula $\frac{2\pi}{|k|}$, where $k = \frac{1}{2}$. This gives a primitive period of 4π units, twice the standard period.

The effect on the graph is to stretch it out along the x-axis, so that one cycle or period of the curve is positioned between -2π and 2π.

ALTERNATIVE SOLUTION: Another approach that is effective in drawing quick sketches of desired graphs is to set down the standard function and then change the units to agree with the given coefficients. This is demonstrated in Figure 4-5G for

$$f(x) = 2 \sec \frac{1}{2} x.$$

Figure 4-5G

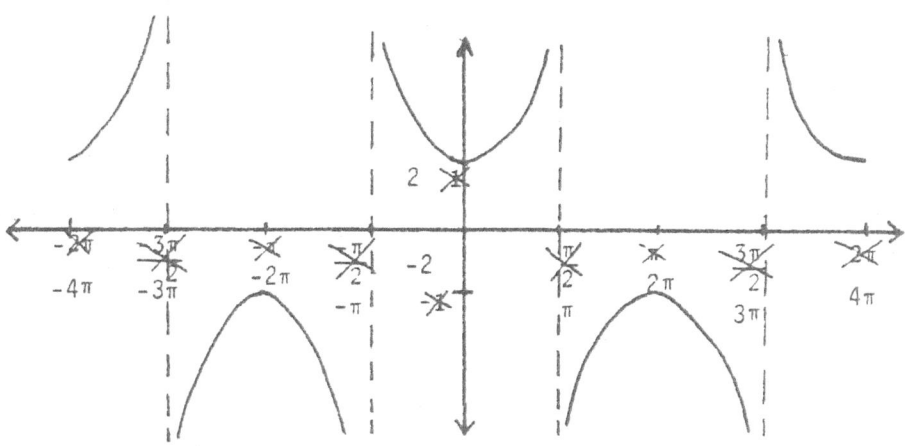

The advantage of this method is the appearance of a larger portion of the graph for functions having longer primitive periods. This technique is used in the following example.

EXAMPLE: Graph h(x) = 3 sin π x.

SOLUTION: The amplitude is 3 and the primitive period is $\frac{2\pi}{|\pi|} = 2$.

First, graph the function y = sin x. Then erase or cross out the units to reflect the coefficients of h.

Figure 4-5H

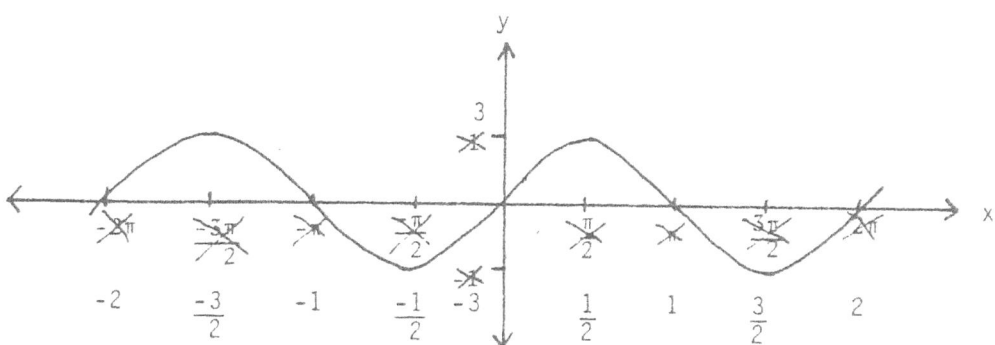

The only disadvantage to this technique is that it may not reflect the true shape of the graph. In the example above, the units along the x and y axis are out of proportion. Figure 4-5I corrects this distortion.

Figure 4-5I

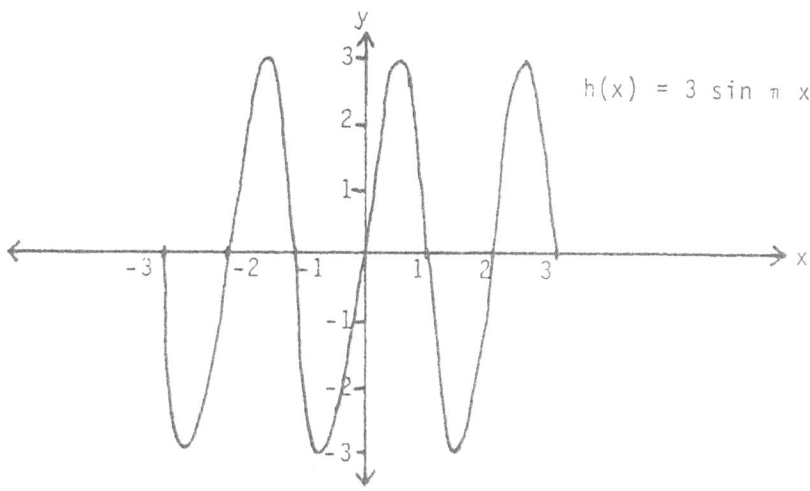

$$h(x) = 3 \sin \pi x$$

Periodic Phenomena in Nature

The example above provides some insight as to how a physicist mathematizes effects in nature that repeat themselves in a regular pattern. Such periodic phenomena as musical tones, alternating electric currents, seismic waves, and heart contractions lend themselves to trigonometric analysis. Sine and cosine waves are particularly effective in providing a mathematical model for these physical effects. In the function above, the coefficient π has the effect of converting the x-axis divisions into integral units. This, in turn, can be interpreted as a time unit axis. Thus, the graph above could represent a natural phenomenon that repeats itself every two seconds. Now consider a function defined by $g(x) = A \sin 120\pi x$. The primitive period of this function is $\frac{2\pi}{|120\pi|} = \frac{1}{60}$. If x is graduated in seconds, this means that 60 periods of the function will occur every second. The formula, therefore, describes the everyday occurrence of household electric current which is generated at 60 cycles per second. In the terminology of physics the reciprocal of the primitive period is known as the <u>frequency</u>. For example, the standard tone for tuning an orchestra has a frequency of 440 Hz. The primitive period of this tone is $\frac{1}{440}$ second.

Periods of Tangent and Cotangent Functions

Since the primitive period of tan and cot is π units, this must be taken into consideration when evaluating factors that alter the periods of these functions.

EXAMPLE: Graph $g(x) = \tan 2x$.

SOLUTION: To determine the period divide π by $|2|$.

$\frac{\pi}{|2|} = \frac{\pi}{2}$. Graph tan x as before and then divide each division point by 2. Note that if $x = \frac{\pi}{8}$, $\tan 2x = \tan 2(\frac{\pi}{8}) = \tan \frac{\pi}{4} = 1$

Figure 4-5J

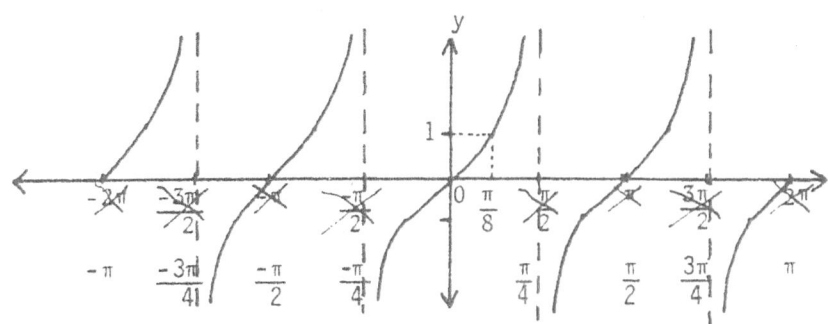

EXAMPLE: Graph $k(x) = 2 \cot \frac{\pi}{3} x$.

SOLUTION: Period $= \frac{\pi}{\left|\frac{\pi}{3}\right|} = 3$. Graph $y = \cot x$ as before and then multiply

each division point by $\frac{3}{\pi}$. Note that if $x = \frac{3}{4}$, then

$2 \cot \frac{\pi}{3} x = 2 \cot \frac{\pi}{3} (\frac{3}{4}) = 2 \cot \frac{\pi}{4} = 2$.

Figure 4-5K

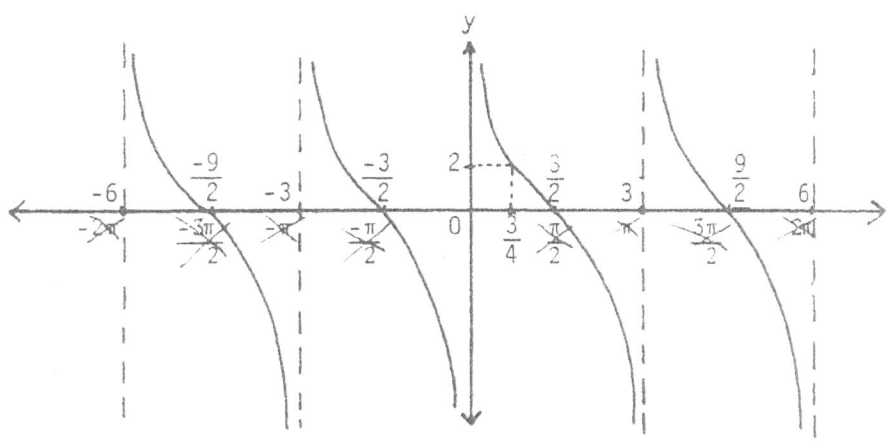

EXERCISE SET 4-5.

Determine the amplitude and period of each function. Graph the function.

1) $f(x) = 3 \cot x$

2) $g(x) = \frac{1}{3} \sin 2x$

3) $h(x) = \frac{2}{5} \cos x$

4) $t(x) = \frac{\pi}{4} \tan \frac{2}{5} x$

5) $k(x) = \frac{3}{2} \sec x$

6) $s(x) = \frac{5}{3} \csc \frac{\pi}{3} x$

7) $r(x) = \frac{8}{3} \tan \frac{1}{3}\pi x$

8) $p(x) = \frac{5}{4} \sin \pi x$

9) $u(x) = 2 \cos 3x$

10) $m(x) = 4 \cot 2\pi x$

Determine the domain and range of the following functions. Graph.

11) $y = 2 \cos 5x$

12) $y = 3 \tan \pi x$

13) $y = 3 \sin 2x$

14) $y = 4 \sec 5x$

15) $y = \frac{1}{2} \cot \frac{3}{4} x$

16) $y = \frac{2}{3} \csc \frac{4}{5} x$

Note: Express the domains and ranges of the above using interval notation.

4-6. SUMMATION GRAPHS AND PHASE SHIFT

Consider the graphs of the quadratic function $y = x^2$ and the linear function $y = x - 1$.

Figure 4-6A

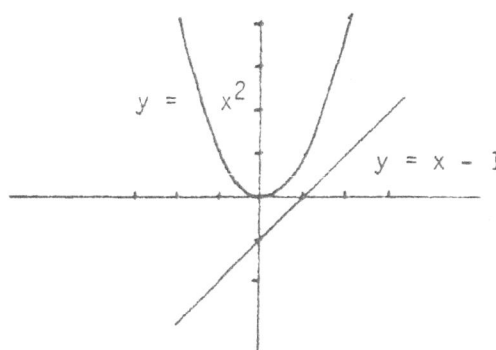

Tables of ordered pairs for the two functions are given below:

x	-2	-1	0	1	2	3
$y = x^2$	4	1	0	1	4	9

x	-2	-1	0	1	2	3
$y = x - 1$	-3	-2	-1	0	1	2

If the ordinates of the two functions are added for the domain values
shown in the tables,

x	-2	-1	0	1	2	3
$y = x^2 + x - 1$	1	-1	-1	1	5	11

ordered pairs are determined for the function $y = x^2 + x - 1$. The resultant
graph may be viewed as a <u>summation</u> of the graphs of the component functions
$y = x^2$ and $y = x - 1$:

Figure 4-6B

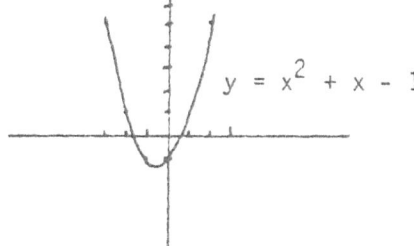

Summation graphs of the trigonometric functions may also be constructed by this method of adding ordinates.

EXAMPLE: Graph the summation function $y = \sin x + \cos x$.

SOLUTION: Select a representative domain for both functions and construct a table of ordered pairs using special domain values.

x	$-\pi$	$-\frac{5\pi}{6}$	$-\frac{3\pi}{4}$	$-\frac{2\pi}{3}$	$-\frac{\pi}{2}$	$-\frac{\pi}{3}$	$-\frac{\pi}{4}$	$-\frac{\pi}{6}$	0	$\frac{\pi}{6}$	$\frac{\pi}{4}$	$\frac{\pi}{3}$	$\frac{\pi}{2}$	$\frac{2\pi}{3}$	$\frac{3\pi}{4}$	$\frac{5\pi}{6}$	π
sin x	0	-.5	-.7	-.9	-1	-.9	-.7	-.5	0	.5	.7	.9	1	.9	.7	.5	0
cos x	-1	-.9	-.7	-.5	0	.5	.7	.9	1	.9	.7	.5	0	-.5	-.7	-.9	-1
y		-1	-1.4	-1.4	-1.4	-1	-.37	0	.37	1	1.4	1.4	1.4	1	.37	0	-.37	-1

Figure 4-6C

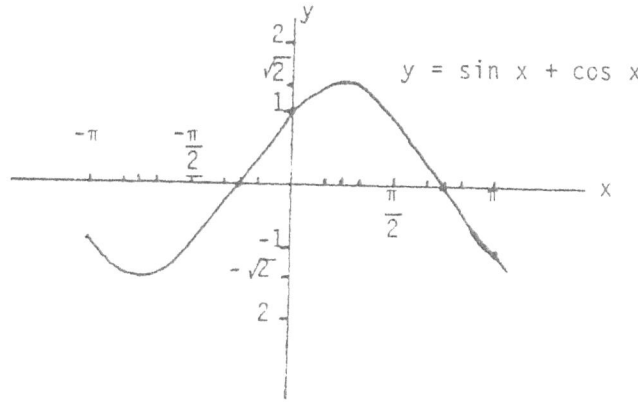

The resultant graph consist of one period of a function that resembles a sine or cosine graph. This summation function has an amplitude of $\sqrt{2}$ and a primitive period of 2π . Later it will be shown how this summation function and similar summations of sine and cosine functions may be expressed as simple sine functions.

The calculator may be used to quickly assemble a sufficient set of ordered pairs to render a good graph of a summation function.

EXAMPLE: Graph $y = 2 \tan x - \cos 2x$.

SOLUTION: Set up a table of x-values from $-\pi$ to π . Use integers and multiples of one-half in addition to quadrantal values.

x	$-\pi$	-3	-2.5	-2	$-\frac{\pi}{2}$	-1.5	-1	-.5	0	.5	1.0	1.5	$\frac{\pi}{2}$	2	2.5	3	π
tan x	0	.14	.75	2.2	∞	-14	-1.6	-.5	0	.5	1.6	14	∞	-2.2	-.75	-.14	0
2 tan x	0	.28	1.5	4.4	∞	-28	-3.2	-1.0	0	1.0	3.2	28	∞	-4.4	-1.5	-.28	0
2x	-2π	-6	-5	-4	$-\pi$	-3	-2	-1	0	1	2	3	π	4	5	6	2π
cos 2x	1	.96	.28	-.7	-1	-.98	-.4	.54	1	.54	-.4	-.98	-1	-.7	.28	.96	1
y	-1	-.7	1.2	5.1	∞	-27	-2.8	-1.5	-1	-.46	3.6	29	∞	-3.7	-1.8	-1.2	-1

Figure 4-6D

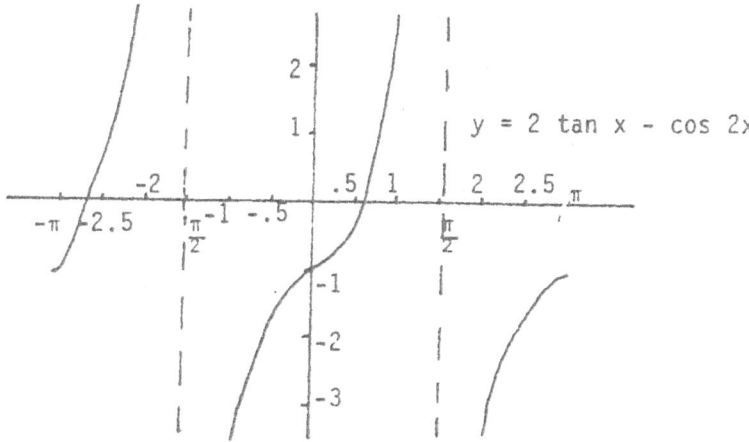

If a quick sketch of a summation graph is desired, draw the component graphs and visually deduce the location of points as in the example below.

EXAMPLE: Graph $f(x) = 2 \cos 3x - \sin 2x$.

SOLUTION: Sketch the graphs of $y = 2 \cos 3x$ and $y = - \sin 2x$ on the same set of axes as shown below. Open a compass to selected ordinates

of y = - sin 2x at points on the domain. If the ordinates are
positive, add to the corresponding ordinates of y = 2 cos 3x .
If the ordinates of - sin 2x are negative, subtract from the corresponding ordinates of y = 2 cos 3x . The compass is a handy
tool for adding ordinates in this manner. Note that although the
primitive period of 2 cos 3x is $\frac{2\pi}{3}$ and the primitive period of
- sin 2x is π , the primitive period of the summation function is
2π which happens to be the lowest common multiple of $\frac{2\pi}{3}$ and π .
This is a general rule for periods of summation functions.

Figure 4-6E

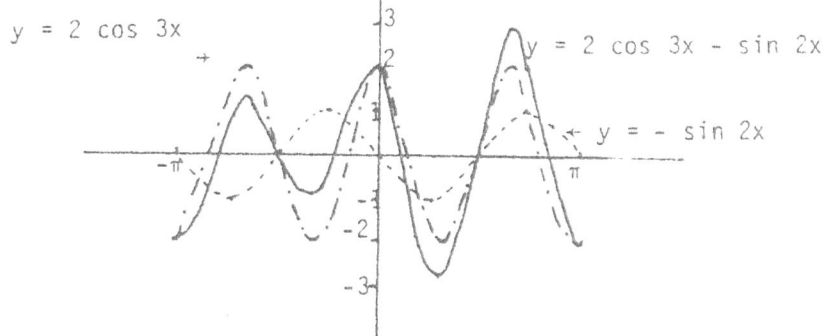

PHASE SHIFT

Earlier it was demonstrated that the sine and cosine graphs, as well as
the secant and cosecant graphs, are identical with the exception of the fact
that they are out of phase with each other by an interval of $\frac{\pi}{2}$ units. These
phase shifts were identified by the complementary relationships, sin $(\frac{\pi}{2} + x)$ =
cos x and csc $(\frac{\pi}{2} + x)$ = sec x . In this section a method of graphing arbitrary
phase shifts will be developed.

EXAMPLE: Graph the function f(x) = cos $(x + \frac{\pi}{4})$.

SOLUTION: A phase shift of $\frac{\pi}{4}$ is identified by the addition of this constant
           to the argument. As a first step in learning to graph phase
           shifts a table will be set up:

x	-2π	$-\frac{7\pi}{4}$	$-\frac{3\pi}{2}$	$-\frac{5\pi}{4}$	$-\pi$	$-\frac{3\pi}{4}$	$-\frac{\pi}{2}$	$-\frac{\pi}{4}$	0	$\frac{\pi}{4}$	$\frac{\pi}{2}$	$\frac{3\pi}{4}$	π
$x + \frac{\pi}{4}$	$-\frac{7\pi}{4}$	$-\frac{3\pi}{2}$	$-\frac{5\pi}{4}$	$-\pi$	$-\frac{3\pi}{4}$	$-\frac{\pi}{2}$	$-\frac{\pi}{4}$	0	$\frac{\pi}{4}$	$\frac{\pi}{2}$	$\frac{3\pi}{4}$	π	$\frac{5\pi}{4}$
$\cos\left(x + \frac{\pi}{4}\right)$	$\frac{1}{\sqrt{2}}$	0	$\frac{-1}{\sqrt{2}}$	-1	$\frac{-1}{\sqrt{2}}$	0	$\frac{1}{\sqrt{2}}$	1	$\frac{1}{\sqrt{2}}$	0	$\frac{-1}{\sqrt{2}}$	-1	$\frac{-1}{\sqrt{2}}$

The ordered pairs $(x, \cos (x + \frac{\pi}{4}))$ are plotted in Figure 4-6F

Figure 4-6F

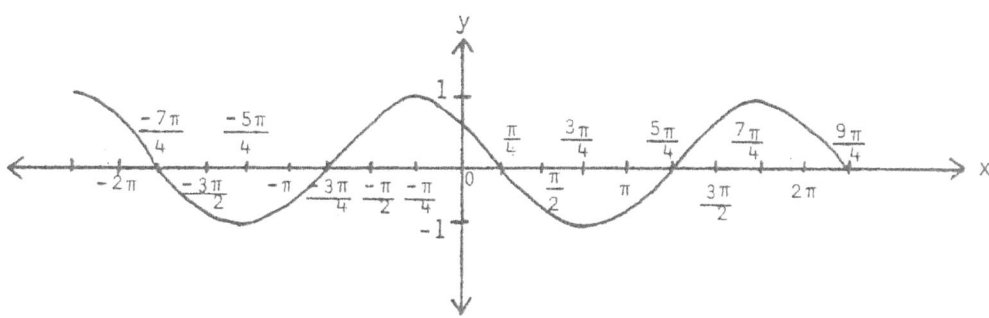

In contrast to cos x, the graph of $\cos (x + \frac{\pi}{4})$ is shifted to the left $\frac{\pi}{4}$ units. It is described as having a <u>lead</u> of $\frac{\pi}{4}$ units. In general, if k > 0, cos (x + k) will have a lead of k units. On the other hand, the function $g(x) = \cos (x - \frac{\pi}{4})$ would be identical to cos x shifted $\frac{\pi}{4}$ units to the right. It would be described as having a <u>lag</u> of $\frac{\pi}{4}$ units (Figure 4-6G).

Figure 4-6G

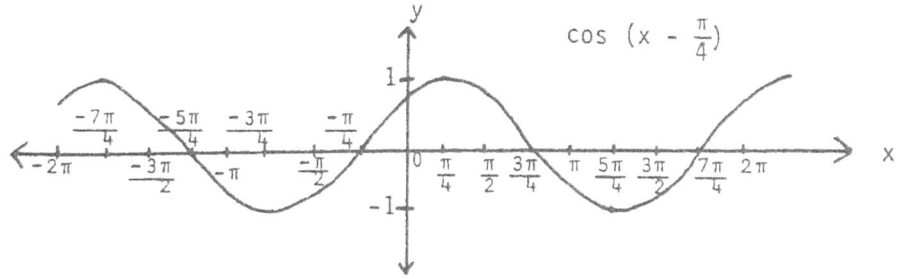

EXAMPLE: Graph $f(x) = \sin\left(x - \frac{\pi}{3}\right)$.

SOLUTION: Instead of constructing a table of ordered pairs, an alternative

graphing method will be described. From the discussion above,

$\sin\left(x - \frac{\pi}{3}\right)$ lags behind the graph of $\sin x$ by $\frac{\pi}{3}$ units. First

graph $\sin x$ without drawing in the y-axis.

Figure 4-6H

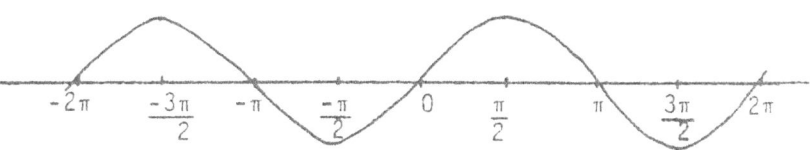

Now, instead of shifting the entire graph $\frac{\pi}{3}$ units to the right, add $\frac{\pi}{3}$

to each of the division points along the x-axis. The altered units will now

appear as in Figure 4-6I.

Figure 4-6I

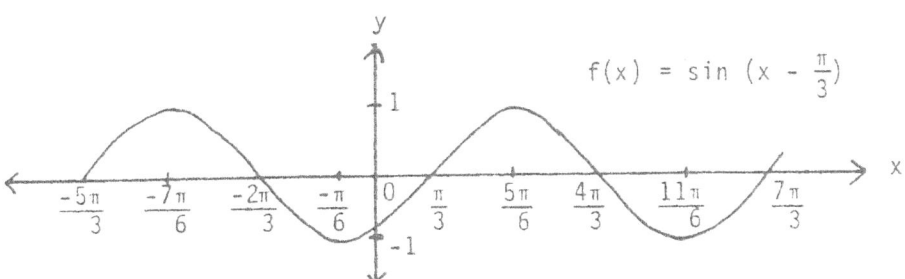

Finally, position the y-axis in its proper intersection at $x = 0$.

Next, this technique will be demonstrated with the tangent function.

EXAMPLE: Graph $f(x) = \tan\left(x + \frac{\pi}{6}\right)$.

SOLUTION: First graph tan x omitting the y-axis.

Figure 4-6J

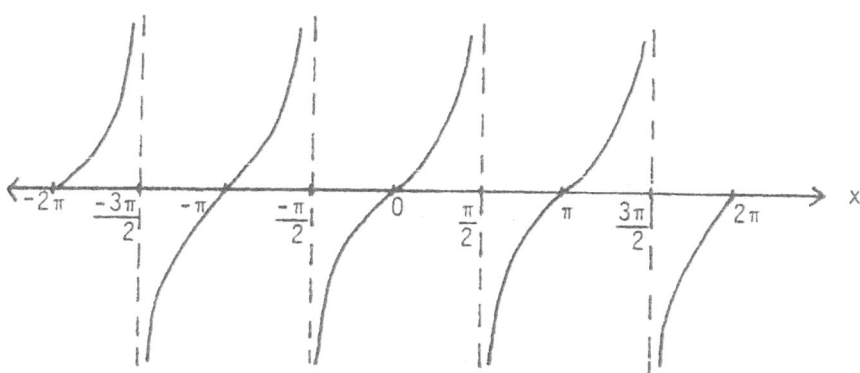

The argument $\left(x + \frac{\pi}{6}\right)$ for f indicates a lead of $\frac{\pi}{6}$ units. Subtract $\frac{\pi}{6}$ units from each division point on the x-axis above.

Figure 4-6K

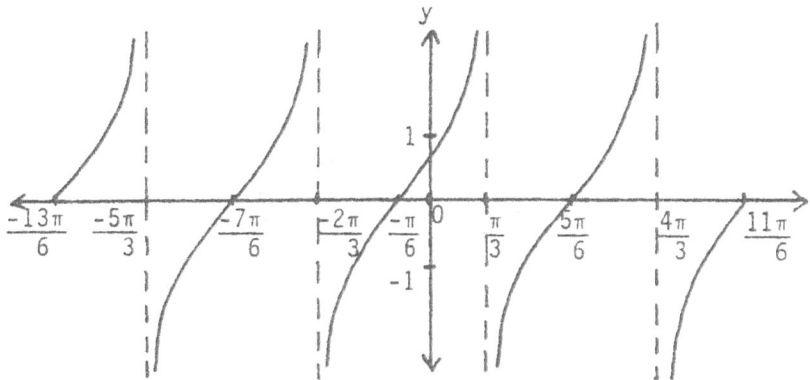

Insert the y-axis in its appropriate position. Note that changing the units not only shifts the graph $\frac{\pi}{6}$ units to the left, but also places the asymptotes in proper positions.

Graphing functions with varying amplitudes, periods, and phase shifts can be simplified by the method of changing units. First, it may be necessary to prepare the functional expression.

EXAMPLE: Graph $g(x) = \frac{2}{3} \cos (2x - \frac{\pi}{6})$.

SOLUTION: The amplitude is $\frac{2}{3}$. Before determining the period and phase shift, it is advisable to factor out 2 from the expression $(2x - \frac{\pi}{6})$. That is, $g(x) = \frac{2}{3} \cos 2 (x - \frac{\pi}{12})$. The factor 2 in front of the parenthesis is used to determine the primitive period as before. The calculation is $\frac{2\pi}{|2|} = \pi$. The phase shift is a lag of $\frac{\pi}{12}$ units, determined by the expression remaining in the parenthesis.

To graph, set up the cos x graph between -2π and 2π, omitting the y-axis. Correct for the period change by dividing each division point by 2.

Figure 4-6L

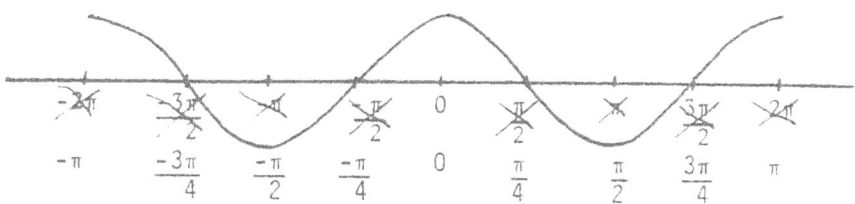

Next, correct for the phase shift by adding $\frac{\pi}{12}$ units to each division point. Insert the y-axis with an amplitude mark of $\pm\frac{2}{3}$.

Figure 4-6M

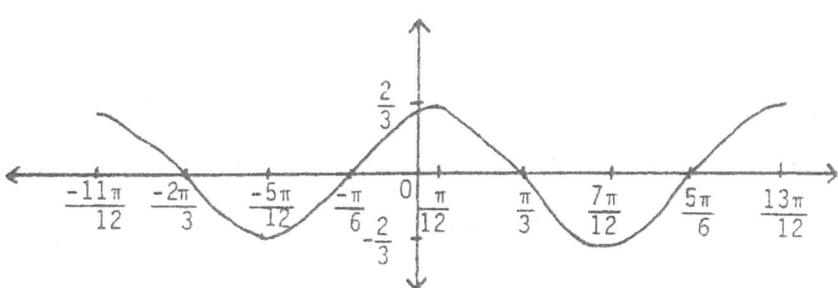

EXERCISE SET 4-6.

In exercises 1 - 10, determine the <u>amplitudes</u>, <u>periods</u>, and <u>phase shifts</u> of the given functions:

1) $f(x) = 2 \cos 3x$

2) $f(x) = -\frac{1}{3} \sin (x - \frac{\pi}{8})$

3) $g(x) = 5 \tan 3x$

4) $h(x) = 2 \sec (3x - \frac{\pi}{4})$

5) $f(x) = -2 \csc \frac{4}{3} x$

6) $g(x) = \pi \cot \pi x$

7) $h(x) = \frac{3}{4} \sin (4x - \pi)$

8) $s(x) = -4 \csc (x - 2)$

9) $t(x) = 10 \cos (\pi - 3x)$

10) $k(x) = \frac{3}{2} \tan (2x - \frac{\pi}{3})$

Graph the following functions:

11) $f(x) = 2 \cos 3x$

12) $g(x) = \frac{1}{2} \sin 4x$

13) $h(x) = - \cot \frac{1}{2} x + 2 \sin x$

14) $k(x) = 3 \sec 2x + 2 \tan x$

15) $t(x) = \frac{3}{2} \csc \frac{1}{3} x - \cos 3x$

16) $p(x) = - \sin \frac{1}{2} x - 2 \cos 2x$

17) $f(x) = 2 \cos (x - \frac{\pi}{3})$

18) $h(x) = -3 \sin (x + \frac{\pi}{4})$

19) $g(x) = 3 \tan \pi x$

20) $q(x) = 4 \sec (2x - \pi)$

21) $p(x) = \frac{5}{4} \sin (2\pi x - \frac{\pi}{2})$

22) $r(x) = -2 \csc (x + \frac{\pi}{4})$

23) $t(x) = 4 \cot (x + \frac{\pi}{3})$

24) $f(x) = 10 \cos 4\pi (x - 3)$

4-7. CHAPTER SUMMARY

1. The following table summarizes the important features of the trigonometric functions.

Function	cos	sin	tan	cot	sec	csc
Domain	Reals	Reals	$x \neq \frac{k\pi}{2}$ $k=\pm1,\pm3$ ...	$x \neq k\pi$ $k=0,\pm1$ ...	$x \neq \frac{k\pi}{2}$ $k=\pm1,\pm3$ ...	$x \neq k\pi$ $k=0,\pm1,\pm2$ ...
Range	$|y| \leq 1$	$|y| \leq 1$	Reals	Reals	$|y| \geq 1$	$|y| \geq 1$
Odd or Even	Even	Odd	Odd	Odd	Even	Odd
Period (Primitive)	2π	2π	π	π	2π	2π
Continuous	Yes	Yes	No	No	No	No
Principal Values +	$0 \leq x \leq \frac{\pi}{2}$	$0 \leq x \leq \frac{\pi}{2}$	$0 \leq x < \frac{\pi}{2}$	$0 < x \leq \frac{\pi}{2}$	$0 \leq x < \frac{\pi}{2}$	$0 < x \leq \frac{\pi}{2}$
Principal Values −	$\frac{\pi}{2} \leq x \leq \pi$	$-\frac{\pi}{2} \leq x \leq 0$	$-\frac{\pi}{2} < x \leq 0$	$\frac{\pi}{2} \leq x < \pi$	$-\pi \leq x < \frac{-\pi}{2}$	$-\pi < x \leq \frac{-\pi}{2}$
Increasing		Sin	Tan			
Decreasing	Cos			Cot		

2. The principal value functions, Sin , Cos , Tan , Cot , Sec , Csc , are one-to-one functions with domains as indicated in the table above.

3. Given $f(x) = A \sin k (x + t)$ or $g(x) = A \cos k (x + t)$, $|A|$ = amplitude, $\frac{2\pi}{|k|}$ = period, and t = phase shift. If $t > 0$, the phase shift is a <u>lead</u>. If $t < 0$, the phase shift is a <u>lag</u>.

4. Given $f(x) = A \tan k (x + t)$ or $g(x) = A \cot k (x + t)$, the factor A has the effect of changing the rate of increase or decrease of the functions. To determine the period, evaluate $\frac{\pi}{|k|}$. The quantity t gives the phase shift as before.

5. Given $f(x) = A \sec k (x + t)$ or $g(x) = A \csc k (x + t)$. Period $= \frac{2\pi}{|k|}$ and t = phase shift. $|A|$ = minimum positive value in the range and $-|A|$ = maximum negative value in the range.

EXERCISE SET 4-7.

Refer to the graphs of the trigonometric functions to determine whether the following statements are <u>true</u> or <u>false</u>:

1) The sin increases on the domain $\frac{3\pi}{2} < x < 2\pi$.

2) Sin and Tan increase on the same domain.

3) Cot and Cos increase on the same domain.

4) Sec and csc intersect at least once in each quadrant.

5) Cot and Cos have the same domain.

6) When sin increases, cos decreases and vice versa.

7) All of the standard trigonometric functions have a period of 2π.

8) If $t > 0$, $\cos (x + t)$ leads $\cos x$.

9) The amplitude of $3 \sin 2x$ is 2.

10) The period of $2 \sin 4x$ is $\frac{\pi}{2}$.

11) tan and -tan are both odd functions.

12) sec and tan are undefined for the same values of x.

13) csc and cos never intersect.

14) csc and cot intersect at least once in each quadrant.

15) tan and cot always intersect at the midpoint of each quadrant.

16) The amplitude factor has no effect on the ranges of the csc functions.

17) The period factor has no effect on the ranges of the trigonometric functions.

18) $\sin x = \tan x$ only at integral multiples of π.

19) $\sin x$ is never equal to $\cot x$.

20) The primitive period of $\tan \pi x$ is 1.

Graph the following functions. Show appropriate units on each axis. Position the y-axis correctly for any phase shift.

21) $f(x) = \frac{2}{3} \cos 5x$

22) $g(x) = -3 \sin \frac{5}{4} x$

23) $h(x) = \frac{1}{2} \tan \frac{1}{2} x$

24) $k(x) = -\frac{3}{2} \sec (x + \frac{\pi}{4})$

25) $s(x) = 4 \cot (2x - \frac{4\pi}{3})$

26) $m(x) = -2 \csc (3x - \pi)$

27) $n(x) = 10 \cos 3\pi x$

28) $p(x) = \frac{4}{3} \sin (5x - 3\pi)$

29) Show graphically that $\tan (\pi - x) = -\tan x$.

30) Show graphically that $\cos (\frac{3\pi}{2} + x) = \sin x$.

Use tables or calculators to find x:

31) $\sin x = .3257$

32) $\cos \frac{3\pi}{8} = x$

33) $\cos x = 1.743$

34) $\tan x = -2.437$

35) $\tan .3758 = x$

36) $\cot x = 1.873$

37) $\tan x = .8540$

38) $\csc x = 7.892$

39) $\csc (-2.5) = x$

40) $\sec x = 2.631$

41) $\sec x = -3.482$

42) $\sec x = -2.7$

43) $\sec (-\frac{5\pi}{8}) = x$

44) $\cos x = -.8534$

45) $\sin (-.1) = x$

46) $\cot x = -2.423$

47) $\cot 1.87 = x$

48) $\sin x = -.395$

Find the smallest positive value of x for which:

49) $\sin x = -.5382$

50) $\cos x = .2$

51) $\tan x = -3.427$

52) $\cot x = -.2876$

53) $\sec x = -9$

54) $\csc x = -1.427$

55) $\cos x = -.4937$

56) $\tan (x + \pi) = -.2789$

CHAPTER 5

INVERSE RELATIONS AND FUNCTIONS

5-1. THE INVERSES OF ALGEBRAIC FUNCTIONS
REVIEW OF THE EXPONENTIAL AND LOGARITHMIC FUNCTIONS

Given a set of ordered pairs,

$$A = \{(-2,5), (3,2), (1,-1), (0,2), (4,-3)\}$$

the inverse of A, symbolized by A^{-1}, is given by

$$A^{-1} = \{(5,-2), (2,3), (-1,1), (2,0), (-3,4)\}.$$

A^{-1} is obtained from A by interchanging the components of each ordered pair. Thus, each x-component in A becomes a y-component in A^{-1} and the corresponding y-components in A become the x-components in A^{-1}.

Since functions are also defined as sets of ordered pairs, the same procedure is applied in forming their inverses.

EXAMPLE: Find the inverse of f(x) = 2x - 3.

SOLUTION: f can be defined as $\{(x,y) : y = 2x - 3\}$.

The inverse f^{-1} is obtained by interchanging x and y in the defining equation for f, y = 2x - 3. Thus, f^{-1} is defined as $\{(x,y) : x = 2y - 3\}$. The interchange of variables has the same effect as interchanging the components of each ordered pair. For example, (2,1) belongs to f while (1,2) belongs to f^{-1}.

Although x = 2y - 3 would serve adequately as the defining equation for f^{-1}, it is standard practice to express y in terms of x for purposes of functional notation.

Since x = 2y - 3

then x + 3 = 2y

and $y = \dfrac{x + 3}{2}$.

Therefore, $f^{-1}(x) = \dfrac{x + 3}{2}$.

In preparation for graphing, tables of ordered pairs for both f and f^{-1} are constructed:

$$y = f(x) = 2x - 3$$

x	0	1	2	3	4	-1	-2
y	-3	-1	1	3	5	-5	-7

$$y = f^{-1}(x) = \frac{x + 3}{2}$$

x	-3	-1	1	3	5	-5	-7
y	0	1	2	3	4	-1	-2

Note how the second table is derived from the first.

Figure 5-1A

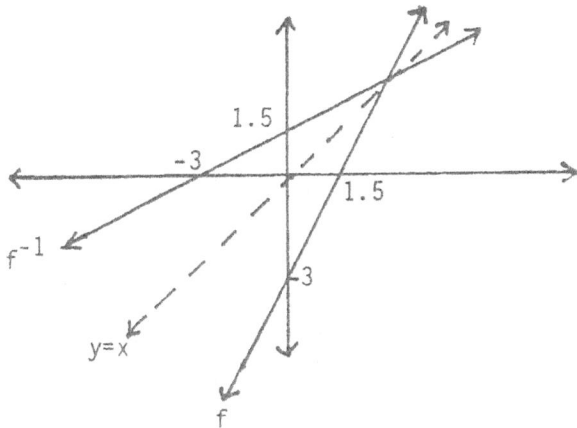

Both f and f^{-1} graph as straight lines that are symmetric with respect to the line y = x. This symmetry is characteristic of every set of ordered pairs and its inverse set. If the graphs intersect, they will do so at some point or points on the line of symmetry. The corresponding ordered pairs define points that are equidistant from the line y = x. For example, (0,-3) and (-3,0) are equidistant from the axis of symmetry, y = x.

If a function is a one-to-one mapping like f above, its inverse will also be a function. In the next example, a parabolic function and its inverse are graphed. Since the parabolic function is a many-to-one mapping, its inverse (a one-to-many mapping) is <u>not</u> a function.

Figure 5-1B

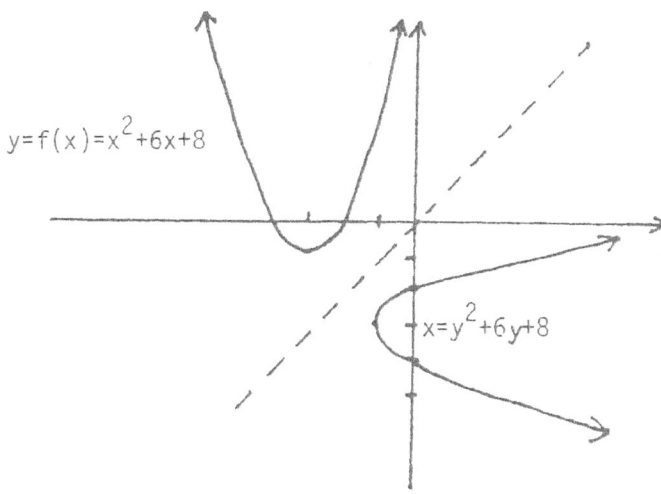

The fact that the inverse of the parabolic function is not a function is underscored when the defining equation is solved explicitly for y in terms of x:

$$x = y^2 + 6y + 8$$

$$x - 8 = y^2 + 6y$$

$$x - 8 + 9 = y^2 + 6y + 9 \qquad \text{(completing the square)}$$

$$x + 1 = (y + 3)^2 \qquad \text{(factoring the square trinomial)}$$

$$\pm\sqrt{x + 1} = y + 3 \qquad \text{(taking square roots of both sides)}$$

$$-3 \pm \sqrt{x + 1} = y \qquad \text{(solving for y)}$$

The meaning of this last equation is that for each $x \geq -1$, $y = -3 + \sqrt{x + 1}$ <u>or</u> $y = -3 - \sqrt{x + 1}$. With two possibilities of y for each value of x, it is obvious that the inverse relation is a one-to-many mapping.

Since all of the standard trigonometric functions are many-to-one mappings, it will be seen in section 5-2 that their inverses are one-to-many mappings and are not, therefore, functions. However, by restricting domains to principal values, inverse trigonometric functions can be defined.

Returning for the moment to Figure 5-1B, it can be seen that the horizontal axis of symmetry, $y = -3$, divides the graph of the inverse into two branches, each of which is the graph of a function. The upper branch is given by $\{(x,y) : y = -3 + \sqrt{x + 1}\}$ and the lower branch by $\{(x,y) : y = -3 - \sqrt{x + 1}\}$. Let $f_1^{-1}(x) = -3 + \sqrt{x + 1}$ and $f_2^{-1}(x) = -3 - \sqrt{x + 1}$. Then the inverse of $f(x) = x^2 + 6x + 8$ may be viewed as $f_1^{-1} \cup f_2^{-1}$, the union of the two inverse functions.

EXAMPLE: Graph $g(x) = |x| + 2$ and its inverse on the same set of axes. Express the inverse as the union of two functions.

SOLUTION: Set up a table of ordered pairs for g and derive a corresponding set for the inverse.

Figure 5-1C

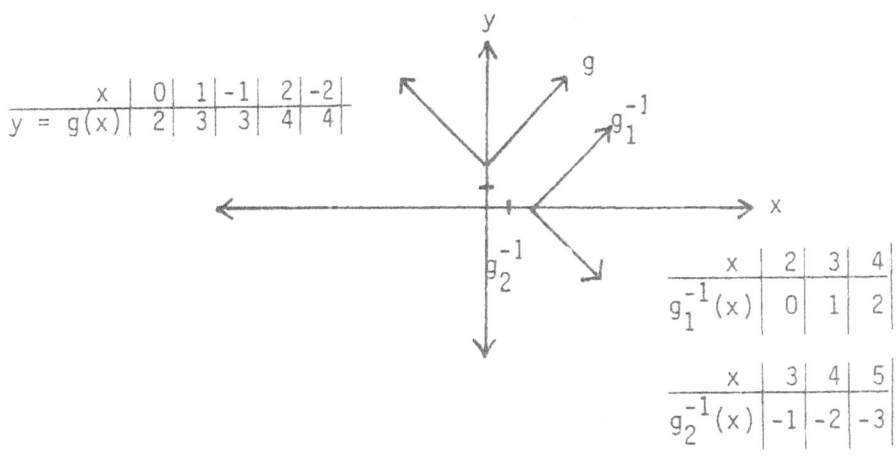

x	0	1	-1	2	-2
y = g(x)	2	3	3	4	4

x	2	3	4
$g_1^{-1}(x)$	0	1	2

x	3	4	5
$g_2^{-1}(x)$	-1	-2	-3

The inverse of $g = g_1^{-1} \cup g_2^{-1}$, where

$$g_1^{-1} = \{(x,y) : x \geq 2 \text{ and } y = x - 2\}$$

$$\text{and } g_2^{-1} = \{(x,y) : x > 2 \text{ and } y = -x + 2\}$$

EXAMPLE: Show algebraically that the inverse of $g(x) = |x| + 2$ can be expressed by the defining equations $y = x - 2$ and $y = -x + 2$.

SOLUTION: g is given by $y = |x| + 2$. Interchanging variables, the <u>inverse of g</u> is given by $x = |y| + 2$. Hence, $|y| = x - 2$.

<table>
<tr><td>CASE I</td><td>CASE II</td></tr>
</table>

CASE I	CASE II
$\|y\| = y$, if $y \geq 0$ (definition of $\|y\|$)	$\|y\| = -y$, if $y < 0$ (definition of $\|y\|$)
Therefore, by substitution,	Therefore, by substitution,
if $\|y\| = x - 2$, we have	if $\|y\| = x - 2$, we have
$y = x - 2$	$-y = x - 2$, and multiplying by -1,
	$y = -x + 2$

EXPONENTIAL AND LOGARITHMIC FUNCTIONS. ·A REVIEW.

In preparation for studying the inverses of the trigonometric functions it is useful to review the exponential and logarithmic functions. The reason for this is that these are examples of non-algebraic functions and, therefore, require special treatment in the handling of their defining equations. To begin, the graphs of $y = 2^x$ and $y = \log_2 x$ are shown on the same set of axes.

Figure 5-1D

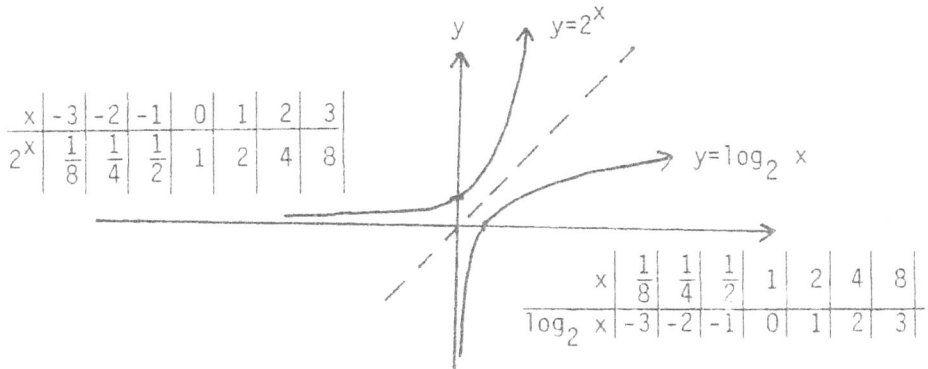

The corresponding ordered pairs are derived by interchanging x and y components. The graphs are symmetric with respect to the line y = x. It is apparent that 2^x and $\log_2 x$ are inverses of each other. Since both are one-to-one mappings, they are both functions. The table below summarizes the features of both functions:

FUNCTION	$f(x) = 2^x$	$f^{-1}(x) = \log_2 x$
Continuity	Continuous	Continuous
Increasing or Decreasing	Increasing	Increasing
Domain	All reals	$\{x : x > 0\}$
Range	$\{y : y > 0\}$	All reals
Asymptotes	X - axis	Y - axis
Intercepts	(0,1)	(1,0)

The defining equation for the exponential function above is $y = 2^x$. To obtain the inverse, the variables are interchanged. That is, the inverse of $y = 2^x$ is $x = 2^y$. There is no way that algebraic techniques may be applied to $x = 2^y$ to obtain a defining equation for y in terms of x for this inverse relation. Instead we take note of the fact that $x = 2^y$ means "y is the exponential power that the base 2 must be raised to in order to obtain x ." The logarithmic notation was invented to symbolize this definition. Hence, the defining equations $x = 2^y$ and $y = \log_2 x$ are equivalent. Since it is frequently easier to evaluate the exponential expression, this equivalence is commonly used.

EXAMPLE: Evaluate $\log_2 16$.

SOLUTION: Let $y = \log_2 16$. The equivalent exponential equation is $16 = 2^y$.
           Since $16 = 2^4$, $2^4 = 2^y$. Equating exponents y = 4. Hence, $\log_2 16 = 4$.

EXAMPLE: Find the base, b, if $\log_b \frac{1}{25} = -2$.

SOLUTION: Let $y = \log_b \frac{1}{25}$. Then $b^y = \frac{1}{25}$. But $y = -2$ is given. Hence,

$b^{-2} = \frac{1}{25}$. Since $\frac{1}{25} = \frac{1}{5^2} = 5^{-2}$, we have $b^{-2} = 5^{-2}$.

Therefore, $b = 5$.

EXAMPLE: Find x, if $\log_3 x = -2$.

SOLUTION: Let $y = \log_3 x$. Then $3^y = x$. But $y = -2$ is given.
Hence $3^{-2} = x$. Therefore, $x = \frac{1}{9}$.

The ability to go back and forth in handling the notation between a function and its inverse is critical to an understanding of the trigonometric functions. The exponential and logarithmic functions provide good practice for this skill.

COMPOSITION OF FUNCTIONS.

Functional notation has a built-in mechanism for directing substitutions of given values.

EXAMPLE: If $f(x) = 2x - 3$, find $f(-4)$.

SOLUTION: $f(-4) = 2(-4) - 3 = -8 - 3 = -11$.

The substitution can be a literal constant.

EXAMPLE: If $g(x) = 2x^2 - 8x + 7$, find $g(c)$.

SOLUTION: $g(c) = 2c^2 - 8c + 7$.

The substitution can be another function.

EXAMPLE: If $f(x) = x - 5$ and $g(x) = 2x + 7$, find $f(g(x))$.

SOLUTION: $f(g(x)) = f(2x + 7) = (2x + 7) - 5 = 2x + 2$.

All of the substitutions in the examples above proceed according to the rule that wherever the variable x appears in the defining expression for f, it is to be replaced by the given value. Thus, in the first example, -4 replaced x in the expression 2x - 3. In the second example, c replaced x in the expression $2x^2 - 3x + 7$. Finally, in the last example (2x + 7) replaced x in the expression x - 5.

This process is not restricted to algebraic functions.

EXAMPLE: If $f(x) = 2^x$, find f(-3).

SOLUTION: $f(-3) = 2^{-3} = \frac{1}{2^3} = \frac{1}{8}$.

EXAMPLE: If $f(x) = \log_2 x$, find f(16).

SOLUTION: $f(16) = \log_2 16 = 4$.

EXAMPLE: If $f(x) = 2^x$, and g(x) = x - 3, find f(g(x)).

SOLUTION: $f(g(x)) = 2^{g(x)} = 2^{x-3}$.

EXAMPLE: If $f(x) = \log_3 x$ and g(x) = 4x + 5, find f(g(x)).

SOLUTION: $f(g(x)) = \log_3 g(x) = \log_3 (4x + 5)$, if 4x + 5 > 0 .

Expressions of the form f(g(x)) are known as <u>compositions of functions</u>. The ordering of the composition is extremely important. f(g(x)) may be read as "f composed with g" or "f composition g". In the examples above, g(x) was the value to be substituted in the defining expression for f. If the ordering is reversed, i. e., g(f(x)), the results will in general differ.

EXAMPLE: If $f(x) = 2x^2 - 3x + 2$ and g(x) = x + 3, find f(g(x)) and g(f(x)).

SOLUTION: $f(g(x)) = 2(g(x))^2 - 3(g(x)) + 2 = 2(x + 3)^2 - 3(x + 3) + 2$

$$= 2(x^2 + 6x + 9) - 3x - 9 + 2$$

$$= 2x^2 + 12x + 18 - 3x - 9 + 2$$

$$= 2x^2 + 9x + 11$$

$g(f(x)) = f(x) + 3 = (2x^2 - 3x + 2) + 3 = 2x^2 - 3x + 5$.

We see that, in general, f(g(x)) ≠ g(f(x)). That is, the composition of functions is <u>not</u> a commutative operation. Of particular interest to the present study, however, is the fact that <u>the composition of inverse functions is</u> <u>commutative and always yields the identity function y = x.</u>

EXAMPLE: If $f(x) = 2x - 3$ and $f^{-1}(x) = \frac{x + 3}{2}$, show that $f(f^{-1}(x)) = f^{-1}(f(x)) = x$.

SOLUTION: $f(f^{-1}(x)) = 2(f^{-1}(x)) - 3 = 2(\frac{x + 3}{2}) - 3 = (x + 3) - 3 = x$

$f^{-1}(f(x)) = \frac{f(x) + 3}{2} = \frac{(2x - 3) + 3}{2} = x$.

EXAMPLE: Let $f(x) = 2^x$ and $g(x) = \log_2 x$. Show that $f(g(x)) = g(f(x)) = x$.

SOLUTION: (i) $f(g(x)) = 2^{g(x)} = 2^{\log_2 x}$. Let $u = 2^{\log_2 x}$ and $v = \log_2 x$. Then $u = 2^v$. Taking the logarithm of both sides of this equation to base 2, $\log_2 u = v \log_2 2$. Since $\log_2 2 = 1$, $\log_2 u = v = \log_2 x$. Hence, $u = x$ or $2^{\log_2 x} = x$.

(ii) $g(f(x)) = \log_2 f(x) = \log_2 2^x$. Let $t = \log_2 2^x$. Then by definition $2^t = 2^x$. Hence, $t = x$ or $\log_2 2^x = x$. The results of (i) and (ii) show that $f(g(x)) = g(f(x)) = x$, as expected since 2^x and $\log_2 x$ are inverse functions.

The composition of inverse functions is exhibited by the fact that the graphs of inverse functions are symmetric with respect to the identity function $y = x$. In the next section it will be seen how this information plays an important role in constructing the graphs of the trigonometric functions and their inverses.

EXERCISE SET 5-1.

In exercises 1 - 10, graph the given function and its inverse on the same set of axes.

1) $f(x) = x - 5$
2) $g(x) = -2x + 5$
3) $h(x) = 2|x| - 4$
4) $k(x) = |x - 3|$
5) $f(x) = x^2 - 2$
6) $g(x) = -x^3$
7) $p(x) = x^2 - 5x + 6$
8) $r(x) = 3^{-x}$
9) $h(x) = \sqrt{4 + x^2}$
10) $k(x) = \log_2 (x + 3)$

In exercises 11 - 20, find a defining equation for each of the inverses graphed in exercises 1 - 10. State whether or not the inverse is a function.

In exercises 21 - 28, a function f is given. Find the indicated value.

EXAMPLE: $2x - 5$, $f^{-1}(3)$.

SOLUTION: If $f(x) = 2x - 5$, $f^{-1} = \{(x,y) : x = 2y - 5\}$. If $x = 3$, $3 = 2y - 5$.
solving for y, $2y = 8$ and $y = 4$. Hence, $f^{-1}(3) = 4$.

21) $3x + 2$, $f^{-1}(-2)$ 22) $-x^3 + 1$, $f^{-1}(3)$ 23) $\log_2 x$, $f^{-1}(-1)$

24) 3^x, $f^{-1}(3)$ 25) $\sqrt{x - 2}$, $f^{-1}(4)$ 26) $2 + \sqrt{x}$, $f^{-1}(-2)$

27) 2^{x-1}, $f^{-1}(4)$ 28) $\dfrac{x - 5}{4}$, $f^{-1}(0)$

29) Find x, if $\log_2 x = -1$ 30) Find b, if $\log_b 125 = 3$

31) Find x, if $\log_{1/8} x = -\dfrac{2}{3}$ 32) Find $\log_{1/2} 4$

In exercises 33 - 36, express the inverse of f as the union of two functions.
Find defining equations for f_1^{-1} and f_2^{-1}.

33) $f(x) = x^2 - 5x + 4$ 34) $f(x) = |2x + 5|$ 35) $f(x) = 1 - x^2$

36) $f(x) = -\sqrt{x^2 + 36}$

In exercises 37 - 44, find an expression for $f(g(x))$.

37) $f(x) = x - 2$, $g(x) = 2x - 4$ 38) $f(x) = x^2$, $g(x) = 3x + 2$

39) $f(x) = 2x + 5$, $g(x) = x^2 - 3$ 40) $f(x) = x^2 - 3x + 4$, $g(x) = 2x - 1$

41) $f(x) = 3^{x+2}$, $g(x) = x - 3$ 42) $f(x) = 2^x$, $g(x) = 3 \log_2 x$

43) $f(x) = |x - 7|$, $g(x) = 7 - x$ 44) $f(x) = \sqrt{x - 4}$, $g(x) = x^2 + 4x + 8$

In exercises 45-48, show that $f(g(x)) = g(f(x)) = x$.

45) $f(x) = 7x - 2$, $g(x) = \dfrac{x + 2}{7}$ 46) $f(x) = \sqrt[3]{x}$, $g(x) = x^3$

47) $f(x) = \sqrt{x - 2}$, $g(x) = x^2 + 2$ 48) $f(x) = 2^{x-2}$, $g(x) = \log_2 4x$

5-2. THE INVERSES OF THE COSINE AND SINE.

The procedures applied to the functions in section 5-1 will serve to introduce the inverses of the trigonometric functions. We shall begin as before with the cosine function. Let $y = \cos x$ and recall the special values between $-\pi$ and π:

x	$-\pi$	$\frac{-5\pi}{6}$	$\frac{-3\pi}{4}$	$\frac{-2\pi}{3}$	$\frac{-\pi}{2}$	$\frac{-\pi}{3}$	$\frac{-\pi}{4}$	$\frac{-\pi}{6}$	0	$\frac{\pi}{6}$	$\frac{\pi}{4}$	$\frac{\pi}{3}$	$\frac{\pi}{2}$	$\frac{2\pi}{3}$	$\frac{3\pi}{4}$	$\frac{5\pi}{6}$	π
$y = \cos x$	-1	$\frac{-\sqrt{3}}{2}$	$\frac{-1}{\sqrt{2}}$	$\frac{-1}{2}$	0	$\frac{1}{2}$	$\frac{1}{\sqrt{2}}$	$\frac{\sqrt{3}}{2}$	1	$\frac{\sqrt{3}}{2}$	$\frac{1}{\sqrt{2}}$	$\frac{1}{2}$	0	$\frac{-1}{2}$	$\frac{-1}{\sqrt{2}}$	$\frac{-\sqrt{3}}{2}$	-1

For the inverse, let $x = \cos y$. This has the effect of interchanging the ordered pairs in the table above:

$x = \cos y$	-1	$\frac{-\sqrt{3}}{2}$	$\frac{-1}{\sqrt{2}}$	$\frac{-1}{2}$	0	$\frac{1}{2}$	$\frac{1}{\sqrt{2}}$	$\frac{\sqrt{3}}{2}$	1	$\frac{\sqrt{3}}{2}$	$\frac{1}{\sqrt{2}}$	$\frac{1}{2}$	0	$\frac{-1}{2}$	$\frac{-1}{\sqrt{2}}$	$\frac{-\sqrt{3}}{2}$	-1
y	$-\pi$	$\frac{-5\pi}{6}$	$\frac{-3\pi}{4}$	$\frac{-2\pi}{3}$	$\frac{-\pi}{2}$	$\frac{-\pi}{3}$	$\frac{-\pi}{4}$	$\frac{-\pi}{6}$	0	$\frac{\pi}{6}$	$\frac{\pi}{4}$	$\frac{\pi}{3}$	$\frac{\pi}{2}$	$\frac{2\pi}{3}$	$\frac{3\pi}{4}$	$\frac{5\pi}{6}$	π

Next, the graphs of $y = \cos x$ and $x = \cos y$ are placed on the same set of coordinate axes:

Figure 5-2A

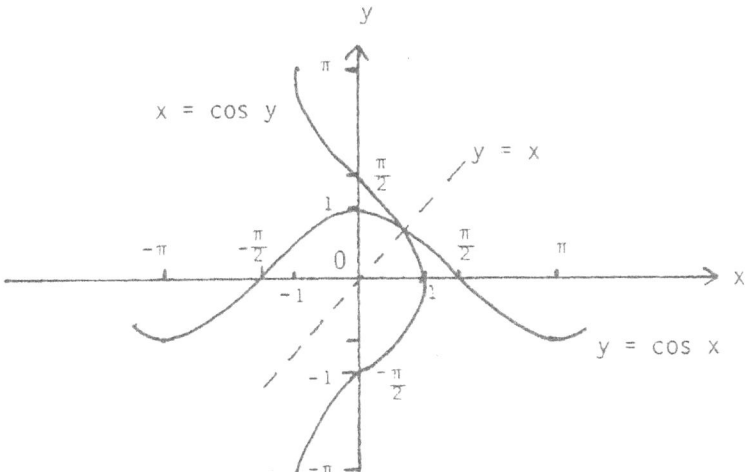

The table below compares the cosine function and its inverse.

Defining Equation	$y = \cos x$	$x = \cos y$
Domain	All reals	$-1 \le x \le 1$
Range	$-1 \le y \le 1$	All reals
Function	Many-to-one	Not a function
Continuity	Continuous	Continuous

The task remaining is to find an equation for the inverse that defines y explicitly in terms of x. Since the trigonometric functions are not algebraic this requires the invention of new notation as was done in defining the logarithmic function. The situation is complicated for the trigonometric functions, in that there are two notational forms used in practice. The more contemporary form makes use of the fact that if $y = f(x)$ is a function, then $y = f^{-1}(x)$ may be acceptable notation for its inverse. Applying this to the function $y = \cos x$, we may express the inverse of the cosine as $y = \cos^{-1} x$. It is to be understood that the raised $^{-1}$ is not to be interpreted as an exponent. That is, $\cos^{-1} x \ne \frac{1}{\cos x}$. Instead, the equivalence for $y = \cos^{-1} x$ is the alternate defining equation $x = \cos y$.

EXAMPLE: Find y if $y = \cos^{-1} \frac{1}{2}$.

SOLUTION: If $y = \cos^{-1} \frac{1}{2}$ then $\frac{1}{2} = \cos y$. From the second table above, if $x = \frac{1}{2}$, $y = \frac{-\pi}{3}$ or $y = \frac{\pi}{3}$. However, since the range of the inverse is the set of all reals and the cosine is periodic with a period of 2π, there is an infinity of solutions. This may be expressed as follows: $y = \pm \frac{\pi}{3} + 2k\pi$, $k = 0, \pm 1, \pm 2, \ldots \ldots$

The other notation that is commonly used derives from our earlier understanding of the trigonometric functions as circular functions. Imagine a real number line wrapped around a unit circle as in Figure 5-2B.

Figure 5-2B

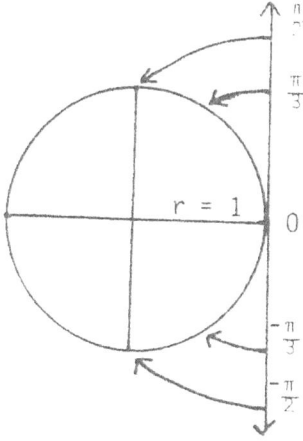

The positive numbers are wrapped in a counterclockwise direction and the negative numbers in a clockwise direction. Since the real number set is infinite, the wrapping would result in a many-to-one mapping of points on the number line to points on the circle. For example, all odd multiples of π, both positive and negative, would map into the same point at the opposite end of the diameter from 0.

Now, consider the wrapping complete and picture as before a central angle of $\frac{\pi}{3}$ radians in standard position (Figure 5-2C).

Figure 5-2C

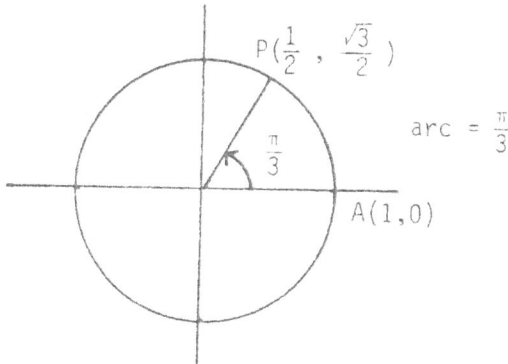

The cosine of the central angle $\frac{\pi}{3}$ is given by the x coordinate $\frac{1}{2}$ as we learned earlier. However, on the <u>unit circle</u>, the arc subtended by an angle has the same measure, so it is just as correct to say the cosine of the arc $\frac{\pi}{3}$ is equal to $\frac{1}{2}$. In other words, if the unit circle is our reference, $\cos\frac{\pi}{3} = \frac{1}{2}$, whether we are thinking of $\frac{\pi}{3}$ radians or an arc of $\frac{\pi}{3}$ units. The wrapping function described above tells us that the cosine of all arcs of the form $\frac{\pi}{3} + 2k\pi$, $k = 0, \pm1, \pm2,\ldots$ will be equal to $\frac{1}{2}$, since all of these arcs will map into the point $P(\frac{1}{2}, \frac{\sqrt{3}}{2})$. The same conclusions apply, of course, to arcs of the form $-\frac{\pi}{3} + 2k\pi$, $k = 0, \pm1, \pm2,\ldots$ as illustrated in Figure 5-2D, all terminating at the point $(\frac{1}{2}, \frac{-\sqrt{3}}{2})$.

Figure 5-2D

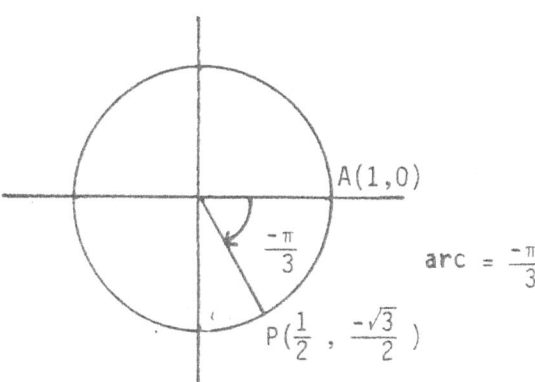

With the discussion above as background, consider the statement $\cos\frac{\pi}{3} = \frac{1}{2}$ again. This statement is a specific substitution in the defining equation $y = \cos x$, where $x = \frac{\pi}{3}$ and $y = \frac{1}{2}$. The corresponding statement for the inverse can be expressed in words as "$\frac{\pi}{3}$ is an arc whose cosine equals $\frac{1}{2}$." In symbols this inverse statement is written:

$$\frac{\pi}{3} = \arccos\frac{1}{2}$$

This statement is a specific substitution in the defining equation for the inverse:

$$y = \arccos x, \text{ where } x = \frac{1}{2} \text{ and } y = \frac{\pi}{3}.$$

Since the inverse relation is a one-to-many mapping, y is not unique. For example, $\arccos \frac{1}{2} = -\frac{\pi}{3}$ also. In fact, $\arccos \frac{1}{2} = \pm \frac{\pi}{3} + 2k\pi$, $k = 0, \pm 1, \pm 2, \ldots$ as revealed by the wrapping function.

The two notations $y = \arccos x$ and $y = \cos^{-1} x$ are equivalent to the defining equation $x = \cos y$.

EXAMPLE: Find $\arccos \frac{-\sqrt{3}}{2}$.

SOLUTION: Let $y = \arccos \frac{-\sqrt{3}}{2}$. Then $\cos y = \frac{-\sqrt{3}}{2}$. The table of inverse values gives $y = \frac{-5\pi}{6}$ or $y = \frac{5\pi}{6}$. The wrapping function yields $y = \pm \frac{5\pi}{6} + 2k\pi$, $k = 0, \pm 1, \pm 2, \ldots$.

EXAMPLE: Show that $\frac{7\pi}{6} = \arccos \frac{-\sqrt{3}}{2}$ from the general solution $\arccos \frac{-\sqrt{3}}{2} = \pm \frac{5\pi}{6} + 2k\pi$, $k = 0, \pm 1, \pm 2, \ldots$.

SOLUTION: If $k = 1$, $\arccos \frac{-\sqrt{3}}{2} = \frac{-5\pi}{6} + 2\pi = \frac{-5\pi}{6} + \frac{12\pi}{6} = \frac{7\pi}{6}$.

PRINCIPAL VALUES AND THE ARCCOS X (OR COS^{-1} X).

In Chapter 4, the function Cos was defined:

$$\text{Cos} = \{(x, y) : 0 \leq x \leq \pi, \text{ and } y = \cos x\}.$$

The principal values defining the domain of Cos produced a one-to-one function. The inverse of a one-to-one function is also a one-to-one function defined as follows:

$$\text{Arccos} = \text{Cos}^{-1} = \{(x, y) : -1 \leq x \leq 1, 0 \leq y \leq \pi, \text{ and } y = \cos^{-1} x\}$$

The graph of Arccos appears as follows:

Figure 5-2E

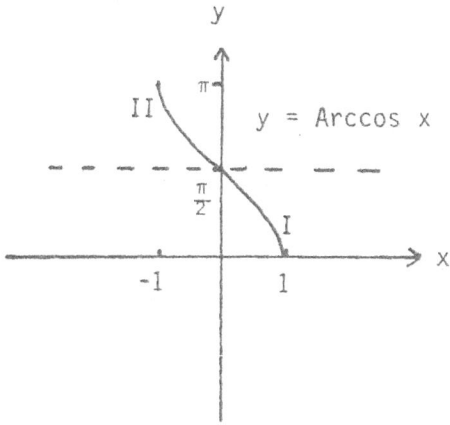

The graph shows that the Arccos function is decreasing over its domain. The first and second quadrants are now related to divisions of the range. We are now in a position to understand why the calculator is equipped with keys labeled INV (for inverse) or ARC (for Arccos, Arcsin, and Arctan functions).

EXAMPLE: Find Arccos .3742.

SOLUTION: Calculator. Radian mode. Input .3742. Press INV or ARC key. Press cos key. Read 1.19 rounded. Therefore, Arccos .3742 = 1.19.

DISCUSSION: Refer to Figure 5-2F.

Figure 5-2F

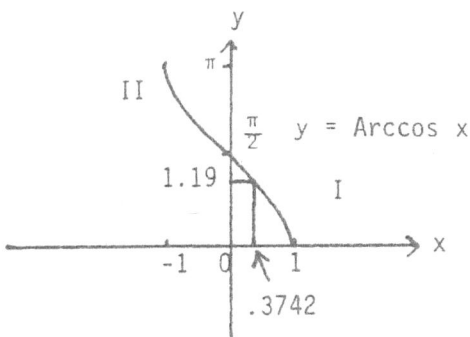

The input .3742 is from the domain of the Arccos function. The calculator returns the value 1.19 from the range, located between 0 and $\frac{\pi}{2}$ as expected from the graph.

The perceptive student will have noted that this problem is identical to that solved earlier in the form:

$$\text{find } \theta, \text{ if } \cos \theta = .3742 \text{ and } 0 < \theta < \frac{\pi}{2}.$$

EXAMPLE: Find Cos^{-1} (-.8496).

SOLUTION: Calculator. Radian mode. Input -.8496. Press INV or ARC key.
Press cos key. Read 2.59. Hence, Cos^{-1} (-.8496) = 2.59. As
expected, negative domain values yield range values between $\frac{\pi}{2}$ and π.

THE INVERSE OF THE SINE FUNCTION.

Figure 5-2G shows the sine function and its inverse.

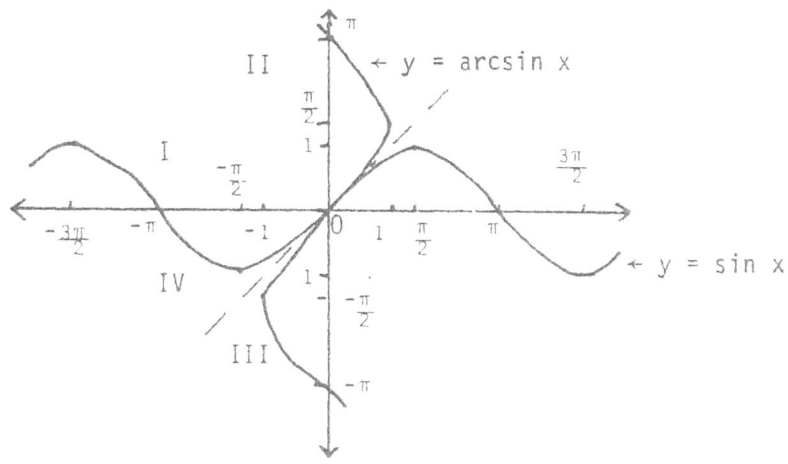

The table below compares the sine function and its inverse.

Defining Equations	$y = \sin x$	$y = \sin^{-1} x$ (or arcsin x)
Domain	All reals	$-1 \le x \le 1$
Range	$-1 \le y \le 1$	All reals
Function	Many-to-one	Not a function
Continuity	Continuous	Continuous

EXAMPLE: Find arcsin $\dfrac{1}{\sqrt{2}}$.

SOLUTION: If $y = \text{arcsin } \dfrac{1}{\sqrt{2}}$, then $\dfrac{1}{\sqrt{2}} = \sin y$. In the first quadrant, $y = \dfrac{\pi}{4}$

and in the second quadrant $y = \dfrac{3\pi}{4}$. The wrapping function yields

the general solutions: $y = \dfrac{\pi}{4} + 2k\pi$, or $y = \dfrac{3\pi}{4} + 2k\pi$, $k = 0, \pm1, \pm2,..$

The inverse function Sin^{-1} (or Arcsin) may be defined as:

Sin^{-1} $= \{(x,y) : -1 \le x \le 1, -\dfrac{\pi}{2} \le y \le \dfrac{\pi}{2}$ and $y = \sin^{-1} x\}$.

Again it may be seen that the principal values of the sine function become the
range of the Arcsin function.

Figure 5-2H

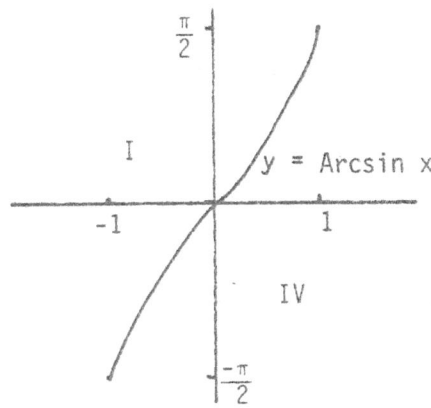

The Arcsin function is increasing over its domain.

EXAMPLE: Find Sin^{-1} .8435.

SOLUTION: Calculator. Radian mode. Input .8435. Press INV or ARC key. Press
sin key. Read 1.00 rounded. Hence Sin^{-1} .8435 = 1.00.

EXAMPLE: Find Arcsin (-.4278).

SOLUTION: Calculator. Radian mode. Input -.4278. Press INV or ARC key. Press
sin key. Read -.44 rounded. Hence Arcsin (-.4278) = -.44.

COMPOSITION OF TRIGONOMETRIC FUNCTIONS AND THEIR INVERSES.

The notation that has been derived in this section allows for the evaluation of expressions that are compositions of trigonometric functions and their inverses (refer to Section 5-1).

EXAMPLE:　Find $\sin \left(\text{Arccos } \frac{\sqrt{3}}{2} \right)$.

SOLUTION:　We read this as follows: "Find the sine of the arc (or angle) whose cosine is equal to $\frac{\sqrt{3}}{2}$." First, the angle whose cosine is $\frac{\sqrt{3}}{2}$ is the principal value $\frac{\pi}{6}$. That is, $\text{Arccos } \frac{\sqrt{3}}{2} = \frac{\pi}{6}$. Next $\sin \frac{\pi}{6} = \frac{1}{2}$. Therefore, $\sin \left(\text{Arccos } \frac{\sqrt{3}}{2} \right) = \sin \frac{\pi}{6} = \frac{1}{2}$.

EXAMPLE:　Find $\cos \left(\text{Sin}^{-1} 0 \right)$.

SOLUTION:　$\text{Sin}^{-1} 0 = 0$ since 0 is the principal value such that $\sin 0 = 0$. Next, $\cos 0 = 1$. Therefore, $\cos \left(\text{Sin}^{-1} 0 \right) = \cos (0) = 1$.

EXAMPLE:　Find $\sin \left(\text{Cos}^{-1} .6834 \right)$.

SOLUTION:　$\text{Cos}^{-1} .6834$ is found by the calculator method to be .82 rounded. Leaving the figure found in the display, press the sin key. Read .7300 to four significant figures. Therefore, $\sin \left(\text{Cos}^{-1} .6834 \right)$ = sin .82 = .7300.

EXAMPLE:　Find $\cos \left(\text{Arcsin} (-.9345) \right)$.

SOLUTION:　Find Arcsin (-.9345) by calculator as before. Read -1.21 rounded. Press cos key. Read .3560 to four significant figures. Therefore, cos (Arcsin (-.9345)) = cos (-1.21) = .3560. The answer is positive as expected for the cosine of an argument in quadrant IV.

The order of composition of the functions may be reversed as in the following:

EXAMPLE:　Find $\text{Arcsin} \left(\cos \frac{\pi}{6} \right)$.

SOLUTION: This is read as "Find the angle whose sine is equal to the cosine of $\frac{\pi}{6}$." First evaluate $\cos \frac{\pi}{6} = \frac{\sqrt{3}}{2}$. Then evaluate Arcsin $\frac{\sqrt{3}}{2} = \frac{\pi}{3}$. Hence, Arcsin $(\cos \frac{\pi}{6})$ = Arcsin $\frac{\sqrt{3}}{2} = \frac{\pi}{3}$.

EXAMPLE: Find Arccos $(\sin \frac{7\pi}{6})$.

SOLUTION: $\sin \frac{7\pi}{6} = -\frac{1}{2}$. Arccos $(-\frac{1}{2}) = \frac{2\pi}{3}$. Therefore, Arccos $(\sin \frac{7\pi}{6})$ Arccos $(-\frac{1}{2}) = \frac{2\pi}{3}$.

EXAMPLE: Find Sin^{-1} (cos 8.92).

SOLUTION: Calculator. Radian mode. Input 8.92. Press cos key. Press INV or ARC key. Press sin key. Read -1.07 rounded.

Hence, Sin^{-1} (cos 8.92) = Sin^{-1} (-.8753) = - 1.07.

Generally it is inappropriate to use the inverse relations $\cos^{-1}$ and $\sin^{-1}$ in compositions of functions.

EXAMPLE: Find $\sin (\cos^{-1} \frac{1}{\sqrt{2}})$.

DISCUSSION: $\cos^{-1} \frac{1}{\sqrt{2}} = \pm \frac{\pi}{4} + 2k\pi$, $k = 0, \pm 1, \pm 2, \ldots$

Since the argument is not unique (single-valued), the solution is indeterminate. If the student should encounter such notation in another textbook, it should be taken for granted that the principal values are implied, i. e., $\cos^{-1} \frac{1}{\sqrt{2}} = \text{Cos}^{-1} \frac{1}{\sqrt{2}}$.

APPLICATIONS OF THE PYTHAGOREAN IDENTITY.

It is possible to obtain exact solutions to certain composition problems by making use of the identity $\sin^2 x + \cos^2 x = 1$.

EXAMPLE: Find $\sin (\text{Arccos} \frac{3}{5})$.

SOLUTION: Let $x = \text{Arccos} \frac{3}{5}$, then $\text{Cos } x = \frac{3}{5}$, $0 \le x \le \frac{\pi}{2}$. Since $\sin^2 x = 1 - \cos^2 x$, $\sin^2 x = 1 - (\frac{3}{5})^2 = 1 - \frac{9}{25} = \frac{16}{25}$.

Therefore, $\sin x = \pm \sqrt{\frac{16}{25}} = \pm \frac{4}{5}$. Since $0 \le x \le \frac{\pi}{2}$, $\sin x = \frac{4}{5}$.

EXAMPLE: Find $\cos \left(\text{Sin}^{-1} \left(\frac{-1}{\sqrt{5}} \right) \right)$.

SOLUTION: Let $x = \text{Sin}^{-1} \left(\frac{-1}{\sqrt{5}} \right)$, then $\text{Sin } x = \frac{-1}{\sqrt{5}}$, $-\frac{\pi}{2} \leq x \leq 0$.

Since $\cos^2 x = 1 - \sin^2 x$, $\cos^2 x = 1 - \left(\frac{-1}{\sqrt{5}} \right)^2 = 1 - \frac{1}{5} = \frac{4}{5}$.

Therefore, $\cos x = \pm \sqrt{\frac{4}{5}} = \pm \frac{2}{\sqrt{5}}$. Since $-\frac{\pi}{2} \leq x \leq 0$, $\cos x = \frac{2}{\sqrt{5}}$.

COMPOSITIONS OF A FUNCTION AND ITS INVERSE.

As noted in Section 5-1, $f(f^{-1}(x)) = f^{-1}(f(x)) = x$. This applies to the trigonometric functions and the inverse trigonometric functions.

EXAMPLE: Find $\cos \left(\text{Arccos } \frac{1}{2} \right)$.

SOLUTION: $\cos \left(\text{Arccos } \frac{1}{2} \right) = \cos \frac{\pi}{3} = \frac{1}{2}$.

EXAMPLE: Find $\text{Sin}^{-1} \left(\sin \left(\frac{-\pi}{4} \right) \right)$.

SOLUTION: $\text{Sin}^{-1} \left(\sin \left(\frac{-\pi}{4} \right) \right) = \text{Sin}^{-1} \left(\frac{-1}{2} \right) = \frac{-\pi}{4}$.

Caution must be observed if the argument of the trigonometric function is not in the range of the inverse trigonometric function.

EXAMPLE: Find $\text{Cos}^{-1} \left(\cos \frac{11\pi}{6} \right)$.

SOLUTION: $\text{Cos}^{-1} \left(\cos \frac{11\pi}{6} \right) = \text{Cos}^{-1} \frac{\sqrt{3}}{2} = \frac{\pi}{6}$.

Since $\frac{11\pi}{6}$ is not in the range of the Cos^{-1} function, the solution must be worked out in detail to obtain the correct result $\frac{\pi}{6}$.

EXERCISE SET 5-2.

Graph the following:

1) $y = \cos^{-1} x$, $-\frac{\pi}{2} \leq y \leq 2\pi$.

2) $y = \arcsin x$, $-\frac{3\pi}{2} \leq y \leq 0$.

3) $y = 2 \text{Cos}^{-1} x$

4) $y = \frac{1}{2} \text{Sin}^{-1} x$

5) $y = \text{Cos}^{-1} 2x$

6) $y = \text{Arcsin } 2x$

7) $y = \text{Cos}^{-1} (-x)$

8) $y = - \text{Sin}^{-1} x$

Find the exact value(s) for the following:

9) arcsin $(-\frac{1}{2})$

10) $\cos^{-1} 0$

11) $\sin^{-1} 2\pi$

12) arccos 5.4

13) arccos (-1)

14) Arcsin $(-\frac{1}{2})$

15) $\cos^{-1} (\frac{-1}{\sqrt{2}})$

16) $\sin^{-1} (\frac{-1}{\sqrt{2}})$

17) Arccos $(\frac{-\sqrt{3}}{2})$

18) $\cos^{-1} \frac{1}{2}$

Find approximate values for the following:

19) $\cos^{-1} .8421$

20) Arcsin .5490

21) $\sin^{-1} .7713$

22) cos .5942

23) sin .8762

24) $\sin^{-1} .8897$

25) Arccos (-.3972)

26) $\cos^{-1} (-\frac{2}{5})$

27) $\sin^{-1} (-.8110)$

28) Arccos $\frac{5}{13}$

Find exact values for the following:

29) sin (Arccos $\frac{\sqrt{3}}{2}$)

30) sin (Arccos $\frac{-1}{\sqrt{2}}$)

31) cos $(\sin^{-1} \frac{1}{2})$

32) cos (Arcsin 0)

33) Arccos (sin $\frac{5\pi}{4}$)

34) $\cos^{-1}$ (sin $\frac{5\pi}{6}$)

35) $\sin^{-1}$ (cos $\frac{4\pi}{3}$)

36) Arcsin (cos π)

37) sin $(\sin^{-1} \frac{3}{5})$

38) cos $(\cos^{-1} \frac{4}{9})$

39) $\sin^{-1}$ (sin $\frac{7\pi}{4}$)

40) cos $(\sin^{-1} (-\frac{1}{2}))$

Use the Pythagorean identity to find exact values for the following:

41) sin (Arccos $\frac{-5}{13}$)

42) sin $(\cos^{-1} \frac{7}{25})$

43) cos $(\sin^{-1} (\frac{-2}{\sqrt{13}}))$

44) cos (Arcsin $(\frac{-8}{17})$)

Find approximate values for the following:

45) $\cos^{-1}$ (sin .8965)

46) Arccos (sin 7.85)

47) $\sin^{-1}$ (cos (-3.56))

48) Arcsin (cos 2.86π)

49) Cos $(\sin^{-1} .7863)$

50) cos (Arcsin (-.4320)

51) sin $(\cos^{-1} .0542)$

52) sin (Arccos (-.8632))

53) $\cos^{-1}$ (cos 17.83)

54) $\cos^{-1}$ (cos (-4.78))

55) Arcsin (sin 5.37)

56) $\sin^{-1}$ (sin (-2.76))

5-3.   INVERSES OF TANGENT AND COTANGENT.

The Arctangent Relation and Arctangent Function

The graph of the arctangent relation is given by:

Figure 5-3A

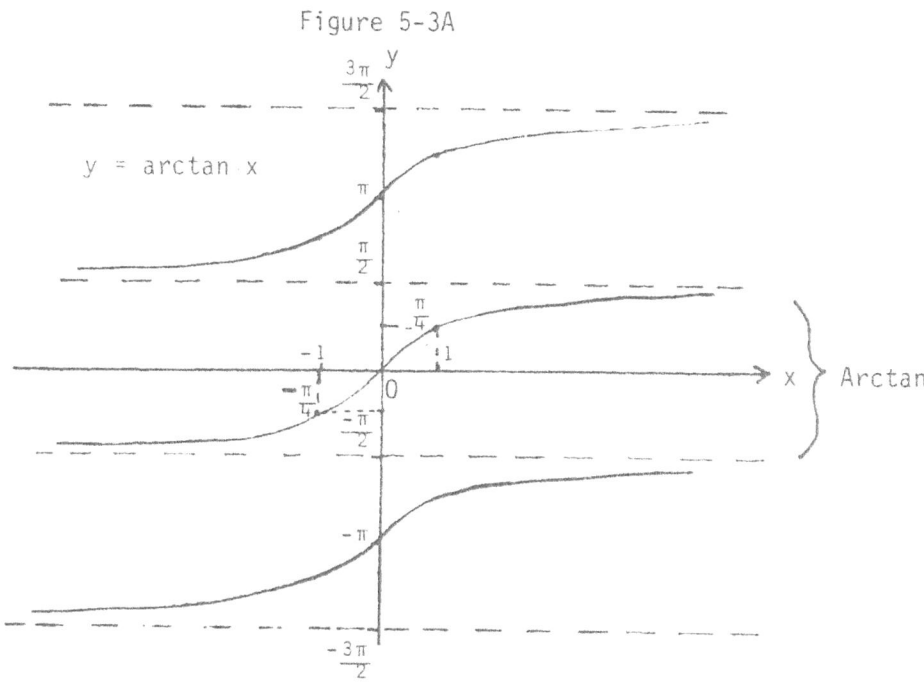

The branches of the relation's graph repeat themselves within the range of all reals.

The limited range of the Arctan function is also shown between the horizontal asymptotes at $y = -\frac{\pi}{2}$ and $y = \frac{\pi}{2}$. The table below compares the Tan function and the Arctan function.

Function	Tan	Arctan (or Tan^{-1})
Domain	$-\frac{\pi}{2} < x < \frac{\pi}{2}$	All reals
Range	All reals	$-\frac{\pi}{2} < y < \frac{\pi}{2}$
Increasing or Decreasing	Increasing	Increasing
Continuity	Continuous	Continuous
Odd or Even	Odd	Odd

Tan^{-1} is defined by:

Tan^{-1} $= \{(x,y) : -\infty < x < \infty, -\frac{\pi}{2} < y < \frac{\pi}{2}$, and $y = tan^{-1} x\}$.

The distinction between the arctan relation and the Arctan function is brought out in the following examples:

EXAMPLE: Find $tan^{-1} 1$ and $Tan^{-1} 1$.

SOLUTION: If $y = tan^{-1} 1$, then $tan\, y = 1$. Since the tangent function is positive in quadrants I and III, $y = \frac{\pi}{4}$ or $y = \frac{5\pi}{4}$. Since the tangent has a primitive period of π units, the general solution may be expressed as $tan^{-1} 1 = \frac{\pi}{4} + k\pi$, where $k = 0, \pm1, \pm2,...$

The principal value for Tan $y = 1$ is $y = \frac{\pi}{4}$. Therefore, $Tan^{-1} 1 = \frac{\pi}{4}$.

EXAMPLE: Find $arctan\, (-\sqrt{3}\,)$ and $Arctan\, (-\sqrt{3}\,)$.

SOLUTION: If $y = arctan\, (-\sqrt{3}\,)$, then $tan\, y = -\sqrt{3}$. The tangent is negative in quadrants II and IV. The special value gives $y = \frac{2\pi}{3}$ or $y = \frac{5\pi}{3}$. As above, the general solution may be written as $arctan\, (-\sqrt{3}) = \frac{2\pi}{3} + k\pi$, $k = 0, \pm1, \pm2,....$

The principal value for Tan $y = -\sqrt{3}$ is $y = -\frac{\pi}{3}$. Therefore, $Arctan\, (-\sqrt{3}\,) = -\frac{\pi}{3}$.

CALCULATOR SOLUTIONS FOR TAN^{-1} X.

The INV (or ARC) key combined with the tan key will yield values in the range of the Arctangent function, $-\frac{\pi}{2} < y < \frac{\pi}{2}$.

EXAMPLE: Find Tan^{-1} .8976.

SOLUTION: Calculator. Radian mode. Input .8976. Press INV or ARC key. Press tan key. Read .73 rounded. Therefore, Tan^{-1} .8976 = . 73.

EXAMPLE: Find Arctan (-3.851).

SOLUTION: Calculator. Radian mode. Input (-3.851). Press INV or ARC key. Press tan key. Read -1.32 rounded. Therefore, Arctan (-3.851) = -1.32.

COMPOSITIONS

The Arctan function may be composed with the sine or cosine functions or the tan function may be composed with the Arcsin or Arccos.

EXAMPLE: Find $\sin (\text{Tan}^{-1} (-1))$.

SOLUTION: $\sin (\text{Tan}^{-1} (-1)) = \sin (- \frac{\pi}{4}) = \frac{-1}{\sqrt{2}}$.

EXAMPLE: Find $\text{Tan}^{-1} (\cos \frac{5\pi}{6})$.

SOLUTION: $\text{Tan}^{-1} (\cos \frac{5\pi}{6}) = \text{Tan}^{-1} (\frac{-\sqrt{3}}{2})$. Care must be exercised before continuing with the solution, for although $\cos \frac{5\pi}{6}$ is a special value, the angle whose tangent is equal to $\frac{-\sqrt{3}}{2}$ is not. The remainder of the solution may be handled by the calculator.

$\text{Tan}^{-1} (\frac{-\sqrt{3}}{2}) \doteq \text{Tan}^{-1} (-.8660) = - .71$ rounded.

EXAMPLE: Find an exact value for $\tan (\text{Sin}^{-1} \frac{4}{5})$.

SOLUTION: $\tan (\text{Sin}^{-1} \frac{4}{5})$ can be evaluated by the Pythagorean identity and the quotient identity $\frac{\sin x}{\cos x} = \tan x$. Let $x = \text{Sin}^{-1} \frac{4}{5}$. Then $\sin x = \frac{4}{5}$

$\cos x = \sqrt{1 - (\frac{4}{5})^2} = \frac{3}{5}$. Therefore, $\tan x = \frac{\frac{4}{5}}{\frac{3}{5}} = \frac{4}{3}$.

Hence $\tan (\text{Sin}^{-1} \frac{4}{5}) = \frac{4}{3}$.

EXAMPLE: Find an approximate value for $\tan (\text{Arccos} (-.5973))$.

SOLUTION: Calculator. Radian mode. Input -.5973. Press INV or ARC key. Press cos key. Read 2.21 rounded. Press tan key. Read -1.343. Hence $\tan (\text{Arccos} (-.5973)) = -1.343$.

THE ARCCOTANGENT RELATION AND COT^{-1} X.

The graph of the arccotangent relation is given by:

Figure 5-3B

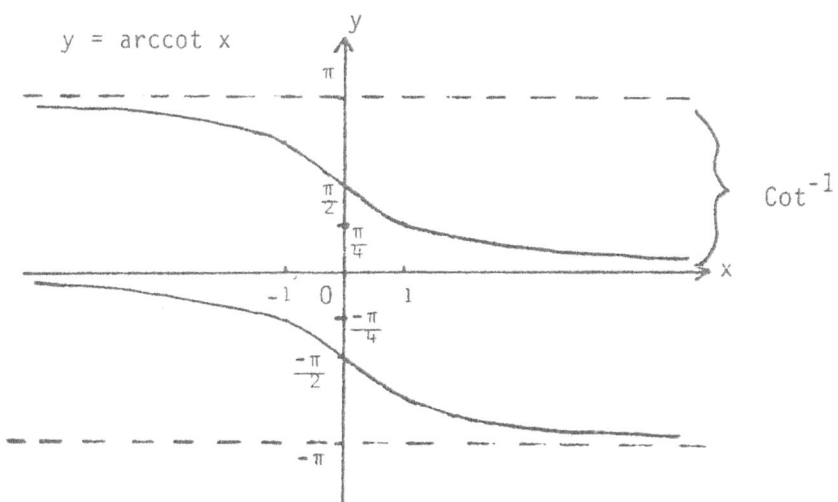

The range of this one-to-many mapping is the set of all reals. The horizontal asymptotes at y = 0 and y = π form the lower and upper bounds for the Arccot function.

Cot and Cot^{-1} are compared below:

Function	Cot	Cot^{-1} (or Arccot)
Domain	0 < x < π	All reals
Range	All reals	0 < y < π
Increasing or Decreasing	Decreasing	Decreasing
Continuity	Continuous	Continuous
Odd or Even	Neither	Neither

Cot^{-1} is defined by:

Cot^{-1} = {(x,y) : - ∞ < x < ∞, 0 < y < π, and y = cot^{-1} x}

The distinction between the arccot relation and the Arccot function is brought out in the following examples:

EXAMPLE: Find arccot $\frac{1}{\sqrt{3}}$ and Arccot $\frac{1}{\sqrt{3}}$.

SOLUTION: Let y = arccot $\frac{1}{\sqrt{3}}$, then cot y = $\frac{1}{\sqrt{3}}$. The cotangent is positive in the first and third quadrants. The special value $\frac{1}{\sqrt{3}}$ gives $y = \frac{\pi}{3}$ in Quadrant I and $y = \frac{4\pi}{3}$ in Quadrant III. With a primitive period of π the general solution may be written as $y = \frac{\pi}{3} + k\pi$, $k = 0, \pm1, \pm2,$. The principal value for Cot $y = \frac{1}{\sqrt{3}}$ is $y = \frac{\pi}{3}$. Therefore, Arccot $\frac{1}{\sqrt{3}} = \frac{\pi}{3}$.

EXAMPLE: Find $\cot^{-1}(-1)$ and $\text{Cot}^{-1}(-1)$.

SOLUTION: Let $y = \cot^{-1}(-1)$, then cot y = -1. The cotangent is negative in the second and fourth quadrants. The special value (-1) gives $y = \frac{3\pi}{4}$ in Quadrant II and $y = \frac{7\pi}{4}$ in Quadrant IV. The general solution may be written as $y = \frac{3\pi}{4} + k\pi$, $k = 0, \pm1, \pm2,\ldots$ $\frac{3\pi}{4}$ is the principal value such that Cot $\frac{3\pi}{4}$ = -1. Therefore, $\text{Cot}^{-1}(-1) = \frac{3\pi}{4}$.

CALCULATOR SOLUTIONS FOR COT^{-1} X.

In Chapter 3 we took note of the fact that values of the cot function are determined by the calculator using the reciprocal relation cot $x = \frac{1}{\tan x}$.

EXAMPLE: Find cot 6.854.

SOLUTION: Radian mode. Input 6.854. Press tan key. Press $\frac{1}{x}$ key. Read cot 6.854 = 1.557. Furthermore, since the principal values for Tan and Cot differ, problems that involved finding principal values of the cotangent between $\frac{\pi}{2}$ and π have to be handled with care.

EXAMPLE: Find x if Cot x = -.8468.

SOLUTION: For negative values of Cot x, $\frac{\pi}{2} < x < \pi$. Calculator. Radian mode Input -.8468. Press $\frac{1}{x}$ key. Press INV or ARC key. Press tan key. Read -.87 rounded. The solution requires an addition of π to place x between $\frac{\pi}{2}$ and π. Add π to the display value. Read 2.27 rounded.

EXAMPLE: Find Cot^{-1} (-.8468).

SOLUTION: This is the same problem as the example above but written with inverse function notation. The calculator solution is identical. As a comparison the steps in the solution are shown in both notations:

$\underline{Cot\ x = -.8468}$	$\underline{Cot^{-1}\ (-.8468)}$
$Cot\ x = -.8468 = \dfrac{1}{Tan\ (x - \pi)}$	$Cot^{-1}\ (-.8468) = Tan^{-1}\ (\dfrac{1}{-.8468}) + \pi$
$Tan\ (x - \pi) = \dfrac{1}{-.8468} = -1.18$	$= Tan^{-1}\ (-1.18) + \pi$
$x - \pi = -.87$	$= -.87 + \pi$
$x = \pi - .87 = 2.27$	$= 2.27$

A comparison of the graphs of the solution is shown below:

Figure 5-3C

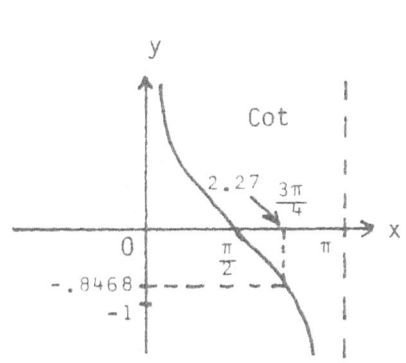

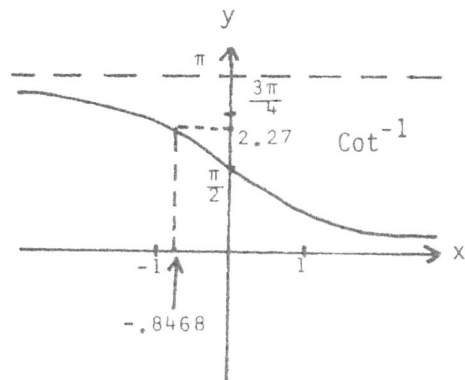

In the next problem the solution will be shown in inverse notation only and followed by the calculator steps.

EXAMPLE: Find Arccot (-4.674).

SOLUTION: Arccot (-4.674) = Arctan $(\frac{1}{-4.674})$ + π

$$= \text{Arctan } (-.2139) + \pi = -.21 + \pi = 2.93.$$

Calculator. Radian mode. Input -4.674. Press $\frac{1}{x}$ key. Press INV key. Press tan key. Add π. Read 2.93.

Of course, if the argument is positive there is no need to add π since $\text{Cot}^{-1} x = \text{Tan}^{-1} \frac{1}{x}$ if x > 0.

EXAMPLE: Find Cot^{-1} 3.82.

SOLUTION: Cot^{-1} 3.82 = $\text{Tan}^{-1} \frac{1}{3.82}$ = Tan^{-1} .2618 = .2560 or .26.

Calculator. Radian mode. Input 3.82. Press $\frac{1}{x}$ key. Press INV key. Press tan key. Read .26 rounded. Figure 5-3D shows the graph of this solution.

Figure 5-3D

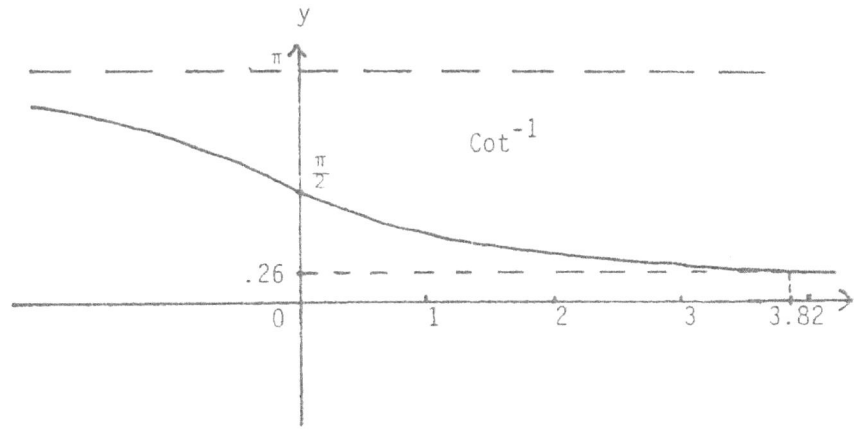

COMPOSITIONS.

Compositions involving tan or cot and their inverse functions are simply handled using the reciprocal identity.

EXAMPLE: Find $\tan(\text{Cot}^{-1}\sqrt{3})$.

SOLUTION: By the reciprocal identity the tangent of the angle whose Cotangent is $\sqrt{3}$ is equal to $\frac{1}{\sqrt{3}}$.   Therefore, $\tan(\text{Cot}^{-1}\sqrt{3}) = \frac{1}{\sqrt{3}}$.

In more abstract mathematical terms the solution above makes use of the composition $f(f^{-1}(x)) = x$ as follows: $\tan(\text{Cot}^{-1}\sqrt{3}) = \tan(\text{Tan}^{-1}\frac{1}{\sqrt{3}}) = \frac{1}{\sqrt{3}}$

Another example will reinforce this approach:

EXAMPLE: Find $\cot(\text{Tan}^{-1}(\frac{-2}{7}))$.

SOLUTION: $\cot(\text{Tan}^{-1}(\frac{-2}{7})) = \cot(\text{Cot}^{-1}(\frac{-7}{2})) = \frac{-7}{2}$.

If the composition is reversed, the range of the inverse function must be kept in mind.

EXAMPLE: Find $\text{Arccot}(\tan\frac{5\pi}{4})$.

SOLUTION: $\text{Arccot}(\tan\frac{5\pi}{4}) = \text{Arccot}(1) = \frac{\pi}{4}$.

Compositions involving decimal approximations also use the reciprocal identity with the help of the calculator.

EXAMPLE: Find $\cot(\text{Tan}^{-1}.3421)$.

SOLUTION: $\cot(\text{Tan}^{-1}.3421) = \cot(\text{Cot}^{-1}\frac{1}{.3421}) = \cot(\text{Cot}^{-1}2.923) = 2.923$.

EXAMPLE: Find $\text{Cot}^{-1}(\tan 6.53)$.

SOLUTION: $\text{Cot}^{-1}(\tan 6.53) = \text{Cot}^{-1}(.2520) = \text{Tan}^{-1}(\frac{1}{.2520})$

$= \text{Tan}^{-1}(3.969) = 1.32$.

Calculator steps: Radian mode. Input 6.53. Press tan key. Press $\frac{1}{x}$ key. Press INV key. Press tan key. Read 1.32.

EXAMPLE: Find $\text{Cot}^{-1}(\tan(-4.6))$.

SOLUTION: $\text{Cot}^{-1}(\tan(-4.6)) = \text{Cot}^{-1}(-8.860) = \text{Tan}^{-1}(\frac{1}{-8.860}) + \pi$

$= \text{Tan}^{-1}(-.1129) + \pi = 3.03$.

The solution requires the identity $Cot^{-1} x = Tan^{-1} (\frac{1}{x}) + \pi$ if $x < 0$.

Calculator steps: Radian mode. Input -4.6. Press tan key. Press $\frac{1}{x}$ key. Press INV key. Press tan key. Add π. Read 3.03.

Compositions with the sine and cosine and their inverse functions are also possible:

EXAMPLE: Find $\cos (Cot^{-1} (-4.68))$.

SOLUTION: Let $y = Cot^{-1} (-4.68)$. Then $\frac{\pi}{2} < y < \pi$. Calculator, Radian mode. Input -4.68. Press $\frac{1}{x}$ key. Press INV or ARC key. Press tan key. Add π. Press cos key. Read -.9779. The solution steps are:

$$Cos (Cot^{-1} (-4.68)) = \cos (Tan^{-1} (\frac{1}{-4.68}) + \pi)$$
$$= \cos (Tan^{-1} (-.2137) + \pi) = \cos 2.93 = -.9779.$$

ALTERNATE SOLUTION: Let $\theta = Cot^{-1} (-4.68)$. Since $\frac{\pi}{2} < \theta < \pi$ and $\cot \theta = -4.68$, then draw a circle as in Chapter 3 for the trigonometric ratios.

Figure 5-3E

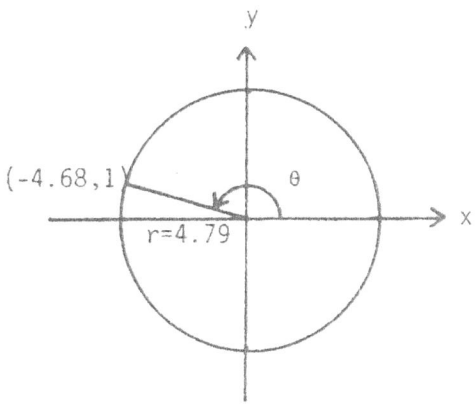

Since $\cot \theta = \frac{x}{y} = -4.68 = \frac{-4.68}{1}$, let $x = -4.68$ and $y = 1$.

Then $r = \sqrt{(-4.68)^2 + 1^2} = 4.79$.

Therefore, $\cos \theta = \frac{x}{r} = \frac{-4.68}{4.79} = -.9779$.

The alternate solution is particularly useful when the argument is a simple fraction and exact values are desired.

EXAMPLE: Find $\sin \left(\text{Tan}^{-1} \left(\frac{-3}{4} \right) \right)$.

SOLUTION: Let $\theta = \text{Tan}^{-1} \left(\frac{-3}{4} \right)$. Then $-\frac{\pi}{2} < \theta < 0$. Draw a sketch.

Figure 5-3F

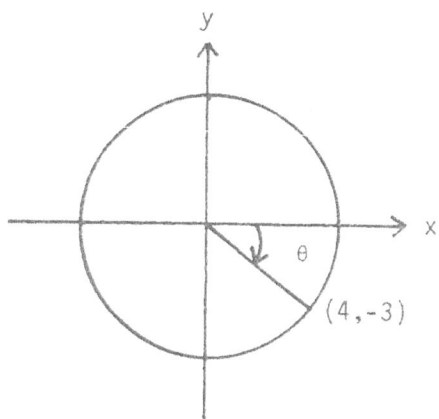

Since $\tan \theta = \frac{y}{x} = \frac{-3}{4}$, $r = \sqrt{(-3)^2 + 4^2} = 5$.

Therefore, $\sin \left(\text{Tan}^{-1} \left(\frac{-3}{4} \right) \right) = \sin \theta = \frac{y}{r} = \frac{-3}{5}$.

COMPOSITIONS AND RESTRICTED DOMAINS.

Compositions of the form $f(g(x))$ are defined if and only if values from the range of g are included in the domain of f. All of the compositions cited previously as examples have obeyed this restriction. The following are examples which do not.

EXAMPLE: $\text{Sin} \left(\text{Cot}^{-1} \left(-\sqrt{3} \right) \right) = \text{Sin} \frac{5\pi}{6}$ which is undefined since the domain of Sin $= \{ x : -\frac{\pi}{2} \le x \le \frac{\pi}{2} \}$.

EXAMPLE: $\cot \left(\text{Sin}^{-1} 0 \right) = \cot 0$ which is undefined since the domain of cot $= \{ x : x \neq k\pi, k = 0, \pm 1, \pm 2, \ldots \}$.

There is nothing to restrict the composition of a function with itself provided its range is a subset of its domain.

EXAMPLE: $\quad \text{Cot}^{-1} (\text{Cot}^{-1} \sqrt{3}) = \text{Cot}^{-1} (\frac{\pi}{6}) = \text{Cot}^{-1} (.5236)$

$$= \text{Tan}^{-1} (1.910) = 1.09.$$

EXERCISE SET 5-3.

Graph the following:

1) $y = \text{Arctan } x, \ 0 \le y < \frac{\pi}{2}$ $\qquad$ 2) $y = \text{arctan } x, \ -\pi < y \le 0.$

3) $y = \text{arccot } x, \ -\pi < y < 0$ $\qquad$ 4) $y = \cot^{-1} x, \ 0 < y < 2\pi$

5) $y = \text{Tan}^{-1} x + \frac{\pi}{2}$ $\qquad\qquad$ 6) $y = \text{Cot}^{-1} x - \frac{\pi}{2}$

Find the exact value(s) of the following:

7) $\text{Arctan } (-1)$ $\qquad$ 8) $\text{Cot}^{-1} \frac{1}{\sqrt{3}}$ $\qquad$ 9) $\tan^{-1} 0$ $\qquad$ 10) $\text{Tan}^{-1} (-\sqrt{3})$

11) $\cot^{-1} (-\sqrt{3})$ $\qquad$ 12) $\text{Cot}^{-1} 0$ $\qquad$ 13) $\tan (\text{Arccot } \frac{1}{2})$

14) $\tan (\text{Cot}^{-1} \frac{2}{3})$ $\qquad$ 15) $\cot (\text{Tan}^{-1} (\frac{-3}{2}))$ $\qquad$ 16) $\tan (\text{Tan}^{-1} \frac{4}{9})$

17) $\cot (\text{Cot}^{-1} (\frac{-3}{7}))$ $\qquad$ 18) $\text{Tan}^{-1} (\tan \frac{\pi}{8})$ $\qquad$ 19) $\text{Cot}^{-1} (\text{Cot } \frac{4\pi}{5})$

20) $\text{Arctan } (\tan \frac{3\pi}{4})$ $\qquad$ 21) $\text{Arctan } (\tan (\frac{-5\pi}{4}))$ $\quad$ 22) $\text{Arccot } (\cot (\frac{-\pi}{3}))$

23) $\text{Cot}^{-1} (\cot \frac{7\pi}{6})$ $\qquad$ 24) $\sin (\text{Tan}^{-1} 1)$ $\qquad$ 25) $\cos (\text{Tan}^{-1} \frac{1}{\sqrt{3}})$

26) $\cos (\text{Cot}^{-1} (\frac{-1}{\sqrt{3}}))$ $\qquad$ 27) $\text{Cot}^{-1} (\sin \pi)$ $\qquad$ 28) $\text{Tan}^{-1} (\cos \pi)$

29) $\sin (\text{Arccot } \frac{2}{3})$ $\qquad$ 30) $\cos (\text{Tan}^{-1} \frac{1}{4})$ $\qquad$ 31) $\cot (\text{Sin}^{-1} (\frac{-3}{7}))$

32) $\tan (\text{Cos}^{-1} (\frac{-8}{17}))$ $\qquad$ 33) $\sin (\text{Cot}^{-1} (\frac{-5}{12}))$ $\quad$ 34) $\cos (\text{Arccot } (\frac{-24}{7}))$

35) $\cos (\text{Tan}^{-1} (\frac{-4}{3}))$ $\qquad$ 36) $\cot (\text{Cos}^{-1} \frac{12}{13})$

Find approximate values for the following:

37) $\text{Cot}^{-1} 3.764$ $\qquad$ 38) $\text{Arccot } (-1.594)$ $\qquad$ 39) $\text{Arctan } (-.8764)$

40) $\text{Tan}^{-1} .5437$ $\qquad$ 41) $\cot^{-1} (2.235)$ $\qquad$ 42) $\tan^{-1} (-4.513)$

43) $\cot^{-1} (-1.015)$ $\qquad$ 44) $\text{arctan } (5.462)$ $\qquad$ 45) $\text{Arctan } (-.0132)$

46) $\text{Cot}^{-1} (\tan 2.35)$ $\qquad$ 47) $\text{Arccot } (\tan (-1.13))$ $\qquad$ 48) $\text{Tan}^{-1} (\cot 3.8)$

49) Arctan (tan 5.4)

50) Cot^{-1} (cot (-.14))

51) Tan^{-1} (Sin (-.4))

52) Tan (Sin^{-1} .9724)

53) Arcsin (tan 3.64)

54) Cos^{-1} (tan (-2.843))

55) Tan^{-1} (Tan^{-1} .6724)

56) Arctan (Tan^{-1} $\sqrt{2}$)

57) sin (Cot^{-1} (-2.843))

58) tan (Cos^{-1} .5382)

59) cot (Sin^{-1} .8878)

60) Tan (Cos^{-1} (-.8555))

5-4.   INVERSES OF SECANT AND COSECANT. (OPTIONAL).

The Arcsecant Relation.

Figure 5-4A shows the arcsecant relation for range values between -2π and
2π. This should be compared with the secant function illustrated in Figure 4-4A.

Figure 5-4A

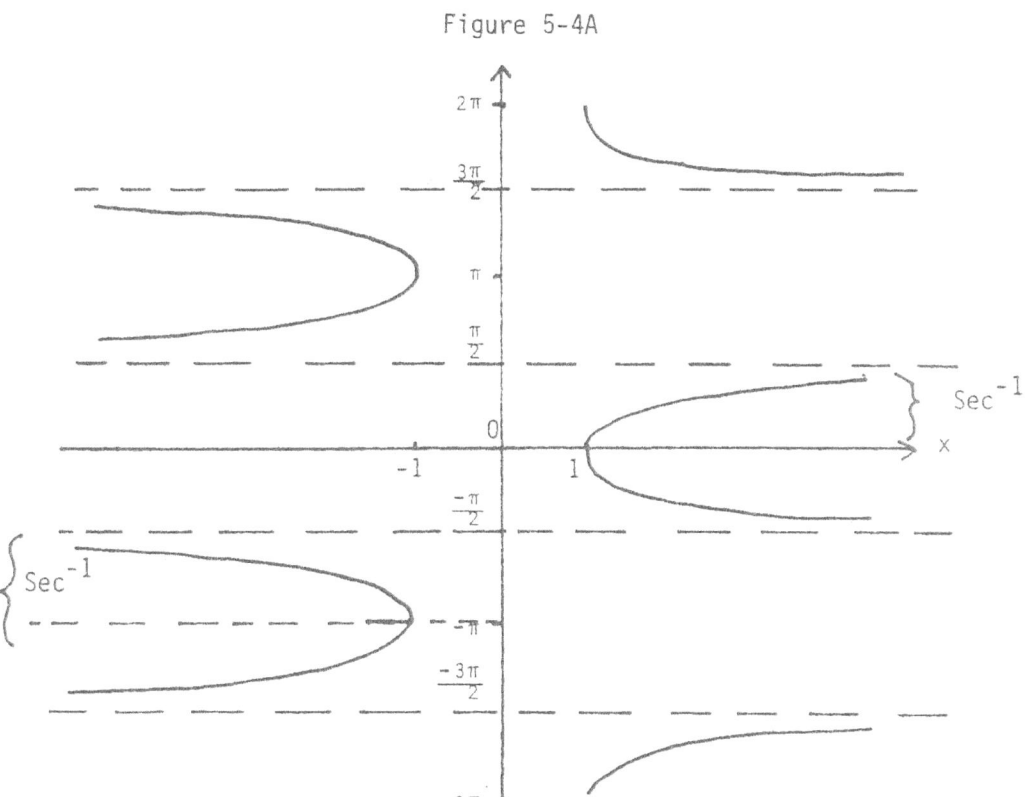

In conformity with the principal values of Sec , the bracketed portions
of the graph show the positive and negative branches of the Arcsecant function
(Sec^{-1}). The table below compares Sec and Sec^{-1} .

Function	Sec	Sec^{-1}
Domain	$\{x : 0 \le x < \frac{\pi}{2}\}$ U $\{x : -\pi \le x < -\frac{\pi}{2}\}$	$\{x : x \ge 1\}$ U $\{x : x \le -1\}$
Range	$\{y : y \ge 1\}$ U $\{y : y \le -1\}$	$\{y : 0 \le y < \frac{\pi}{2}\}$ U $\{y : -\pi \le y < -\frac{\pi}{2}\}$
Increasing	On $\{x : 0 \le x < \frac{\pi}{2}\}$	On $\{x : x \ge 1\}$
Decreasing	On $\{x : -\pi \le x < -\frac{\pi}{2}\}$	On $\{x : x \le -1\}$
Continuity	Discontinuous	Discontinuous
Odd or Even	Neither (sec is even)	Neither

Sec^{-1} is defined by:

Sec^{-1} = $\{(x,y) : |x| \ge 1, 0 \le y < \frac{\pi}{2}$ or $-\pi \le y < -\frac{\pi}{2}$, and $y = \sec^{-1} x\}$.

Solutions to problems involving Sec^{-1} x usually depend upon the reciprocal

identity sec x = $\frac{1}{\cos x}$.

EXAMPLE: Find Sec^{-1} $\sqrt{2}$.

SOLUTION: Let y = Sec^{-1} $\sqrt{2}$. Then Sec y = $\sqrt{2}$. Therefore, Cos y = $\frac{1}{\sqrt{2}}$ and

$y = \frac{\pi}{4}$.

EXAMPLE: Find Sec^{-1} 3.421.

SOLUTION: Sec^{-1} 3.421 = Cos^{-1} ($\frac{1}{3.421}$) = Cos^{-1} (.2923) = 1.27.

Care must be exercised since $\frac{\pi}{2} \le$ Cos^{-1} x $\le \pi$ and $-\pi \le$ Sec^{-1} x < $-\frac{\pi}{2}$

for negative reciprocals of x.

EXAMPLE: Find Arcsec (-2.594).

SOLUTION: The solution follows the same pattern on the calculator as given in

Section 4-4. Input 2.594. Press $\frac{1}{x}$ key. Press Inv key. Press cos key.

Subtract π. Read Arcsec(-2.594) = -1.97. Using inverse notation:

$$\text{Arcsec } (-2.594) = \text{Arccos } (0.3855) - \pi = 1.175 - \pi = -1.97 \ .$$

Compositions are also possible within the restrictions of the domains of the functions involved.

EXAMPLE: Find $\text{Sec}^{-1} (\sin \frac{2\pi}{3})$.

SOLUTION: $\text{Sec}^{-1} (\sin \frac{2\pi}{3}) = \text{Sec}^{-1} (\frac{\sqrt{3}}{2})$. Since $\frac{\sqrt{3}}{2} < 1$ and the domain of

$\text{Sec}^{-1} = \{x : |x| \geq 1\}$, there is no solution. In general, no

compositions of the form $\text{Sec}^{-1} (\sin x)$ or $\text{Sec}^{-1} (\cos x)$ can be formed

unless $\sin x = \pm 1$ or $\cos x = \pm 1$.

EXAMPLE: Find $\text{Sec}^{-1} (\tan 1.3)$.

SOLUTION: $\text{Sec}^{-1} (\tan 1.3) = \text{Sec}^{-1} (3.602) = \text{Cos}^{-1} (.2776) = 1.29$.

EXAMPLE: Find $\text{Sec}^{-1} (\tan .43)$.

SOLUTION: $\text{Sec}^{-1} (\tan .43) = \text{Sec}^{-1} (.4586)$. No solution.

If x is negative, $\text{Sec}^{-1} x$ is only defined for $x \leq -1$.

EXAMPLE: Find $\text{Sec}^{-1} (\cot 3.09)$.

SOLUTION: $\text{Sec}^{-1} (\cot 3.09) = \text{Sec}^{-1} (\frac{1}{\tan 3.09}) = \text{Sec}^{-1} (-19.37)$

$$= \text{Cos}^{-1}(.0516) - \pi = 1.52 - \pi = -1.62 \ .$$

THE ARCCOSECANT RELATION.

The graph of the arccosecant relation is given by:

Figure 5-4B

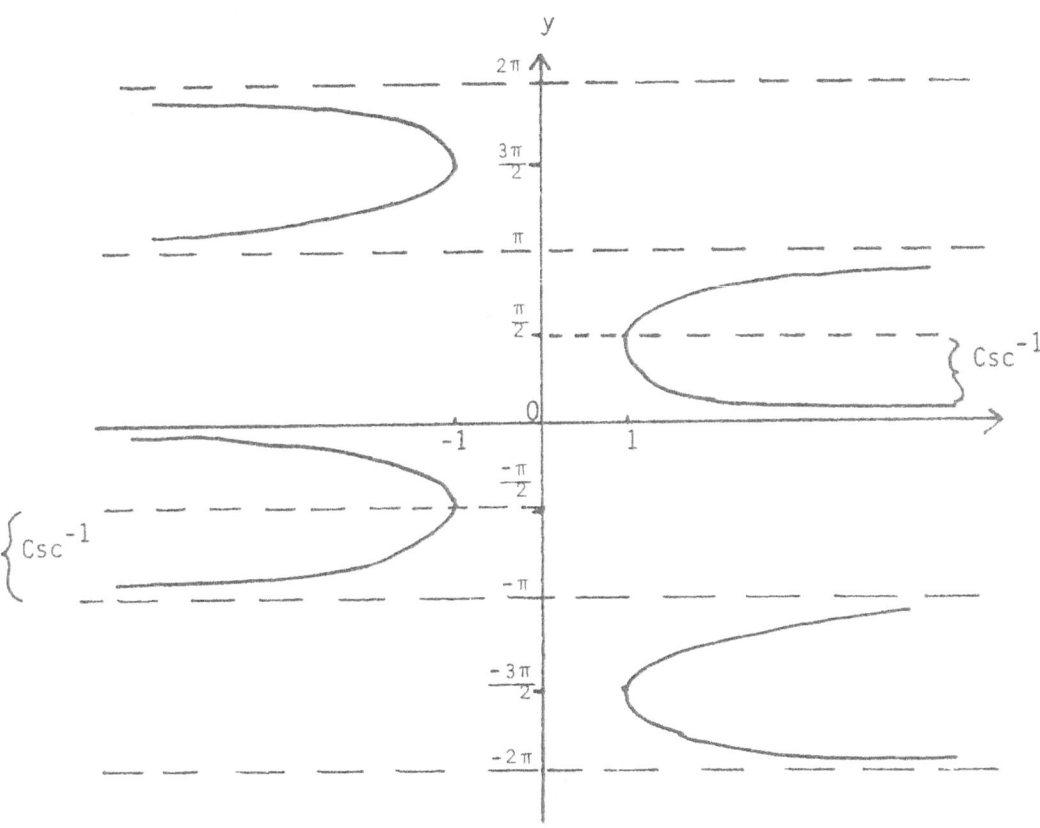

The table below compares Csc   and Csc^{-1} :

Function	Csc	Csc^{-1}
Domain	$\{x: 0 < x \leq \frac{\pi}{2}\} \cup \{x: -\pi < x \leq -\frac{\pi}{2}\}$	$\{x: x \geq 1\} \cup \{x: x \leq -1\}$
Range	$\{y: y \geq 1\} \cup \{y: y \leq -1\}$	$\{y: 0 < y \leq \frac{\pi}{2}\} \cup$ $\{y: -\pi < y \leq -\frac{\pi}{2}\}$
Increasing	On $\{x: -\pi < x \leq -\frac{\pi}{2}\}$	On $\{x: x \leq -1\}$
Decreasing	On $\{x: 0 < x \leq \frac{\pi}{2}\}$	On $\{x: x \geq 1\}$
Continuity	Discontinuous	Discontinuous
Odd or Even	Neither (csc is odd)	Neither

208

Csc^{-1} is defined by:

$\text{Csc}^{-1} = \{(x,y) : |x| \geq 1,\ 0 < y \leq \frac{\pi}{2}\ \text{or}\ -\pi < y \leq -\frac{\pi}{2}\,,\ \text{and } y = \csc^{-1} x\}.$

Problems involving $\text{Csc}^{-1} x$ usually depend on the reciprocal identity $\csc x = \frac{1}{\sin x}$ for their solution.

EXAMPLE: Find $\text{Csc}^{-1} 1.532$.

SOLUTION: $\text{Csc}^{-1} 1.532 = \text{Sin}^{-1} (\frac{1}{1.532}) = .71.$

EXAMPLE: Find $\text{Csc}^{-1} (-4.863)$.

SOLUTION: Refer to Figure 4-4D and its accompanying explanation. The calculator steps may now be translated in terms of inverse notation:

$\text{Csc}^{-1} (-4.863) = \text{Sin}^{-1} (\frac{1}{4.863}) - \pi = -2.93.$

The rule to follow then for finding the Arccosecant of a negative argument is to subtract π from the Arcsin of its positive reciprocal.

Another example will illustrate the rule graphically:

EXAMPLE: Find Arccsc (-1.835).

SOLUTION: Arccsc $(-1.835) = \text{Arcsin} (\frac{1}{1.835}) - \pi = .5763 - \pi = -2.57$ rounded.

Calculator steps: Radian mode. Input 1.835. Press $\frac{1}{x}$ key.

Press INV key. Press sin key. Subtract π. Read -2.57.

Figure 5-4C

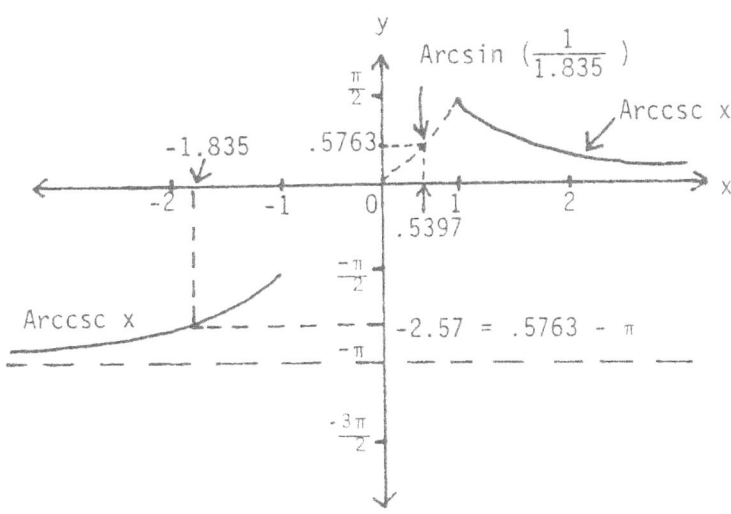

A few examples of compositions will complete this section:

EXAMPLE: Find $\text{Csc}^{-1} (\tan 1.35)$.

SOLUTION: $\text{Csc}^{-1} (\tan 1.35) = \text{Csc}^{-1} (4.455) = \text{Sin}^{-1} (.2245) = .23$.

EXAMPLE: Find $\text{Csc}^{-1} (\cot 1.35)$.

SOLUTION: $\text{Csc}^{-1} (\cot 1.35) = \text{Csc}^{-1} (\frac{1}{\tan 1.35}) = \text{Csc}^{-1} (.2245)$ which is undefined.

EXAMPLE: Find $\cot (\text{Csc}^{-1} 1.35)$.

SOLUTION: $\cot (\text{Csc}^{-1} 1.35) = \cot (\text{Sin}^{-1} .7407) = \cot .8342$

$$= \frac{1}{\tan .8342} = \frac{1}{1.103} = .9069.$$

Calculator steps: Radian mode. Input 1.35. Press $\frac{1}{x}$ key. Press INV key. Press sin key. Press tan key. Press $\frac{1}{x}$ key. Read .9069 rounded.

ALTERNATE SOLUTION: Circle model for trigonometric ratios. Let $\theta = \text{Csc}^{-1} 1.35$.

Then $\csc \theta = 1.35 = \frac{r}{y} = \frac{1.35}{1}$. $x^2 + y^2 = r^2$. Substituting,

$x^2 + (1)^2 = (1.35)^2$, $x = \sqrt{(1.35)^2 - 1} = \sqrt{.8225} = .9069$.

$\cot \theta = \frac{x}{y} = \frac{.9069}{1} = .9069$.

EXAMPLE: Find $\tan (\text{Csc}^{-1} (-3.276))$.

SOLUTION: $\tan (\text{Csc}^{-1} (-3.276)) = \tan (\text{Sin}^{-1} (\frac{1}{3.276}) - \pi)$

$= \tan (\text{Sin}^{-1} (.3053) - \pi) = \tan (.3102 - \pi) = \tan (-2.831) = .3205$.

ALTERNATE SOLUTION: Circle model for trigonometric ratios.

Let $\theta = \text{Csc}^{-1} (-3.276)$. Then $\csc \theta = -3.276$ and θ is in Quadrant III.

$\csc \theta = -3.276 = \frac{r}{y} = \frac{3.276}{-1}$. $x^2 + y^2 = r^2$. Substituting,

$x^2 + (-1)^2 = (3.276)^2$. $x = -\sqrt{(3.276)^2 - 1} = -\sqrt{9.732} = -3.120$

$\tan \theta = \frac{y}{x} = \frac{-1}{-3.120} = .3205$.

EXERCISE SET 5-4.

Graph the following:

1) $y = \text{arcsec } x, \; -\pi < y < \pi$ 2) $y = \text{arccsc } x, \; -\pi < y < \pi$

3) $y = \text{Arcsec } x, \; -\pi < y < \pi$ 4) $y = \text{Arccsc } x, \; -\pi < y < \pi$

5) $y = \text{Sec}^{-1} x + \dfrac{\pi}{2}$ 6) $y = \text{Csc}^{-1} x + \dfrac{\pi}{2}$

Find exact values for the following:

7) $\text{Sec}^{-1} \sqrt{2}$ 8) $\text{Csc}^{-1} \sqrt{2}$ 9) $\text{Arcsec } \left(\dfrac{-2}{\sqrt{3}} \right)$

10) $\text{Arccsc } \left(\dfrac{-2}{\sqrt{3}} \right)$ 11) $\text{Csc}^{-1} \dfrac{2}{\sqrt{3}}$ 12) $\text{Sec}^{-1} 2$

13) $\tan (\text{Csc}^{-1} (-2))$ 14) $\sec (\text{Csc}^{-1} (-1))$ 15) $\csc (\text{Sec}^{-1} 1)$

16) $\cot \left(\text{Sec}^{-1} \left(\dfrac{-2}{\sqrt{3}} \right) \right)$ 17) $\tan (\text{Sec}^{-1} (-1))$ 18) $\cot (\text{Csc}^{-1} 2)$

19) $\text{Csc}^{-1} \left(\tan \dfrac{\pi}{4} \right)$ 20) $\text{Csc}^{-1} \left(\cot \dfrac{-\pi}{4} \right)$ 21) $\text{Arccsc } \left(\tan \dfrac{5\pi}{4} \right)$

22) $\text{Sec}^{-1} \left(\cot \left(\dfrac{-5\pi}{4} \right) \right)$ 23) $\text{Arcsec } \left(\sec \dfrac{2\pi}{3} \right)$ 24) $\text{Arccsc } \left(\csc \dfrac{7\pi}{6} \right)$

Find approximate values for the following:

25) $\text{Csc}^{-1} \left(\tan \dfrac{5\pi}{6} \right)$ 26) $\text{Csc}^{-1} \left(\tan \dfrac{2\pi}{3} \right)$ 27) $\text{Sec}^{-1} \left(\tan \left(\dfrac{-\pi}{3} \right) \right)$

28) $\sin \left(\text{Csc}^{-1} \dfrac{5}{3} \right)$ 29) $\cos (\text{Csc}^{-1} 2.763)$ 30) $\tan (\text{Sec}^{-1} (-1.543))$

31) $\cot (\text{Csc}^{-1} (-1.118))$ 32) $\text{Cot}^{-1} (\sec 3.56)$

33) $\text{Sec}^{-1} (\tan 3.56)$ 34) $\text{Arcsec } (\cot (-3.56))$

35) $\text{Csc}^{-1} (\sec 2.89)$ 36) $\text{Sec}^{-1} (\csc 4.37)$

5-5.   THE INVERSE FUNCTIONS (RELATIONS) AND VARIATIONS IN
       AMPLITUDES, PERIODS, AND PHASE SHIFTS.

Amplitude

A comparison of $y = 2 \sin x$ and $y = 2 \sin^{-1} x$ will demonstrate the effects of a change in coefficient for the function and its inverse.

Figure 5-5A

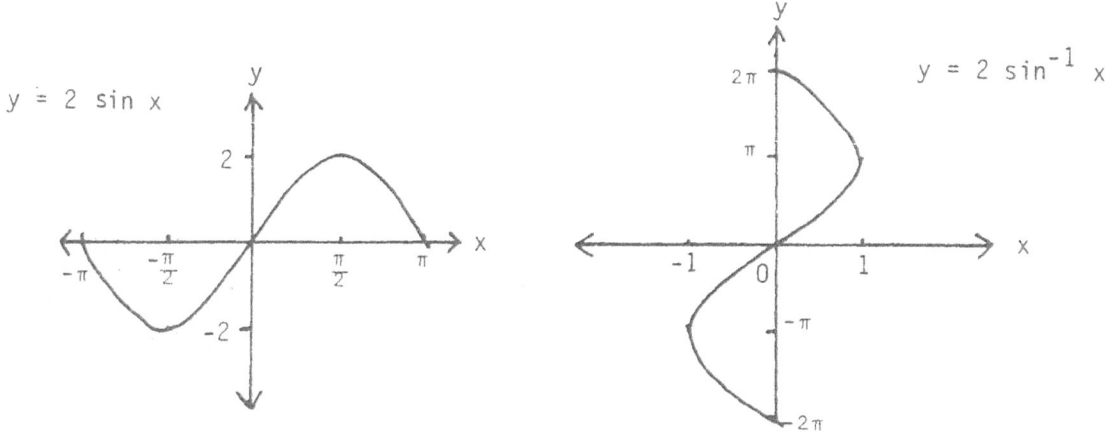

If the "amplitude" coefficient is greater than 1, the graph will be "stretched" along the y-axis.   Conversely, if the coefficient is between 0 and 1, the graph will contract along the y-axis.

Figure 5-5B

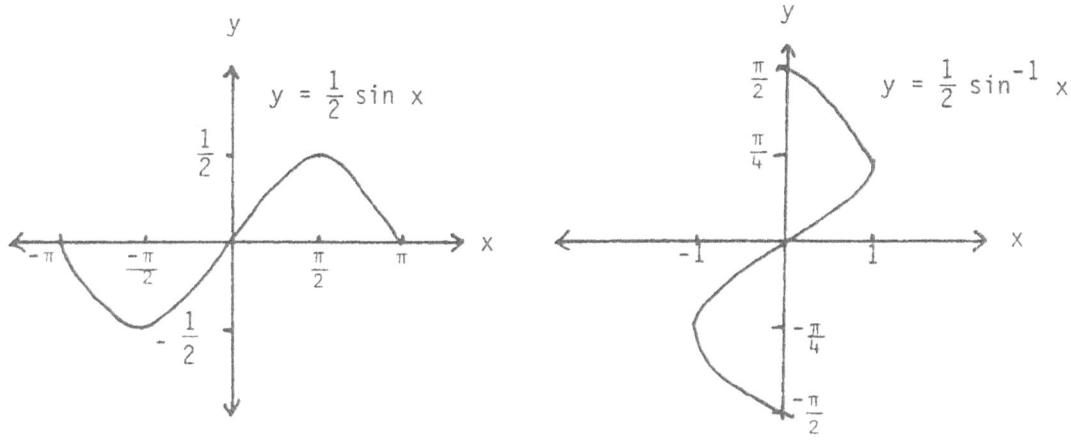

If the "amplitude" coefficient is negative there will result a reflection with respect to the x-axis for the function and a reflection with respect to the x-axis for its inverse.

Figure 5-5C

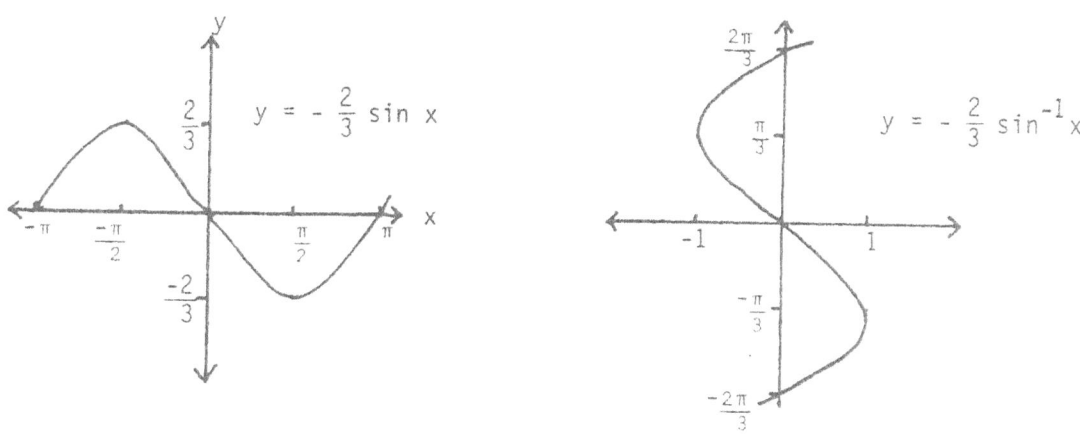

Graphs of the principal value functions and their inverse functions are similarly affected. A comparison of $y = 2 \tan x$ and $y = 2 \tan^{-1} x$ is shown.

Figure 5-5D

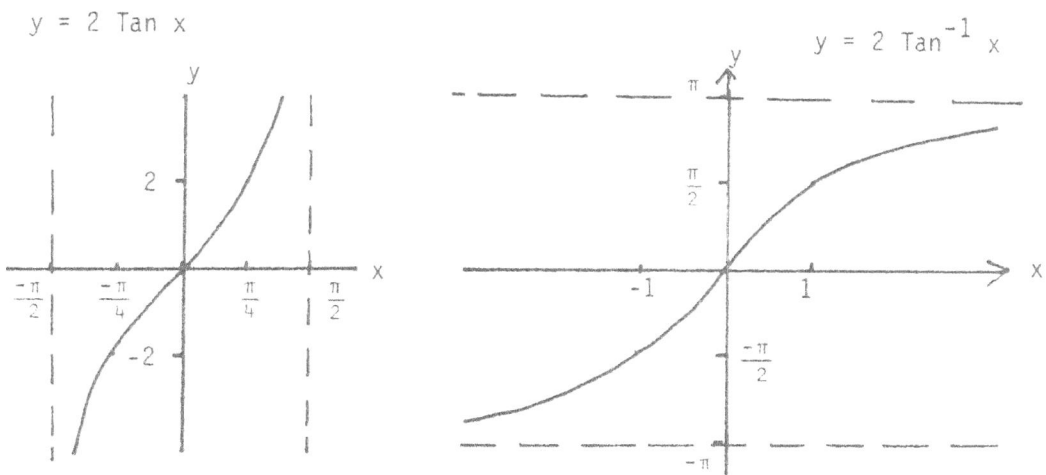

Note that the change in coefficient for $y = 2 \tan^{-1} x$ has the effect of doubling its range in comparison to $y = \tan^{-1} x$.

<u>Period</u>

A comparison of $y = \sin 2x$ and $y = \sin^{-1} 2x$ will show the effects of changing the argument's coefficient.

Figure 5-5E

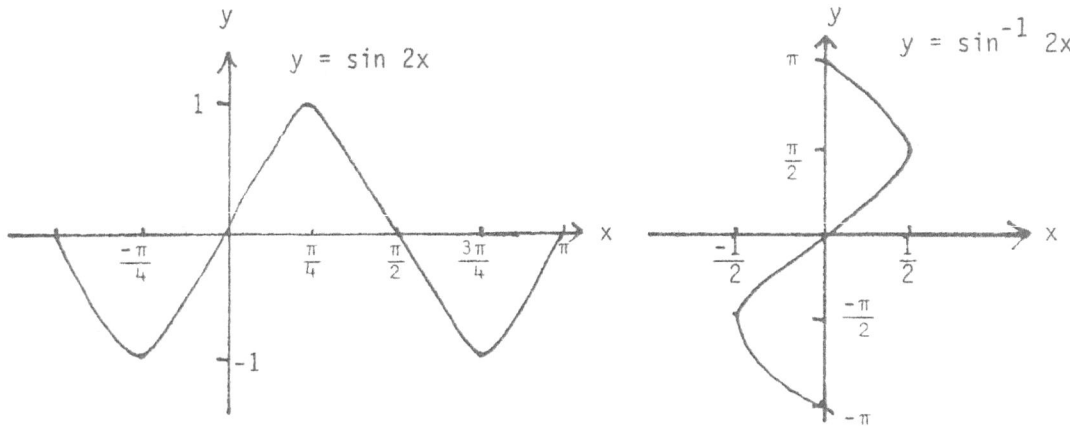

In each case, the units along the x-axis are <u>divided</u> by the coefficient 2. For the inverse relation this results in a contraction of the domain.

Consider now the dual effects of a change in both coefficients in a comparison of $y = 3 \cos \frac{1}{2} x$ and $y = 3 \cos^{-1} \frac{1}{2} x$.

Figure 5-5F

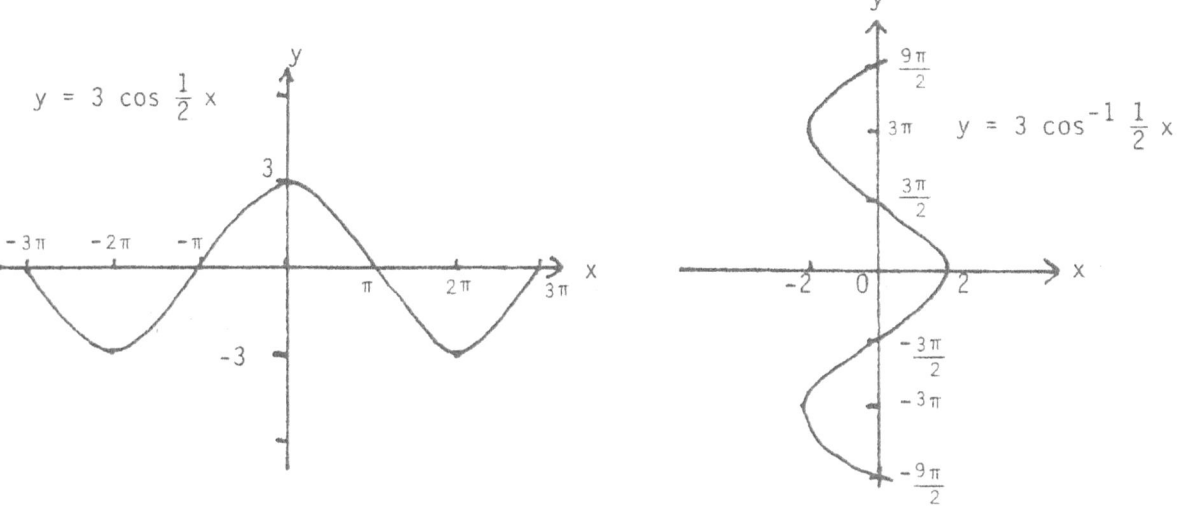

In both cases the units along the x-axis are divided by $\frac{1}{2}$ (multiplied by 2) and the units along the y-axis are multiplied by 3 as compared to the standard function y = cos x and its inverse.

Phase Shift

The effects of adding a constant to the argument of a trigonometric function were described in section 4-6. The graphing techniques will be carried over to the inverses. First, recall how y = sin $(x + \frac{\pi}{3})$ was graphed. The standard function is sketched without placement of the y-axis.

Figure 5-5G

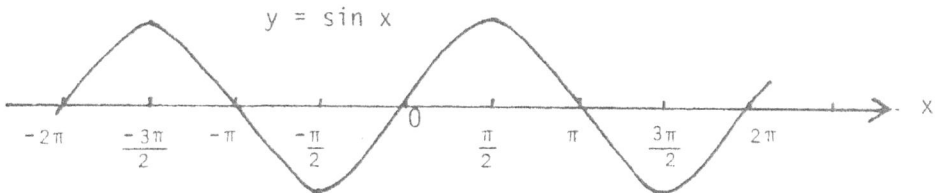

Since the addition of $\frac{\pi}{3}$ to x represents a lead, subtract $\frac{\pi}{3}$ from each x division point. Then place the y-axis at the new zero position.

Figure 5-5H

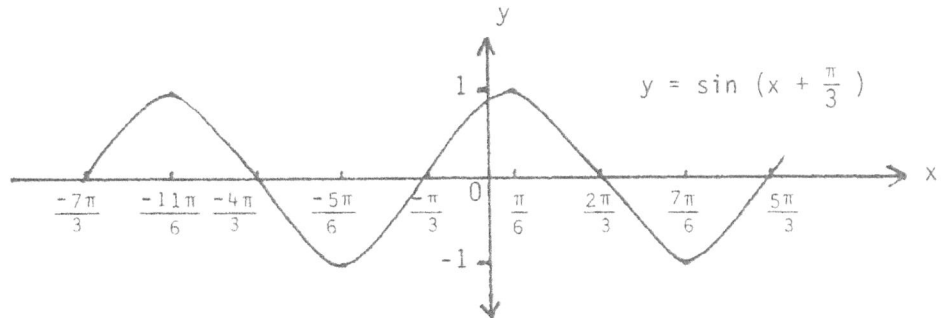

The technique above is now applied to $y = \sin^{-1}(x + \frac{\pi}{3})$. Sketch the standard inverse relation $y = \sin^{-1} x$ without the y-axis.

Figure 5-5I

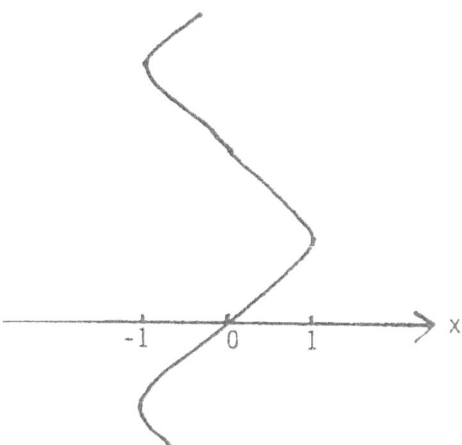

Subtract $\frac{\pi}{3}$ from the x division points and place the y-axis with its unchanged values as indicated:

Figure 5-5J

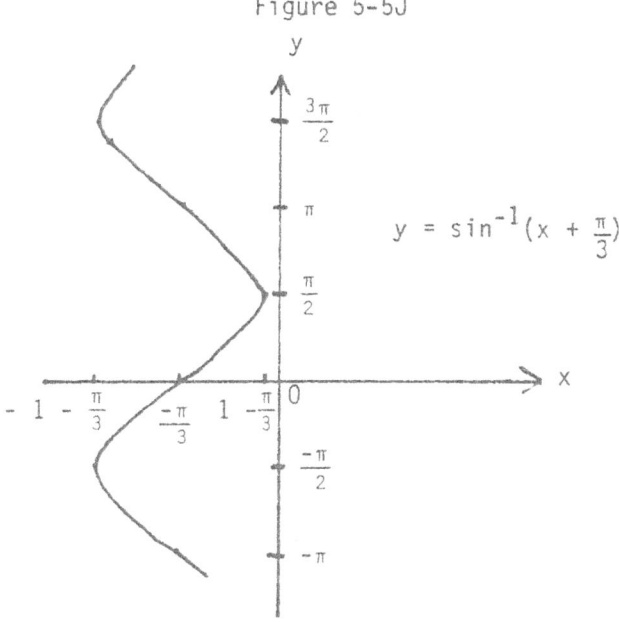

$$y = \sin^{-1}(x + \frac{\pi}{3})$$

As expected, the entire graph is translated $\frac{\pi}{3}$ units to the left.

<u>FUNCTIONS WITH PHASE SHIFTS AND THEIR INVERSES:</u>

A visual comparison of $y = \sin (x + \frac{\pi}{3})$, Figure 5-5H and $y = \sin^{-1} (x + \frac{\pi}{3})$, Figure 5-5J, will confirm that they are <u>not</u> inverses of each other. If we remember that the graphs of a function and its inverse are symmetric with respect to the line $y = x$, it will be apparent that the inverse of $y = \sin (x + \frac{\pi}{3})$ is $y = \sin^{-1} x - \frac{\pi}{3}$. The graph of this inverse is shown in Figure 5-5K. Note that the graph of $y = \sin^{-1} x$ is shifted downward by $\frac{\pi}{3}$ units.

Figure 5-5K

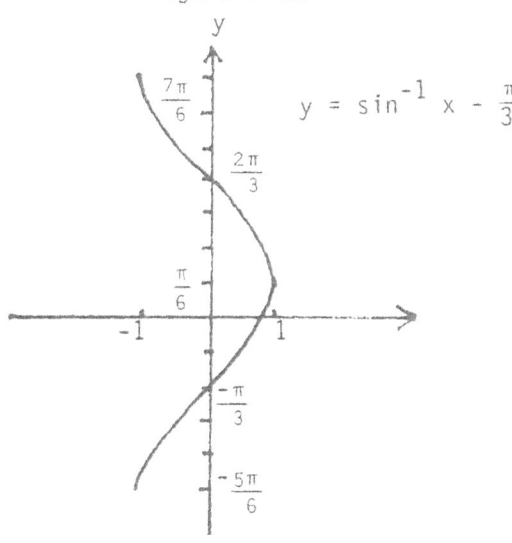

EXAMPLE: Prove the inverse of $y = \sin (x + \frac{\pi}{3})$ is $y = \sin^{-1} x - \frac{\pi}{3}$.

SOLUTION: If $y = \sin^{-1} x - \frac{\pi}{3}$, $y + \frac{\pi}{3} = \sin^{-1} x$.

$y + \frac{\pi}{3} = \sin^{-1} x$ is defined by $x = \sin (y + \frac{\pi}{3})$.

Interchanging x and y for the inverse, $y = \sin (x + \frac{\pi}{3})$.

EXAMPLE: Find the inverse of $y = \text{Cos} (x - \frac{\pi}{4})$ and sketch its graph.

SOLUTION: Interchange variables for the inverse: $x = \text{Cos} (y - \frac{\pi}{4})$. By definition $y - \frac{\pi}{4} = \text{Cos}^{-1} x$. Hence, $y = \text{Cos}^{-1} x + \frac{\pi}{4}$. The graph is sketched below.

Figure 5-5L

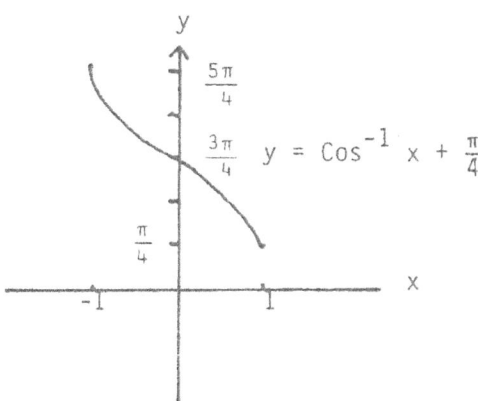

$$y = \cos^{-1} x + \frac{\pi}{4}$$

An example illustrating all of the effects covered in this section will conclude this exposition.

EXAMPLE: Find the inverse of $y = 3 \tan (2x - \frac{\pi}{3})$ and sketch its graph.

SOLTUION: The inverse is given by $x = 3 \tan (2y - \frac{\pi}{3})$. Then $\frac{x}{3} = \tan (2y - \frac{\pi}{3})$ and by definition, $2y - \frac{\pi}{3} = \tan^{-1} \frac{x}{3}$. Solving for y,

$2y = \tan^{-1} \frac{x}{3} + \frac{\pi}{3}$ and $y = \frac{1}{2} \tan^{-1} \frac{x}{3} + \frac{\pi}{6}$.

To graph, begin with a sketch of $\tan^{-1} x$.

Figure 5-5M

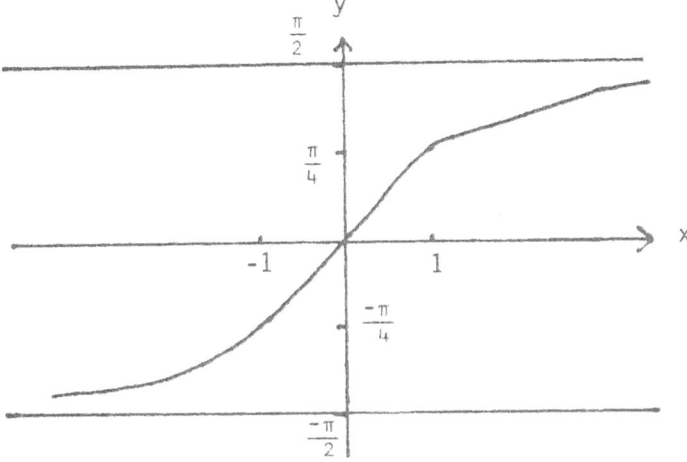

The coefficient, $\frac{1}{2}$, contracts the y values to $\frac{1}{2}$ of their indicated values. The coefficient of the argument, $\frac{1}{3}$, stretches the graph along the x-axis by a multiple of 3. Finally, the addition of $\frac{\pi}{6}$ shifts the graph $\frac{\pi}{6}$ units in the positive y direction. The results are shown below:

Figure 5-5N

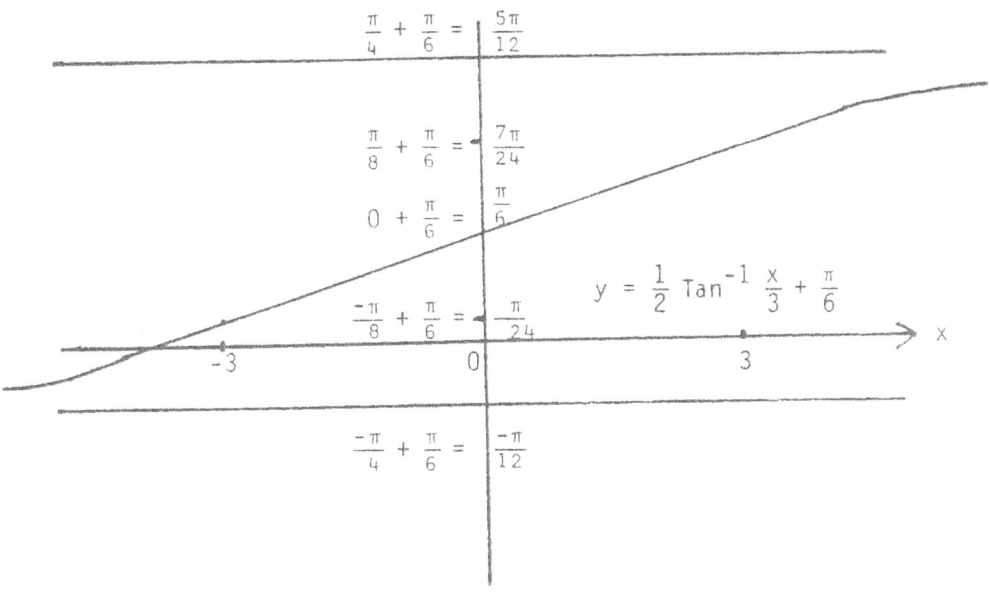

$$\frac{\pi}{4} + \frac{\pi}{6} = \frac{5\pi}{12}$$

$$\frac{\pi}{8} + \frac{\pi}{6} = \frac{7\pi}{24}$$

$$0 + \frac{\pi}{6} = \frac{\pi}{6}$$

$$y = \frac{1}{2} \, \text{Tan}^{-1} \frac{x}{3} + \frac{\pi}{6}$$

$$\frac{-\pi}{8} + \frac{\pi}{6} = \frac{\pi}{24}$$

$$\frac{-\pi}{4} + \frac{\pi}{6} = \frac{-\pi}{12}$$

EXERCISE SET 5-5.

Sketch the graphs of the following:

1) $y = \frac{1}{2} \, \text{Cos}^{-1} x$

2) $y = \frac{2}{3} \, \text{Arccos } x$

3) $y = - \, \text{Sin}^{-1} x$

4) $y = \frac{-3}{2} \, \text{Arcsin } x$

5) $y = 2 \, \text{Tan}^{-1} 2x$

6) $y = \frac{1}{3} \, \text{Arctan } \frac{1}{2} x$

7) $y = \pi \, \text{Cot}^{-1} x$

8) $y = \frac{3}{2} \, \text{Arcsin } \pi x$

9) $y = \frac{2\pi}{3} \, \text{Arccos } (x - 1)$

10) $y = 3 \, \text{Tan}^{-1} (x + 1)$

11) $y = 2 \, \text{Sin}^{-1} x - \frac{\pi}{3}$

12) $y = \frac{1}{2} \, \text{Cos}^{-1} (x + 2) + \frac{\pi}{4}$

Find the inverse of the following:

13) $y = \cos (x - \frac{\pi}{4})$

14) $y = 2 \sin (x + \frac{\pi}{3})$

15) $y = \frac{1}{2} \tan (x - \frac{\pi}{6})$

16) $y = -3 \, \text{Cot} (x + \frac{\pi}{2})$

17) $y = 2 \, \text{Cos}^{-1} \frac{x}{3} - \frac{\pi}{4}$

18) $y = \frac{1}{2} \, \text{Tan}^{-1} \frac{x}{4} + \frac{\pi}{3}$

5-6. <u>TRIGONOMETRIC EQUATIONS</u>.

Throughout the earlier sections of this text simple trigonometric equations have been solved without focusing special attention on the fact. A few examples will bear this out.

<u>EXAMPLE</u>: Find x, $0 \le x \le 2\pi$, if $\sin x = \frac{\sqrt{3}}{2}$.

<u>SOLUTION</u>: If $\sin x = \frac{\sqrt{3}}{2}$ then $x = \sin^{-1} \frac{\sqrt{3}}{2} = \frac{\pi}{3}$ or $\frac{2\pi}{3}$.

<u>EXAMPLE</u>: Solve for x, $\mathrm{Sin}^{-1} x = \frac{-\pi}{3}$.

<u>SOLUTION</u>: If $\mathrm{Sin}^{-1} x = \frac{-\pi}{3}$ then $x = \mathrm{Sin}(\frac{-\pi}{3}) = \frac{-\sqrt{3}}{2}$.

<u>EXAMPLE</u>: Find $\mathrm{Tan}^{-1} 1$.

<u>SOLUTION</u>: Let $x = \mathrm{Tan}^{-1} 1$. Then $\mathrm{Tan}\, x = 1$. Hence $x = \frac{\pi}{4}$.

In this example x was introduced to show the problem could be formulated as an equation.

<u>EXAMPLE</u>: Solve for x : $\tan x = \frac{\sin x}{\cos x}$, $\cos x \ne 0$.

<u>SOLUTION</u>: This equation is an <u>identity</u> and is true for all real numbers for which the equation is defined. The solution set

$$= \{x : x \ne \frac{k\pi}{2}, k = \pm1, \pm3, \ldots\}$$

<u>EXAMPLE</u>: Solve for x, $\sec x = -2.357$.

<u>SOLUTION</u>: Calculator. Radian mode. Input -2.357. Press $\frac{1}{x}$ key. Press INV key. Press cos key. Read 2.01 rounded. The general solution is given by $\{x : x = \pm2.01 + 2k\pi, k = 0, \pm1, \pm2, \ldots\}$

In the remainder of this section trigonometric equations will be solved making use of skills developed in algebra. For many trigonometric equations the solutions are an exact parallel to their algebraic counterparts. This will be demonstrated in a side by side analysis of the steps in each solution.

Trigonometric equations differ from those solved in algebra by the need for an additional step in their solutions. This step requires an evaluation of the inverse function or relation.

FIRST DEGREE EQUATIONS.

EXAMPLE: Solve for x.

Algebra	Steps	Trigonometry
$2x = 1$	Given	$2 \cos x = 1$
$x = \dfrac{1}{2}$	If $ab = c$, $a = \dfrac{c}{b}$	$\cos x = \dfrac{1}{2}$
	Inverse relation	$x = \cos^{-1} \dfrac{1}{2}$
	Evaluation	$x = \pm\dfrac{\pi}{3} + 2k\pi$, $k = 0, \pm1, \pm2,..$

EXAMPLE: Solve for x.

Algebra	Steps	Trigonometry
$\dfrac{x}{2} + 1 = x$	Given	$\dfrac{\cot x}{2} + 1 = \cot x$
$x + 2 = 2x$	If $a = b$, $ac = bc$	$\cot x + 2 = 2 \cot x$
$2 = 2x - x$	If $a = b$, $a-c=b-c$	$2 = 2 \cot x - \cot x$
$2 = x$	Combining terms	$2 = \cot x$
	Inverse relation	$x = \cot^{-1} 2$
	Reciprocal identity	$x = \tan^{-1} .5$
	Evaluation	$x = .46 + k\pi$, $k = 0, \pm1, \pm2...$

EXAMPLE: Solve for x.

Algebra	Steps	Trigonometry
$\dfrac{3}{x + 2} = \dfrac{1}{x - 2}$	Given	$\dfrac{3}{\csc x + 2} = \dfrac{1}{\csc x - 2}$
$3(x-2) = (x+2)\cdot 1$	If $\dfrac{a}{b} = \dfrac{c}{d}$, $ad = bc$	$3(\csc x - 2) = (\csc x + 2)\cdot 1$
$3x - 6 = x + 2$	Distributive Axiom	$3 \csc x - 6 = \csc x + 2$
$3x - x = 2 + 6$	If $a=b$, $a-c = b-c$	$3 \csc x - \csc x = 2 + 6$
$2x = 8$	Combining terms	$2 \csc x = 8$
$x = 4$	If $ac = bc$, $a = b$	$\csc x = 4$
	Inverse function	$x = \csc^{-1} 4$
	Reciprocal identity	$x = \sin^{-1} .25$
	Evaluation	$x = .25$

EXAMPLE: Solve for x.

Algebra	Steps	Trigonometry
$\lvert x - 1 \rvert = 4$	Given	$\lvert \text{Tan } x - 1 \rvert = 4$
$x - 1 = \pm 4$	If $\lvert a \rvert = b$, $a = \pm b$	$\text{Tan } x - 1 = \pm 4$
$x = 1 \pm 4$	If $a = b$, $a+c = b+c$	$\text{Tan } x = 1 \pm 4$
$x = 5$ or $x = -3$	Combining terms	$\text{Tan } x = 5$ or $\text{Tan } x = -3$
	Inverse function	$x = \text{Tan}^{-1} 5$ or $x = \text{Tan}^{-1}(-3)$
	Evaluation	$x = 1.37$ or $x = -1.25$

QUADRATIC EQUATIONS

The three methods learned in algebra for solving quadratic equations are factoring, completion of the square, and the quadratic formula. These techniques are demonstrated below.

EXAMPLE: Solve for x.

Algebra	Steps	Trigonometry
$2x + 1 = 8x^2$	Given	$2 \text{ Cos } x + 1 = 8 \text{ Cos}^2 x$
$8x^2 - 2x - 1 = 0$	Standard form	$8 \text{ Cos}^2 x - 2 \text{ Cos } x - 1 = 0$
$(4x + 1)(2x - 1) = 0$	Factoring	$(4 \text{ Cos } x + 1)(2 \text{ Cos } x - 1) = 0$
$4x+1 = 0 \mid 2x - 1 = 0$	If $ab=0$, $a=0$ or $b=0$	$4 \text{ Cos } x + 1 = 0 \mid 2 \text{ Cos } x - 1 = 0$
$4x = -1 \mid 2x = 1$	If $a=b$, $a-c = b-c$	$4 \text{ Cos } x = -1 \mid 2 \text{ Cos } x = 1$
$x = \frac{-1}{4} \mid x = \frac{1}{2}$	If $ac = bc$, $a = b$	$\text{Cos } x = -\frac{1}{4} \mid \text{Cos } x = \frac{1}{2}$
	Inverse function	$x = \text{Cos}^{-1}(\frac{-1}{4}) \mid x = \text{Cos}^{-1} \frac{1}{2}$
	Evaluation	$x = 1.82 \mid x = \frac{\pi}{3}$

EXAMPLE: Solve for x.

Algebra	Steps	Trigonometry
$x^2 + 2x = 2$	Given	$\cot^2 x + 2\cot x = 2$
$x^2 + 2x + 1 = 2 + 1$	Completing the square	$\cot^2 x + 2\cot x + 1 = 2 + 1$
$(x + 1)^2 = 3$	Factoring	$(\cot x + 1)^2 = 3$
$\lvert x + 1 \rvert = \sqrt{3}$	Taking square roots	$\lvert \cot x + 1 \rvert = \sqrt{3}$
$x + 1 = \pm\sqrt{3}$	If $\lvert a \rvert = b$, $a = \pm b$	$\cot x + 1 = \pm\sqrt{3}$

		Steps		
$x+1 = \sqrt{3}$	$x+1 = -\sqrt{3}$	Cases	$\cot x +1= \sqrt{3}$	$\cot x +1=-\sqrt{3}$
$x=-1+\sqrt{3}$	$x=-1-\sqrt{3}$	If $a=b$, $a-c=b-c$	$\cot x = -1+\sqrt{3}$	$\cot x=-1-\sqrt{3}$
		Inverse function	$x=\cot^{-1}(-1+\sqrt{3})$	$x=\cot^{-1}(-1-\sqrt{3})$
		Approximation	$x=\cot^{-1}(.7321)$	$x=\cot^{-1}(-2.732)$
		Reciprocal identity	$x=\tan^{-1}(1.366)$	$x=\tan^{-1}(-.3660)+\pi$
		Evaluation	$x = .94$	$x = 2.79$

In preparation for the next example, a review of the quadratic formula is in order. Given an equation in the form $ax^2 + bx + c = 0$, where $a \neq 0$, the complete solution is given by $x = \dfrac{-b \pm\sqrt{b^2 - 4ac}}{2a}$

EXAMPLE: Solve for x.

Algebra	Steps	Trigonometry
$2x^2 + 4x = 1$	Given	$2\sin^2 x + 4\sin x = 1$
$2x^2 + 4x - 1 = 0$	Standard form	$2\sin^2 x + 4\sin x - 1 = 0$
$a=2$, $b=4$, $c=-1$	Coefficients defined	$a=2$, $b=4$, $c=-1$
$x=\dfrac{-4\pm\sqrt{(4)^2-4(2)(-1)}}{2(2)}$	Formula substitution	$\sin x = \dfrac{-4\pm\sqrt{(4)^2-4(2)(-1)}}{2(2)}$
$x = \dfrac{-4\pm\sqrt{24}}{4}$	Simplifying	$\sin x = \dfrac{-4 \pm \sqrt{24}}{4}$

		Steps		
$x = \dfrac{-4+\sqrt{24}}{4}$	$x = \dfrac{-4-\sqrt{24}}{4}$	Cases	$\sin x = \dfrac{-4+\sqrt{24}}{4}$	$\sin x = \dfrac{-4-\sqrt{24}}{4}$
$x = \dfrac{-4+2\sqrt{6}}{4}$	$x = \dfrac{-4-2\sqrt{6}}{4}$	Simplifying the radical	$\sin x = \dfrac{-4+2\sqrt{6}}{4}$	$\sin x = \dfrac{-4-2\sqrt{6}}{4}$

$x = \dfrac{-2+\sqrt{6}}{2}$ | $x = \dfrac{-2-\sqrt{6}}{2}$	Dividing out a common factor	$Sin\ x = \dfrac{-2+\sqrt{6}}{2}$	$Sin\ x = \dfrac{-2-\sqrt{6}}{2}$
	Approximation	$Sin\ x = .2247$	$Sin\ x = -2.22$
	Inverse function	$x = Sin^{-1} .2247$	No solution
	Evaluation	$x = .23$	

EQUATIONS OF HIGHER DEGREE

Equations of the third, fourth, etc., degree may be solvable by factoring. In the following examples the student should supply the reasons for the indicated steps.

EXAMPLE: Solve for x.

Algebra

Trigonometry

$x^4 - 3x^2 + 2 = 0$ $Csc^4 x - 3\ Csc^2 x + 2 = 0$

$(x^2 - 1)(x^2 - 2) = 0$ $(Csc^2 x - 1)(Csc^2 x - 2) = 0$

$x^2-1 = 0$	$x^2-2 = 0$	$Csc^2 x - 1 = 0$	$Csc^2 x - 2 = 0$
$x^2 = 1$	$x^2 = 2$	$Csc^2 x = 1$	$Csc^2 x = 2$
$x = \pm 1$	$x = \pm\sqrt{2}$	$Csc\ x = \pm 1$	$Csc\ x = \pm\sqrt{2}$
		$x=Csc^{-1} 1$ or $x=Csc^{-1}(-1)$	$x=Csc^{-1} \sqrt{2}$ or $x=Csc^{-1}(-\sqrt{2}\ )$
		$x = \dfrac{\pi}{2}$ or $x = -\dfrac{\pi}{2}$	$x = \dfrac{\pi}{4}$ or $x = \dfrac{-3\pi}{4}$

EXAMPLE: Solve for x.

Algebra

Trigonometry

$x^3 - 3x^2 + x = 0$ $Sin^3 x - 3\ Sin^2 x + Sin\ x = 0$

$x(x^2 - 3x + 1) = 0$ $Sin\ x\ (Sin^2 x - 3\ Sin\ x + 1) = 0$

$x=0$	$x^2-3x+1 = 0$	$Sin\ x = 0$	$Sin^2 x - 3\ Sin\ x + 1 = 0$
	$x = \dfrac{3\pm\sqrt{(-3)^2-4(1)(1)}}{2(1)}$		$Sin\ x = \dfrac{3\pm\sqrt{(-3)^2-4(1)(1)}}{2(1)}$
	$x = \dfrac{3\pm\sqrt{5}}{2}$		$Sin\ x = \dfrac{3\pm\sqrt{5}}{2}$
			$Sin\ x = .3820$ | $Sin\ x = 2.618$
			$x = Sin^{-1} .3820$ | No Solution
		$x = 0$	$x = .39$

EQUATIONS WITH RADICALS.

Solving equations containing square roots of expressions requires the squaring of both sides of the equations. Care must be exercised in checking out solutions in the original equation as squaring may introduce extraneous solutions.

EXAMPLE: Solve for x.

Algebra	Step	Trigonometry
$3\sqrt{x+2} = x+4$	Given	$3\sqrt{\cot x + 2} = \cot x + 4$
$9(x+2) = (x+4)^2$	Squaring	$9(\cot x + 2) = (\cot x + 4)^2$
$9x+18 = x^2+8x+16$	Distributive Axiom	$9\cot x + 18 = \cot^2 x + 8\cot x + 16$
$x^2 - x - 2 = 0$	Standard form	$\cot^2 x - \cot x - 2 = 0$
$(x-2)(x+1) = 0$	Factoring	$(\cot x - 2)(\cot x + 1) = 0$

$x-2 = 0$ $\mid$ $x+1 = 0$	If ab=0, a=0 or b=0	$\cot x -2 = 0$	$\mid$	$\cot x +1 = 0$
$x = 2$ $\mid$ $x = -1$	If a+b=0, a=-b	$\cot x = 2$	$\mid$	$\cot x = -1$
	Inverse function	$x = \cot^{-1} 2$	$\mid$	$x = \cot^{-1}(1)$
	Evaluation	$x = .46$	$\mid$	$x = \frac{3\pi}{4}$

In the following example squaring occurs twice to remove radicals. Checking reveals the presence of an extraneous solution.

Algebra	Step	Trigonometry
$\sqrt{2x-3} = 2-\sqrt{x-1}$	Given	$\sqrt{2\sec x-3} = 2-\sqrt{\sec x -1}$
$2x-3=4-4\sqrt{x-1} + (x-1)$	Squaring	$2\sec x -3 = 4 - 4\sqrt{\sec x -1} + (\sec x -1)$
$x-6 = -4\sqrt{x-1}$	Simplifying	$\sec x - 6 = -4\sqrt{\sec x -1}$
$x^2-12x+36 = 16x-16$	Squaring	$\sec^2 x-12 \sec x+36 = 16 \sec x-16$
$x^2 - 28x + 52 = 0$	Standard form	$\sec^2 x-28 \sec x+52 = 0$
$(x-26)(x-2) = 0$	Factoring	$(\sec x-26)(\sec x-2) = 0$

x-26 = 0	x-2 = 0
x = 26	x = 2
Not a solution	x = 2

Sec x -26 = 0	Sec x -2 = 0
Sec x = 26	Sec x = 2
Not a solution	$x = Sec^{-1} 2$
	$x = \frac{\pi}{3}$

EXPONENTIAL EQUATIONS.

Certain exponential equations are easily solved by equating exponents. If $b^x = b^y$ then x = y. The technique requires expressing both sides of the equation as powers of the given base.

EXAMPLE: Solve for x.

Algebra	Step	Trigonometry
$2^x = 8$	Given	$2^{Tan\ x} = 8$
$2^x = 2^3$	$8 = 2^3$	$2^{Tan\ x} = 2^3$
x = 3	If $b^x = b^y$, x=y	Tan x = 3
	Inverse function	$x = Tan^{-1} 3$
	Evaluation	x = 1.25

EXAMPLE: Solve for x. Fill in missing steps.

Algebra	Step	Trigonometry
$2^{3x} = \frac{1}{64}$	Given	$2^{3\ Csc\ x} = \frac{1}{64}$
$2^{3x} = 2^{-6}$	$\frac{1}{64} = 2^{-6}$	$2^{3\ Csc\ x} = 2^{-6}$
3x = -6		3 Csc x = -6
x = -2		Csc x = -2
		$x = Csc^{-1} (-2)$
		$x = -\frac{5\pi}{6}$

A more general method of solving exponential equations requires the use of logarithms. Scientific calculators are equipped to handle both common logarithms (Base 10 : log key) and natural logarithms (Base e $\doteq$ 2.7182818 : ln x key).* The following examples will illustrate their use.

---

* To obtain an approximation to the irrational number e on the calculator, input 1. Press INV key. Press ln x key. Read display.

EXAMPLE: Solve for x. Fill in missing steps.

Algebra	Step	Trigonometry
$4^{5x+3} = 7$	Given	$4^{5\,\text{Sin}\,x+3} = 7$
$(4^{5x})(4^3) = 7$	$a^{m+m} = a^m \cdot a^n$	$(4^{5\,\text{Sin}\,x})(4^3) = 7$
$4^{5x} = \frac{7}{64}$		$4^{5\,\text{Sin}\,x} = \frac{7}{64}$
$5x \log 4 = \log(\frac{7}{64})$	If $a^m=b$, $m \log a = \log b$	$(5\,\text{Sin}\,x)(\log 4) = \log(\frac{7}{64})$
$x = \frac{\log(\frac{7}{64})}{5 \log 4}$		$\text{Sin}\,x = \frac{\log(\frac{7}{64})}{5 \log 4}$
$x = -.319$		$\text{Sin}\,x = -.319$
		$x = \text{Sin}^{-1}(-.319)$
		$x = -.32$

In the following example, the step involving the calculator's ln x key will be indicated.

EXAMPLE: Solve for x. Fill in missing steps.

Algebra	Step	Trigonometry
$e^{2x+1} = 3$	Given	$e^{2\,\text{Cos}\,x+1} = 3$
$(2x+1)(\ln e) = \ln 3$	If $a^m=b$, $m \ln a = \ln b$	$(2\,\text{Cos}\,x+1)(\ln e) = \ln 3$
$2x + 1 = \ln 3$	$\ln e = 1$	$2\,\text{Cos}\,x+1 = \ln 3$
$2x = \ln 3 - 1$		$2\,\text{Cos}\,x = \ln 3 - 1$
$x = \frac{\ln 3 - 1}{2}$		$\text{Cos}\,x = \frac{\ln 3 - 1}{2}$
$x = .0493$	Input 3. Press ln key. Subtract 1. Divide by 2	$\text{Cos}\,x = .0493$
		$x = \text{Cos}^{-1}(.0493)$
		$x = 1.52$

LOGARITHMIC EQUATIONS.

Equations may contain expressions that are compositions of the logarithmic and trigonometric functions. Their solutions depend upon successive applications of the inverse functions.

EXAMPLE: Solve for x. Fill in missing steps.

Algebra	Step	Trigonometry
$\log_2 (x - 3) = 1$		$\log_2 (\text{Tan } x - 3) = 1$
$x - 3 = 2'$	If $\log_b m = n$, $m = b^n$	$\text{Tan } x - 3 = 2'$
$x = 5$		$\text{Tan } x = 5$
		$x = \text{Tan}^{-1} 5 = 1.37$

The laws of logarithms are frequently applied.

EXAMPLE: Solve for x. Fill in missing steps.

Algebra	Step	Trigonometry
$\log_3 x - \log_3 (x-4) = \frac{1}{2}$		$\log_3 \text{Sec } x - \log_3 (\text{Sec } x - 4) = \frac{1}{2}$
$\log_3 \frac{x}{x-4} = \frac{1}{2}$	$\log_b \frac{M}{N} = \log_b M - \log_b N$	$\log_3 \frac{\text{Sec } x}{\text{Sec } x-4} = \frac{1}{2}$
$\frac{x}{x-4} = 3^{1/2} = \sqrt{3}$		$\frac{\text{Sec } x}{\text{Sec } x-4} = 3^{1/2} = \sqrt{3}$
$x = \sqrt{3}\, x - 4\sqrt{3}$		$\text{Sec } x = \sqrt{3} \text{ Sec } x - 4\sqrt{3}$
$4\sqrt{3} = \sqrt{3}\, x - x = (\sqrt{3}-1)x$		$4\sqrt{3} = \sqrt{3} \text{ Sec } x - \text{Sec } x = (\sqrt{3} - 1) \text{ Sec } x$
$x = \dfrac{4\sqrt{3}}{\sqrt{3} - 1}$		$\text{Sec } x = \dfrac{4\sqrt{3}}{\sqrt{3} - 1} \doteq 9.464$
		$x = \text{Sec}^{-1} 9.464 = 1.46$

Since the domain of the logarithmic function is $\{x : x > 0\}$, absolute value expressions are frequently employed to assure solutions.

EXAMPLE: Solve for x.

Algebra	Step	Trigonometry
$\log |x| = -.3576$		$\log |\text{Sin } x| = -.3576$
$|x| = 10^{-.3576}$	$\log x = \log_{10} x$	$|\text{Sin } x| = 10^{-.3576}$
$|x| = .4389$	Calculator: Input $-.3576$. Press INV key. Press log key	$|\text{Sin } x| = .4389$
$x = \pm .4389$		$\text{Sin } x = \pm .4389$
		$x = \text{Sin}^{-1} .4389$ or $\text{Sin}^{-1} (-.4389)$
		$x = .45$ or $x = -.45$

Familiarity with natural logarithms is also essential in preparing for the Calculus.

EXAMPLE: Solve for x.

Algebra	Step	Trigonometry
$\ln x + \ln 2 = 1$		$\ln \text{Cot } x + \ln 2 = 1$
$\ln 2x = 1$	$\ln(MN) = \ln M + \ln N$	$\ln 2 \text{ Cot } x = 1$
$2x = e$		$2 \text{ Cot } x = e$
$x = \dfrac{e}{2}$		$\text{Cot } x = \dfrac{e}{2} \doteq 1.359$
		$x = \text{Cot}^{-1} 1.359 = .63$

Exponential equations involving different bases are solved by converting them to logarithmic equations.

EXAMPLE: Solve for x.

Algebra	Step	Trigonometry
$2^{x-3} = 3^x$		$2^{\text{Sec} x - 3} = 3^{\text{Sec } x}$
$(x-3)\log 2 = x \log 3$		$(\text{Sec} x - 3)\log 2 = (\text{Sec } x)(\log 3)$
$x \log 2 - 3 \log 2 = x \log 3$		$(\text{Sec } x)(\log 2) - 3 \log 2 = (\text{Sec } x)(\log 3)$
$x \log 2 - x \log 3 = 3 \log 2$		$(\text{Sec } x)(\log 2) - (\text{Sec } x)(\log 3) = 3 \log 2$
$x(\log 2 - \log 3) = 3 \log 2$		$\text{Sec } x(\log 2 - \log 3) = 3 \log 2$
$x = \dfrac{3 \log 2}{\log 2 - \log 3}$		$\text{Sec } x = \dfrac{3 \log 2}{\log 2 - \log 3}$
$x = \dfrac{\log 8}{\log \left(\frac{2}{3}\right)}$		$\text{Sec } x = \dfrac{\log 8}{\log \left(\frac{2}{3}\right)} \doteq -5.129$
		$x = \text{Sec}^{-1}(-5.129) = -1.77$

CONDITIONAL EQUATIONS INVOLVING IDENTITIES.

When more than one trigonometric function appears in an equation, the identities are used to transform the equation into an algebraic counterpart.

EXAMPLE: Solve for x. $\cot x = \sin x, \quad 0 < x < \dfrac{\pi}{2}$.

SOLUTION: $\cot x = \sin x$

$\dfrac{\cos x}{\sin x} = \sin x$ identity

$\cos x = \sin^2 x$

$\cos x = 1 - \cos^2 x$ identity

$\cos^2 x + \cos x - 1 = 0$ standard form

$\cos x = \dfrac{-1 \pm \sqrt{(1)^2 - 4(1)(-1)}}{2(1)}$ quadratic formula

$\cos x = \dfrac{-1 \pm \sqrt{5}}{2}$

$\cos x = \dfrac{-1+\sqrt{5}}{2}$	$\cos x = \dfrac{-1-\sqrt{5}}{2}$
$\cos x = .6180$	$\cos x = -1.618$
$x = \cos^{-1} .6180$	no solution
$x = .90$	

EXAMPLE: Solve for x. $0 < x < 2\pi$ and $4\tan^2 x - 3\sec^2 x = 0$

SOLUTION: $4\tan^2 x - 3\sec^2 x = 0$

$4\tan^2 x - 3(\tan^2 x + 1) = 0$ identity

$4\tan^2 x - 3\tan^2 x - 3 = 0$

$\tan^2 x - 3 = 0$

$\tan^2 x = 3$

$\tan x = \pm\sqrt{3}$

$\tan x = \sqrt{3}$	$\tan x = -\sqrt{3}$
$x = \tan^{-1}\sqrt{3}$	$x = \tan^{-1}(-\sqrt{3})$
$x = \dfrac{\pi}{3} , \dfrac{4\pi}{3}$	$x = \dfrac{2\pi}{3} , \dfrac{5\pi}{3}$

Factoring is frequently effective in these kinds of equations.

EXAMPLE: Solve for x. $0 \leq x \leq 2\pi$ and

$2\sin x \cos x + \cos x - 2\sin x - 1 = 0$

SOLUTION: $2 \sin x \cos x + \cos x - 2 \sin x - 1 = 0$

$\cos x (2 \sin x + 1) - 1(2 \sin x + 1) = 0$

$(2 \sin x + 1)(\cos x - 1) = 0$

$2 \sin x + 1 = 0$	$\cos x - 1 = 0$
$\sin x = -\dfrac{1}{2}$	$\cos x = 1$
$x = \dfrac{7\pi}{6}, \dfrac{11\pi}{6}$	$x = 0, 2\pi$

EQUATIONS WITH MULTIPLE ARGUMENTS.

If the coefficient of the argument differs from 1, the number of solutions to an equation will vary according to the given domain of the variable. A comparison of the solutions to $\tan x = 1$, $\tan 2x = 1$, and $\tan \frac{1}{2} x = 1$ will reveal these differences.

EXAMPLE: Solve for x. $0 < x < 2\pi$.

SOLUTION:

$\tan x = 1$	$\tan 2x = 1$	$\tan \frac{1}{2} x = 1$
$x = \tan^{-1} 1 = \dfrac{\pi}{4}, \dfrac{5\pi}{4}$	$2x = \tan^{-1} 1$	$\dfrac{1}{2} x = \tan^{-1} 1$
	$2x = \dfrac{\pi}{4}, \dfrac{5\pi}{4}, \dfrac{9\pi}{4}, \dfrac{13\pi}{4}$	$\dfrac{1}{2} x = \dfrac{\pi}{4}$
	$x = \dfrac{\pi}{8}, \dfrac{5\pi}{8}, \dfrac{9\pi}{8}, \dfrac{13\pi}{8}$	$x = \dfrac{\pi}{2}$

As a general rule, with four quadrants available for the solution set $(0 < x < 2\pi)$, there will be 2k solutions for equations of the form $f(kx) = c$ where f is a trigonometric function.

EXAMPLE: Solve for x, $0 < x < 2\pi$. $2 \sin 3x = -1$

SOLUTION: $2 \sin 3x = -1$

$\sin 3x = -\dfrac{1}{2}$

$3x = \sin^{-1} \left(-\dfrac{1}{2}\right)$

$3x = \dfrac{7\pi}{6}, \dfrac{11\pi}{6}, \dfrac{19\pi}{6}, \dfrac{23\pi}{6}, \dfrac{31\pi}{6}, \dfrac{35\pi}{6}$

$x = \dfrac{7\pi}{18}, \dfrac{11\pi}{18}, \dfrac{19\pi}{18}, \dfrac{23\pi}{18}, \dfrac{31\pi}{18}, \dfrac{35\pi}{18}$

If the principal value function is given, the number of solutions will be reduced to a single value.

EXAMPLE: Solve for x. 2 Sin 3x = -1

SOLUTION: 2 Sin 3x = -1

$$\text{Sin } 3x = -\frac{1}{2}$$

$$3x = \text{Sin}^{-1}\left(-\frac{1}{2}\right)$$

$$3x = -\frac{\pi}{6}$$

$$x = \frac{-\pi}{18}$$

It is also possible that no solution will be available in the given domain of the solution set.

EXAMPLE: Solve for x. $0 < x < 2\pi$. $\sec \frac{1}{3} x = -\sqrt{2}$

SOLUTION: $\sec \frac{1}{3} x = -\sqrt{2}$

$$\frac{1}{3} x = \sec^{-1}(-\sqrt{2})$$

$$\frac{1}{3} x = \frac{3\pi}{4}, \frac{5\pi}{4}$$

$$x = \frac{9\pi}{4}, \frac{15\pi}{4}.$$

Since both solutions are greater than 2π there is no solution in the given domain.

EXERCISE SET 5-6.

Solve the following equations for all values of x in the indicated domain. Leave answers in multiples of π whenever possible. Otherwise round off answers to hundredths.

1) $2 \sin x = \sqrt{3}$

2) $2 \csc x = \frac{-1}{\sqrt{3}}$

3) $\frac{1}{3} \tan x = \frac{-1}{\sqrt{3}}$

4) $\cos x = -.4896, \frac{\pi}{2} < x < \frac{3\pi}{2}$

5) $\cot x = 1.843, -\frac{\pi}{2} < x < \frac{3\pi}{2}$

6) $\sec x = -1.427$

7) $\tan^2 x - 1 = 3$

8) $\sin^2 x = \frac{1}{4}$

9) $\sqrt{3}\tan x = -1, 0 < x < 2\pi$

10) $2 \cos x = -\sqrt{3}, -2\pi < x < 0$

11) $\sqrt{3} \csc x = -2$

12) $\cot x = -1$

13) $\cot x = \sqrt{3}, 0 < x < 2\pi$

14) $3 \tan^2 x - 1 = 0$

15) $4 \cos^2 x - 3 = 9$

16) $3 \csc^2 x - 4 = 0$

17) $\tan^3 x - 3 \tan x = 0$

18) $\cot^2 x - 1 = 2$

19) $2 - \sin^2 x = \frac{7}{4}$

20) $\csc^2 x - 1 = \tan \frac{\pi}{4}, 0 < x < 2\pi$

21) $\tan^2 x - 1 = \sec \frac{\pi}{3}, 0 < x < 2\pi$

22) $\frac{2}{\tan x + 1} = \tan x - 1$

23) $1 - \cos x = \frac{1}{4 + 4 \cos x}$

24) $2 \cos^2 x + \cos x - 1 = 0$

25) $2 \csc^2 x = 3 - \cot^2 x$

26) $2 \tan^2 x = 2 - \sec^2 x$

27) $3 \sin^2 x - \cos^2 x = 0, 0 \le x \le 2\pi$

28) $\frac{\sqrt{3} \tan x}{2 \cos x} = \frac{-1}{\sqrt{3}}, 0 < x < 2\pi$

29) $3 \sin x + 1 = 0$

30) $\frac{\cot x + 2}{4} = \frac{1}{2 \cot x - 1}$

31) $3 \tan^2 x - 4 \tan x - 2 = 0$

32) $|\tan x - 3| = 7$

33) $7 \sin^2 x + 2 \sin x - 1 = 0$

34) $8 \sec^4 x - 22 \sec^2 x + 15 = 0$

35) $2 \cos^3 x - 5 \cos^2 x + \cos x = 0$

36) $\sqrt{2} \sin x = 1, -\pi < x < \pi$

37) $\sqrt{3} \tan x = 2$

38) $3\sqrt{3} \sec x + 1 = 5\sqrt{\sec x} + 1$

39) $\sqrt{\cot x + 3} = 1 - 2 \cot x$

40) $3^{\sin x} = 1$

41) $10^{\cos x+3} = 150$

42) $3^{\cot x-1} = 27$

43) $e^{5 \cos x+1} = 7$

44) $e^{2 \cos x-3} = 3^{\cos x+1}, 0 < x < 2\pi$

45) $\log_2 (\text{Csc } x + 1) = 4$

46) $\log_4 \text{Sin } x - \log_4 (4 \text{ Sin } x + 2) = \frac{3}{2}$

47) $\log |\text{Tan } x + 1| = 2 \log |\text{Tan } x|$

48) $\ln |\text{Sec } x + 2| = 3$

49) $\sin x = \sqrt{3} \cos x$, $0 < x < 2\pi$

50) $1 - 3 \cos x = 2 \sin x$, $\frac{\pi}{2} < x < 2\pi$

51) $2 \text{ Tan } x + 3 \text{ Sec } x = 4$

52) $\sin^2 2x = \frac{1}{2}$, $0 \leq x \leq 2\pi$

53) $\cos^2 3x = \frac{1}{4}$, $0 \leq x \leq 2\pi$

54) $\cos^2 x - 3 - 3 \sin x = 0$, $0 \leq x \leq 2\pi$

55) $\text{Cos } \frac{1}{2} x = \frac{2}{5}$

56) $\cot \frac{1}{3} x = .5432$, $-\pi < x < \pi$

57) $4 \sin x = \tan x$, $0 < x < 2\pi$

58) $5 \cos x = 2 \csc x$, $-\pi < x < \pi$

59) $\frac{\text{Csc } x}{\text{Cot } x} = \sqrt{2}$

60) $\text{Tan } (\ln x + 3) = 4$

5-7. CHAPTER SUMMARY

1. The standard trigonometric functions cos x, sin x, tan x, cot x, sec x, csc x are many-to-one functions. Consequently their inverses are not functions. The expression $\sin^{-1} \frac{1}{\sqrt{2}}$ or arcsin $\frac{1}{\sqrt{2}}$ is many-valued; i.e., arcsin $\frac{1}{\sqrt{2}} = \frac{\pi}{4} + 2k\pi$, or $\frac{3\pi}{4} + 2k\pi$, k = 0, ±1, ±2,... .

2. If the domains of the trigonometric functions are restricted to their principal values then the functions become one-to-one and their inverses are one-to-one functions also. The following table summarizes their properties:

Function	Cos^{-1}	Sin^{-1}	Tan^{-1}	Cot^{-1}	Sec^{-1}	Csc^{-1}
Domain	$-1 \leq x \leq 1$	$-1 \leq x \leq 1$	Reals	Reals	$\|x\| \geq 1$	$\|x\| \geq 1$
Range	$0 \leq y \leq \pi$	$\frac{-\pi}{2} \leq y \leq \frac{\pi}{2}$	$\frac{-\pi}{2} < y < \frac{\pi}{2}$	$0 < y < \pi$	$0 \leq y < \frac{\pi}{2}$, $-\pi \leq y < \frac{-\pi}{2}$	$0 < y \leq \frac{\pi}{2}$, $-\pi < y \leq \frac{-\pi}{2}$
Odd or Even		Odd	Odd			
Continuous	Yes	Yes	Yes	Yes	No	No
Increasing or Decreasing	Decr.	Incr.	Incr.	Decr.	Incr. on $x \geq 1$ Decr. on $x \leq -1$	Incr. on $x \leq -1$ Decr. on $x \geq 1$

3. The scientific electronic calculator is programmed to evaluate $\text{Cos}^{-1}x$, $\text{Sin}^{-1}x$, and $\text{Tan}^{-1}x$ for all values of x in the domains given above. The remaining functions require the reciprocal identities and the addition or subtraction of π for negative values of x. The procedures are summarized as follows:

$\text{Cos}^{-1}x$: Select Degree or Radian mode. Input x. Press INV or ARC key. Press cos key. Read display.

$\text{Sin}^{-1}x$: Same as above except press sin key after INV.

$\text{Tan}^{-1}x$: Same as above except press tan key after INV.

$\text{Cot}^{-1}x$: Select Degree or Radian mode. Input x. Press $\frac{1}{x}$ key. Press INV or ARC key. Press tan key. If $x > 0$ read answer in display. If $x < 0$ add π for answer. $\text{Cot}^{-1}x = \text{Tan}^{-1}\frac{1}{x}$ if $x > 0$ and $\text{Cot}^{-1}x = \text{Tan}^{-1}\frac{1}{x} + \pi$ if $x < 0$.

$\text{Sec}^{-1}x$: Select Degree or Radian mode. Input $|x|$. Press $\frac{1}{x}$ key. Press INV or ARC key. Press cos key. If $x > 1$ read answer in display. If $x < -1$ subtract π for answer. $\text{Sec}^{-1}x = \text{Cos}^{-1}|\frac{1}{x}|$ if $x \geq 1$ and $\text{Sec}^{-1}x = \text{Cos}^{-1}|\frac{1}{x}| - \pi$ if $x \leq -1$.

$\text{Csc}^{-1}x$: Select Degree or Radian mode. Input $|x|$. Press $\frac{1}{x}$ key. Press INV or ARC key. Press sin key. If $x > 1$ read answer in display. If $x < -1$ subtract π for answer. $\text{Csc}^{-1}x = \text{Sin}^{-1}|\frac{1}{x}|$ if $x \geq 1$ and $\text{Csc}^{-1}x = \text{Sin}^{-1}|\frac{1}{x}| - \pi$ if $x \leq -1$.

4. The circle model below identifies the quadrants for the inverse functional values:

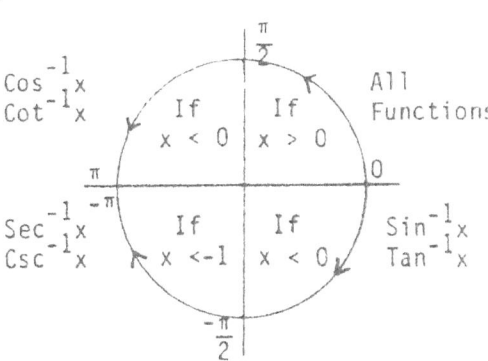

5. The graphs of the inverse trigonometric functions are sketched below:

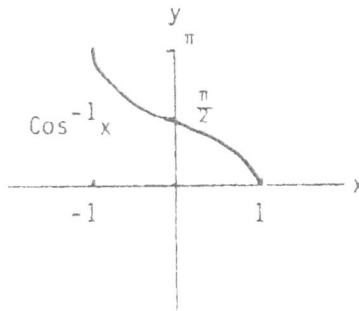

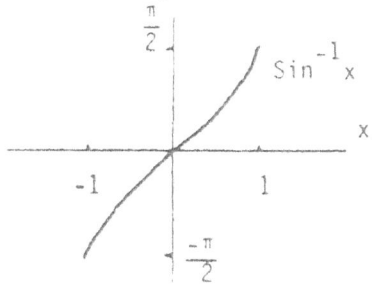

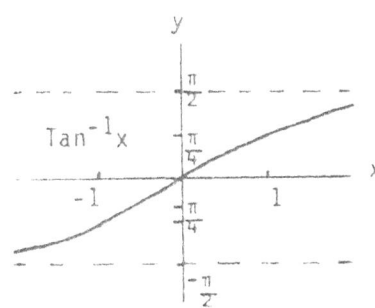

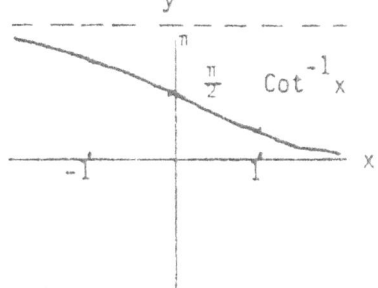

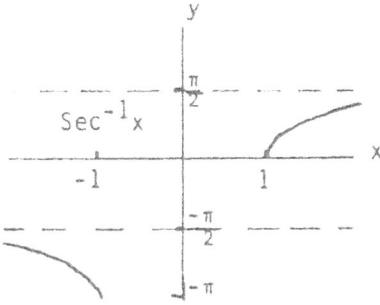

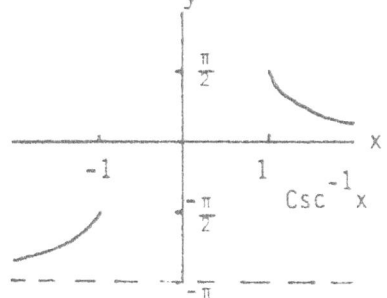

6. The solutions of conditional trigonometric equations follows algebraic rules with the additional step of evaluating inverse trigonometric functions.

<u>EXERCISE SET 5-7.</u>

Find the inverses of the following functions by specifying their defining equations. Determine whether or not the inverses are functions.

1) $f(x) = 3x + 7$

2) $g(x) = 2x^2 + 5x - 4$

3) $h(x) = \sqrt{x}$

4) $k(x) = \log_3 (x + 2)$

5) $m(x) = 2 \cos 4x$

6) $t(x) = -3 \text{ Tan}^{-1} x$

7) $r(x) = \frac{1}{2} \text{ Sec } (x + \frac{\pi}{3})$

8) $s(x) = \pi \text{ Sin } (3x + \pi)$

9) $k(x) = \frac{1}{2} \text{ Arcsin } \frac{x}{3} - \frac{\pi}{4}$

10) $p(x) = 2 \text{ Cot } (\frac{1}{2}x - \frac{\pi}{12})$

Graph the following:

11) $f(x) = 2 \text{ Arcsin } \frac{x}{3}$

12) $g(x) = -2 \text{ Cos}^{-1} x + \pi$

13) $t(x) = \frac{1}{3} \text{ Cot}^{-1} x - \frac{\pi}{4}$

14) $q(x) = \frac{1}{\pi} \text{ Tan}^{-1} x$

15) $p(x) = 2 \text{ Sin}^{-1} x + \text{ Cos}^{-1} x$

Solve the following for x:

16) $\text{Cos}^{-1} x = -.5462$

17) $\text{Tan}^{-1} x = \frac{-\pi}{6}$

18) $x = \text{Sec}^{-1} 5$

19) $x = \text{Tan}^{-1} 3.7 + \text{Cot}^{-1} (-1.89)$

20) $\sin (\text{Cot}^{-1} x) = -.7$

21) $x = \cos (\text{Tan}^{-1} \frac{4}{3})$

22) $\log \text{ Sin } x = -.254$

23) $\text{Tan } (\log x) = .4763$

24) $6 \text{ Sin } x + 9 \text{ Cos}^2 x = 10$

25) $4 \text{ Tan } x = \text{Sec}^2 x$

Find $f(g(x))$ for the following:

26) $f(x) = x^2 - 2x$, $g(x) = 3x - 5$

27) $f(x) = \log x^2$, $g(x) = x - 5$

28) $f(x) = \cos x$, $g(x) = \tan^{-1} x$

29) $f(x) = e^{\tan x}$, $g(x) = \tan^{-1} x$

30) $f(x) = 3$, $g(x) = \sec e^x$

Solve for x; $0 \leq x < 2\pi$

31) $\cos 3x = \dfrac{-1}{2}$

32) $\tan \dfrac{1}{2} x = \sqrt{3}$

CHAPTER 6

ADVANCED FORMULAS

IDENTITIES AND CONDITIONAL EQUATIONS

6-1. <u>THE DISTANCE FORMULA</u>. <u>DERIVATIONS OF THE SUM AND DIFFERENCE FORMULAS</u>.

Let $P_1(x_1,y_1)$ and $P_2(x_2,y_2)$ be two points in the plane whose coordinates are known. The distance between P_1 and P_2 can be determined by application of the Pythagorean Theorem. Figure 6-1A illustrates the situation:

FIGURE 6-1A

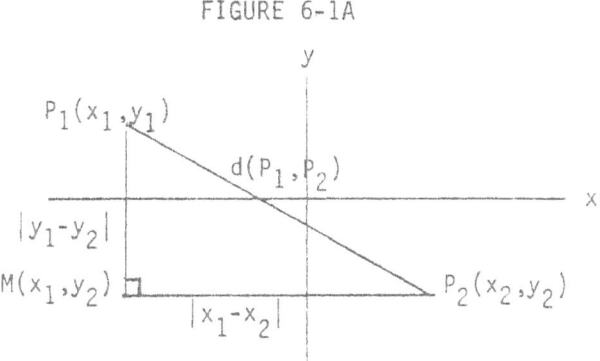

When line segments from P_1 and P_2 are drawn perpendicular to the x and y axes respectively and extended to meet at M, a right triangle is formed with line segment P_1P_2 as the hypotenuse. The ordered pair (x_1,y_2) are the coordinates of M. Side P_1M has length $|y_1-y_2|$ and side MP_2 has length $|x_1-x_2|$. By the Pythagorean Theorem it follows that

$$d(P_1P_2)^2 = |x_1-x_2|^2 + |y_1-y_2|^2 = (x_1-x_2)^2 + (y_1-y_2)^2.$$

Formula 6-1A: $d(P_1P_2) = \sqrt{(x_1-x_2)^2 + (y_1-y_2)^2}$.

Since $(x_1-x_2)^2 = (x_2-x_1)^2$ and $(y_1-y_2)^2 = (y_2-y_1)^2$, the order of the subtractions will not affect the results.

EXAMPLE: Find the distance between P(-3,5) and Q(7,-5).

SOLUTION: $d(PQ) = \sqrt{(-3-7)^2 + (5-(-5))^2} = \sqrt{(-10)^2 + 10^2} = \sqrt{100 + 100}$

$= \sqrt{200} = 10\sqrt{2}$.

Consider now two points located on a circle with radius r and center at the origin.

FIGURE 6-1B

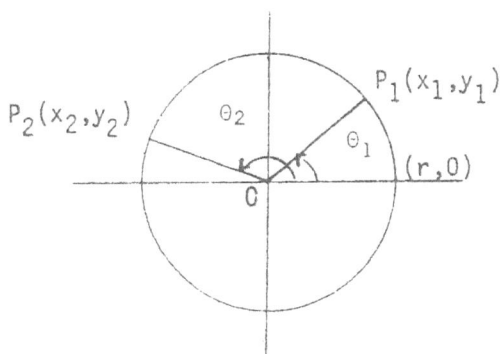

Let θ_1 and θ_2 be the central angles in standard position terminating at P_1 and P_2 respectively. From the ratio definitions we have,

$\cos\theta_1 = \dfrac{x_1}{r}$, $\sin\theta_1 = \dfrac{y_1}{r}$, $\cos\theta_2 = \dfrac{x_2}{r}$, $\sin\theta_2 = \dfrac{y_2}{r}$.

From this it follows that $x_1 = r\cos\theta_1$, $y_1 = r\sin\theta_1$, $x_2 = r\cos\theta_2$, and $y_2 = r\sin\theta_2$. Hence, the ordered pairs $(x_1,y_1) = (r\cos\theta_1 , r\sin\theta_1)$ and $(x_2,y_2) = (r\cos\theta_2 , r\sin\theta_2)$.

To simplify matters let P_1 and P_2 be points on the unit circle and focus your attention on the central angle $\theta_2-\theta_1$ generated in a counterclockwise direction from OP_1 to OP_2.

FIGURE 6-1C

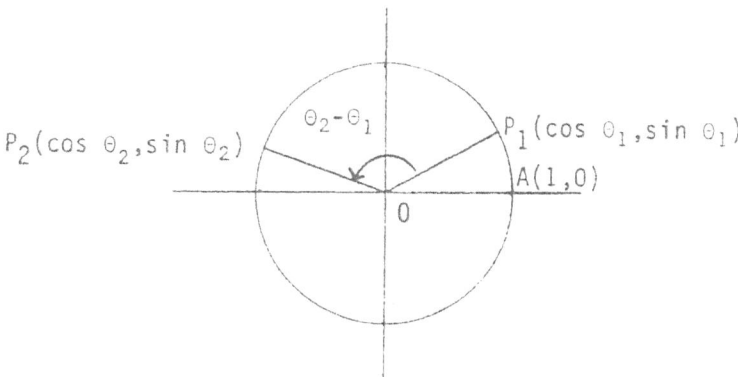

If $\theta_2-\theta_1$ is rotated clockwise into standard position, its terminal side coincides with OQ as shown in 6-1D.

FIGURE 6-1D

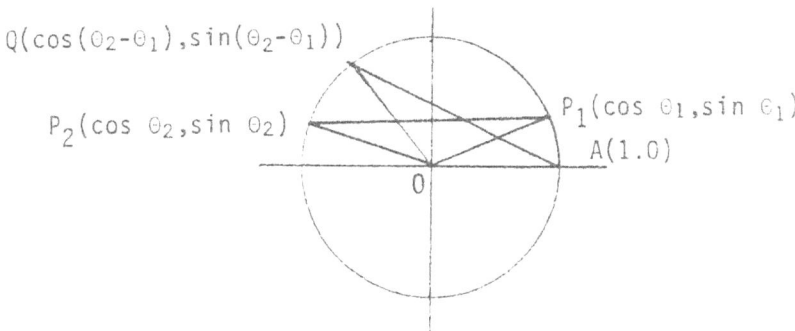

Since the arcs AQ and P_1P_2 are equal, the chords connecting these points are equal as well. Thus $d(AQ) = d(P_1P_2)$. Applying the distance formula:

$$\sqrt{(\cos(\theta_2-\theta_1)-1)^2 + (\sin(\theta_2-\theta_1)-0)^2} = \sqrt{(\cos\theta_2-\cos\theta_1)^2 + (\sin\theta_2-\sin\theta_1)^2}$$

Squaring both sides and expanding the binomials:

$$[\cos^2(\theta_2-\theta_1) - 2\cos(\theta_2-\theta_1) + 1] + \sin^2(\theta_2-\theta_1) =$$

$$[\cos^2\theta_2 - 2\cos\theta_2\cos\theta_1 + \cos^2\theta_1] + [\sin^2\theta_2 - 2\sin\theta_2\sin\theta_1 + \sin^2\theta_1]$$

Since $\cos^2 (\theta_2-\theta_1) + \sin^2 (\theta_2-\theta_1) = 1$, $\cos^2 \theta_2 + \sin^2 \theta_2 = 1$, and $\cos^2 \theta_1 + \sin^2 \theta_1 = 1$, substituting yields

$$2 - 2 \cos (\theta_2-\theta_1) = 2 - 2 \cos \theta_2 \cos \theta_1 - 2 \sin \theta_2 \sin \theta_1$$

Subtracting 2 from each side:

$$- 2 \cos (\theta_2-\theta_1) = - 2 \cos \theta_2 \cos \theta_1 - 2 \sin \theta_2 \sin \theta_1$$

Finally, dividing both sides by -2 yields the formula:

Formula 6-1B
$$\boxed{\cos (\theta_2-\theta_1) = \cos \theta_2 \cos \theta_1 + \sin \theta_2 \sin \theta_1}$$

EXAMPLE: Derive the $\cos 15^\circ$ from the sines and cosines of 60° and 45°.

SOLUTION: $\cos 15^\circ = \cos (60^\circ - 45^\circ) = \cos 60^\circ \cos 45^\circ + \sin 60^\circ \sin 45^\circ$

$$= \frac{1}{2} \cdot \frac{1}{\sqrt{2}} + \frac{\sqrt{3}}{2} \cdot \frac{1}{\sqrt{2}}$$

$$= \frac{1}{2\sqrt{2}} + \frac{\sqrt{3}}{2\sqrt{2}} = \frac{1+\sqrt{3}}{2\sqrt{2}}$$

$$= \frac{\sqrt{2} + \sqrt{6}}{4} \quad \text{rationalizing the denominator.}$$

Since $\cos (-\theta) = \cos \theta$, it follows that $\cos (\theta_2-\theta_1) = \cos (\theta_1-\theta_2)$. The formula for the cosine of the difference of two angles may be used to derive other important formulas.

Let x and y be placeholders for real numbers, then Formula 6-1B becomes $\cos (x-y) = \cos x \cos y + \sin x \sin y$. The following will now be derived:

Formula 6-1C: $\cos (x+y) = \cos x \cos y - \sin x \sin y$

Formula 6-1D: $\sin (x+y) = \sin x \cos y + \cos x \sin y$

Formula 6-1E: $\sin (x-y) = \sin x \cos y - \cos x \sin y$

Derivations	Steps	Reasons

6-1C: $\cos(x+y) = \cos(x-(-y))$ $\qquad$ $a + b = a - (-b)$

$\qquad = \cos x \cos(-y) + \sin x \sin(-y)$ $\qquad$ 6-1B

$\qquad = \cos x \cos y + (\sin x)(-\sin y)$ $\qquad$ $\cos(-y) = \cos y$ and $\sin(-y) = -\sin y$

$\qquad = \cos x \cos y - \sin x \sin y$ $\qquad$ $a(-b) = -ab$

6-1D: $\sin(x+y) = \cos\left[\frac{\pi}{2} - (x+y)\right]$ $\qquad$ $\cos\left(\frac{\pi}{2} - \theta\right) = \sin \theta$

$\qquad = \cos\left[\left(\frac{\pi}{2} - x\right) - y\right]$ $\qquad$ $a-(b+c) = a-b-c$

$\qquad = \cos\left(\frac{\pi}{2}-x\right)\cos y + \sin\left(\frac{\pi}{2}-x\right)\sin y$ $\qquad$ 6-1B

$\qquad = \sin x \cos y + \cos x \sin y$ $\qquad$ $\cos\left(\frac{\pi}{2}-\theta\right) = \sin \theta$ and

$\qquad\qquad\qquad\qquad\qquad\qquad\qquad\qquad$ $\sin\left(\frac{\pi}{2}-\theta\right) = \cos \theta$

6-1E: $\sin(x-y) = \sin(x+(-y))$ $\qquad$ $a-b = a+(-b)$

$\qquad = \sin x \cos(-y) + \cos x \sin(-y)$ $\qquad$ 6-1D

$\qquad = \sin x \cos y + (\cos x)(-\sin y)$ $\qquad$ $\cos(-\theta) = \cos \theta,\ \sin(-\theta) = -\sin \theta$

$\qquad = \sin x \cos y - \cos x \sin y$ $\qquad$ $a(-b) = -ab$

If x and y are replaced by degree measures the formulas above still hold.

EXAMPLE: Using special values and formulas 6-1B-E, find

$\qquad$ a) $\sin 105^{\circ}$ $\qquad$ b) $\cos 75^{\circ}$ $\qquad$ c) $\sin 195^{\circ}$

SOLUTIONS: a) $\sin 105^{\circ} = \sin(60^{\circ} + 45^{\circ})$

$\qquad = \sin 60^{\circ} \cos 45^{\circ} + \cos 60^{\circ} \sin 45^{\circ}$

$\qquad = \frac{\sqrt{3}}{2} \cdot \frac{1}{\sqrt{2}} + \frac{1}{2} \cdot \frac{1}{\sqrt{2}} = \frac{\sqrt{3}+1}{2\sqrt{2}} = \frac{\sqrt{6}+\sqrt{2}}{4}$

b) $\cos 75^0 = \cos(45^0 + 30^0) = \cos 45^0 \cos 30^0 - \sin 45^0 \sin 30^0$

$= \dfrac{1}{\sqrt{2}} \cdot \dfrac{\sqrt{3}}{2} - \dfrac{1}{\sqrt{2}} \cdot \dfrac{1}{2} = \dfrac{\sqrt{3}-1}{2\sqrt{2}} = \dfrac{\sqrt{6}-\sqrt{2}}{4}$

c) $\sin 195^0 = \sin(180^0 + 15^0) = \sin 180^0 \cos 15^0 + \cos 180^0 \sin 15^0$

$= 0 + (-1)\sin 15^0 = -\sin 15^0 = -\cos 75^0$ by the complementary rule.

From an earlier example (b)above), $\cos 75^0 = \dfrac{\sqrt{6}-\sqrt{2}}{4}$.

Therefore, $\sin 195^0 = \dfrac{\sqrt{2} - \sqrt{6}}{4}$.

EXAMPLE: If $\sin x = \dfrac{3}{5}$, $0 < x < \dfrac{\pi}{2}$ and $\cos y = \dfrac{-5}{13}$,

$\dfrac{\pi}{2} < y < \pi$, find $\sin(x-y)$.

SOLUTION: By 6-1E, $\sin(x-y) = \sin x \cos y - \cos x \sin y$.

Substituting the given values,

$\sin(x-y) = (\dfrac{3}{5})(\dfrac{-5}{13}) - \cos x \sin y = \dfrac{-15}{65} - \cos x \sin y$.

The Pythagorean identity $\sin^2 x + \cos^2 x = 1$ and $\sin^2 y + \cos^2 y = 1$

yields $\cos x = \dfrac{4}{5}$ and $\sin y = \dfrac{12}{13}$.

Hence, $\sin(x-y) = \dfrac{-15}{65} - (\dfrac{4}{5})(\dfrac{12}{13}) = \dfrac{-15}{65} - \dfrac{48}{65} = \dfrac{-63}{65}$

Tan (x+y) and Tan (x-y)

The sum and difference formulas for the tangent function may be derived from the quotient identity:

$\tan(x+y) = \dfrac{\sin(x+y)}{\cos(x+y)} = \dfrac{\sin x \cos y + \cos x \sin y}{\cos x \cos y - \sin x \sin y}$

Dividing numerator and denominator by $\cos x \cos y$:

$\tan(x+y) = \dfrac{\dfrac{\sin x \cos y}{\cos x \cos y} + \dfrac{\cos x \sin y}{\cos x \cos y}}{\dfrac{\cos x \cos y}{\cos x \cos y} - \dfrac{\sin x \sin y}{\cos x \cos y}} = \dfrac{\dfrac{\sin x}{\cos x} + \dfrac{\sin y}{\cos y}}{1 - \dfrac{\sin x}{\cos x} \cdot \dfrac{\sin y}{\cos y}} = \dfrac{\tan x + \tan y}{1 - \tan x \tan y}$

$$\tan (x-y) = \tan (x+(-y)) = \frac{\tan x + \tan (-y)}{1 - \tan x \tan (-y)}$$

$$= \frac{\tan x - \tan y}{1 + \tan x \tan y} \quad , \quad x,y \neq \frac{k\pi}{2} \; , \; k = \pm 1, \pm 3, \ldots$$

Summarizing,

Formula 6-1F: $\tan (x+y) = \dfrac{\tan x + \tan y}{1 - \tan x \tan y}$

Formula 6-1G: $\tan (x-y) = \dfrac{\tan x - \tan y}{1 + \tan x \tan y}$

EXAMPLE: Find $\tan 75^{\circ}$ using special values.

SOLUTION: $\tan 75^{\circ} = \tan (45^{\circ} + 30^{\circ}) = \dfrac{\tan 45^{\circ} + \tan 30^{\circ}}{1 - \tan 45^{\circ} \tan 30^{\circ}}$

$$= \frac{1 + \frac{1}{\sqrt{3}}}{1 - 1 \cdot \frac{1}{\sqrt{3}}} = \frac{\frac{\sqrt{3}+1}{\sqrt{3}}}{\frac{\sqrt{3}-1}{\sqrt{3}}} = \frac{\sqrt{3}+1}{\sqrt{3}-1}$$

Rationalizing the denominator,

$$\tan 75^{\circ} = \frac{(\sqrt{3}+1)}{(\sqrt{3}-1)} \cdot \frac{(\sqrt{3}+1)}{(\sqrt{3}+1)} = \frac{4+2\sqrt{3}}{3-1} = \frac{4+2\sqrt{3}}{2} = 2+\sqrt{3}$$

EXAMPLE: If $\sin x = \dfrac{-7}{25}$, $\pi < x < \dfrac{3\pi}{2}$ and

$\cos y = \dfrac{12}{13}$, $\dfrac{3\pi}{2} < y < 2\pi$, find $\tan (y-x)$

SOLUTION: $\tan (y-x) = \dfrac{\tan y - \tan x}{1 + \tan y \tan x}$

By the ratio definitions, $\tan x = \dfrac{7}{24}$ and $\tan y = \dfrac{-5}{12}$

Hence, $\tan (y-x) = \dfrac{\frac{-5}{12} - \frac{7}{24}}{1 + (\frac{-5}{12})(\frac{7}{24})} = \dfrac{\frac{-17}{24}}{\frac{253}{288}} = \dfrac{-17}{24} \cdot \dfrac{288}{253} = \dfrac{-17}{1} \cdot \dfrac{12}{253} = \dfrac{-204}{253}$

EXAMPLE: Prove the identity $\dfrac{\sin(x+y) + \sin(x-y)}{\cos y} = 2 \sin x$

SOLUTION: $\dfrac{\sin(x+y) + \sin(x-y)}{\cos y}$

$$= \frac{(\sin x \cos y + \cos x \sin y) + (\sin x \cos y - \cos x \sin y)}{\cos y}$$

$$= \frac{2 \sin x \cos y}{\cos y}$$

$$= 2 \sin x$$

Sum and Difference formulas may be used to derive basic reduction formulas:

EXAMPLE: Show that $\sin (180^{\circ} + \theta) = -\sin \theta$.

SOLUTION: $\sin (180^{\circ} + \theta) = \sin 180^{\circ} \cos \theta + \cos 180^{\circ} \sin \theta$

$$= 0 \cdot \cos \theta + (-1) \sin \theta = -\sin \theta.$$

EXAMPLE: Show that $\cos (\frac{3\pi}{2} + x) = \sin x$.

SOLUTION: $\cos (\frac{3\pi}{2} + x) = \cos \frac{3\pi}{2} \cos x - \sin \frac{3\pi}{2} \sin x$

$$= 0 \cdot \cos x - (-1) \sin x = \sin x.$$

EXAMPLE: Express $\sin (\frac{\pi}{3} + x)$ in terms of functions of x.

SOLUTION: $\sin (\frac{\pi}{3} + x) = \sin \frac{\pi}{3} \cos x + \cos \frac{\pi}{3} \sin x$

$$= \frac{\sqrt{3}}{2} \cos x + \frac{1}{2} \sin x$$

$$= \frac{\sqrt{3} \cos x + \sin x}{2}$$

EXERCISE SET 6-1.

Find the distance between the given points in the plane:

1) $(-7,2),(10,4)$

2) $(5,6),(-1,-2)$

3) $(11.27,3.8),(-7,6.2)$

4) $(-1.32,4.65),(5.82,-6.34)$

5) $(2\pi,3),(-\sqrt{2},4)$

6) $(\cos \frac{\pi}{3},\sin \frac{\pi}{3}),(\cos \frac{4\pi}{3},\sin \frac{4\pi}{3})$

7) $(3 \cos 7, 3 \sin 7),(4 \cos 1, 4 \sin 1)$

8) Find the midpoint of the line segment joining $(-3,5)$ and $(7,2)$.

9) Derive Formula 6-1C in the same manner as 6-1B was derived making use of the distance formula.

Given the sines and cosines of 30^0, 45^0, and 60^0, find the values of the following:

10) $\sin 15^0$ 11) $\sin 75^0$ 12) $\tan 105^0$ 13) $\cos 105^0$

Prove the following identities:

14) $\tan (x + \frac{\pi}{4}) = \dfrac{\cos x + \sin x}{\cos x - \sin x}$

15) $\dfrac{\sin (x+y)}{\sin (x-y)} = \dfrac{\tan x + \tan y}{\tan x - \tan y}$

16) $2 \sin (45^0 + x) \sin (45^0 - x) = \cos^2 x - \sin^2 x$

17) $\tan x - \tan y = \dfrac{\sin (x-y)}{\cos x \cos y}$

18) $\sin (\frac{\pi}{2} + x) - \cos (\frac{\pi}{2} + x) = \sin x + \cos x$

19) $\sin (x+y) + \sin (x-y) = 2 \sin x \cos y$

20) $\sec (x+y) = \dfrac{\sec x \sec y}{1-\tan x \tan y}$

Express the following as a function of one angle:

Example: $\sin 40^0 \cos 30^0 + \cos 40^0 \sin 30^0 = \sin (40^0 + 30^0) = \sin 70^0$

21) $\cos 85^0 \cos 33^0 - \sin 85^0 \sin 33^0$

22) $\dfrac{\tan 25^0 + \tan 38^0}{1-\tan 25^0 \tan 38^0}$

23) $\sin 95^0 \cos 24^0 - \cos 95^0 \sin 24^0$

24) $\sin \frac{\pi}{3} \cos \frac{5\pi}{4} + \sin \frac{5\pi}{4} \cos \frac{\pi}{3}$

Express the following in terms of functions of x:

25) $\cos (x - \frac{\pi}{6})$ 26) $\tan (\frac{5\pi}{4} + x)$ 27) $\sin (x + \frac{\pi}{4})$

Use the sum and difference formulas to derive the following:

28) $\cos (\frac{3\pi}{2} - x) = -\sin x$

29) $\sin (360^{\circ} - x) = -\sin x$

30) $\tan (\pi + x) = \tan x$

Find the value of the following:

31) A and B are quadrant I angles. $\cos A = \frac{5}{13}$, $\sin B = \frac{3}{5}$.

Find $\sin (A+B)$.

32) $\cos A = \frac{-7}{25}$, A is in Quadrant II, $\cos B = \frac{3}{5}$, B is in Quadrant IV.

Find $\cos (B-A)$. Note: $\frac{\pi}{2} < A < \pi$ and $\frac{3\pi}{2} < B < 2\pi$.

33) $\tan x = \frac{4}{3}$, $0 < x < \frac{\pi}{2}$ and $\tan y = \frac{-24}{7}$, $\frac{\pi}{2} < y < \pi$.

Find $\tan (x-y)$.

34) $\sin x = \frac{1}{\sqrt{5}}$, $\cos y = \frac{1}{2}$ and both x and y are in Quadrant I.

Find $\cot (x+y)$.

6-2. DOUBLE ANGLE AND HALF ANGLE FORMULAS.

Formulas for functions of arguments that are twice as large ("double angle") or half as large ("half angle") as a given argument are derived directly from the formulas of section 6-1. Since they have such widespread applications in advanced mathematics, the student is well-advised to memorize them.

DOUBLE ANGLE FORMULAS

FORMULA 6-2A   DERIVATION

$\sin 2x = 2 \sin x \cos x$

$$\sin 2x = \sin (x + x)$$
$$= \sin x \cos x + \cos x \sin x \quad (6-1D)$$
$$= 2 \sin x \cos x$$

EXAMPLE: Given $\sin 60^{0} = \dfrac{\sqrt{3}}{2}$ and $\cos 60^{0} = 1/2$, find $\sin 120^{0}$.

SOLUTION: $\sin 120^{0} = 2 \sin 60^{0} \cos 60^{0} = 2 \left(\dfrac{\sqrt{3}}{2} \right) \left(\dfrac{1}{2} \right) = \dfrac{\sqrt{3}}{2}$

There are three alternate forms for $\cos 2x$:

FORMULA 6-2B   DERIVATION

$\cos 2x = \cos^{2} x - \sin^{2} x$

$$\cos 2x = \cos (x + x)$$
$$= \cos x \cos x - \sin x \sin x \quad (6-1C)$$
$$= \cos^{2} x - \sin^{2} x$$

FORMULA 6-2B'   DERIVATION

$\cos 2x = 2 \cos^{2} x - 1$

$$\cos 2x = \cos^{2} x - \sin^{2} x \quad (6-2B)$$
$$= \cos^{2} x - (1 - \cos^{2} x)$$

by the Pythagorean Identity
$$= \cos^{2} x - 1 + \cos^{2} x$$
$$= 2 \cos^{2} x - 1$$

FORMULA 6-2B"   DERIVATION

$\cos 2x = 1 - 2 \sin^{2} x$

Left to the student (See Exercise 1).

EXAMPLE: Show that $\cos 90^0 = 0$ given that $\cos 45^0 = \frac{1}{\sqrt{2}} = \sin 45^0$.

SOLUTION: $\cos 90^0 = \cos^2 45^0 - \sin^2 45^0$ by 6-2B.

$$= (\frac{1}{\sqrt{2}})^2 - (\frac{1}{\sqrt{2}})^2$$

$$= \frac{1}{2} - \frac{1}{2} = 0$$

FORMULA 6-2C

$$\tan 2x = \frac{2 \tan x}{1 - \tan^2 x}$$

DERIVATION

$$\tan 2x = \frac{\tan x + \tan x}{1 - \tan x \tan x} \quad (6-1F)$$

$$= \frac{2 \tan x}{1 - \tan^2 x}$$

EXAMPLE: If $\sin x = \frac{-3}{5}$ and $\pi < x < \frac{3\pi}{2}$, find $\sin 2x$, $\cos 2x$, and $\tan 2x$.

SOLUTION: By the Pythagorean Identity, if $\sin x = \frac{-3}{5}$ and $\pi \leq x < \frac{3\pi}{2}$,

$\cos x = \frac{-4}{5}$. Therefore, $\sin 2x = 2 \sin x \cos x = 2 (\frac{-3}{5})(\frac{-4}{5}) = \frac{24}{25}$.

$\cos 2x = 2 \cos^2 x - 1 = 2 (\frac{-4}{5})^2 - 1 = 2 (\frac{16}{25}) - 1 = (\frac{32}{25}) - 1 = \frac{7}{25}$.

$$\tan 2x = \frac{2 \tan x}{1 - \tan^2 x} = \frac{2 (\frac{3}{4})}{1 - (\frac{3}{4})^2} = \frac{\frac{6}{4}}{1 - (\frac{9}{16})} = \frac{\frac{6}{4}}{\frac{7}{16}} = \frac{24}{7}.$$

$\tan x$ is obtained by the quotient identity $\tan x = \frac{\sin x}{\cos x}$.

As a check, $\tan 2x = \frac{\sin 2x}{\cos 2x} = \frac{\frac{24}{25}}{\frac{7}{25}} = \frac{24}{7}$.

HALF ANGLE FORMULAS

Formula 6-2B' may be used to derive an expression for $\cos \frac{x}{2}$.

FORMULA 6-2D

$$\cos \frac{x}{2} = \pm\sqrt{\frac{1 + \cos x}{2}}$$

DERIVATION

$\cos 2x = 2 \cos^2 x - 1$ (6-2B').

$\cos x = 2 \cos^2 \frac{x}{2} - 1$, the same formula where x is the double angle.

$$1 + \cos x = 2 \cos^2 \frac{x}{2} \text{ , adding 1}$$

$$\frac{1 + \cos x}{2} = \cos^2 \frac{x}{2} \text{ , dividing by 2}$$

$$\pm \sqrt{\frac{1 + \cos x}{2}} = \cos \frac{x}{2} \text{ , taking square roots}$$

$$\cos \frac{x}{2} = \pm \sqrt{\frac{1 + \cos x}{2}} \text{ , symmetric law}$$

Note: The choice of sign will depend upon the quadrant in which $\frac{x}{2}$ terminates.

EXAMPLE: Find an exact value for $\cos 15^O$.

SOLUTION: $\cos 15^O = \sqrt{\frac{1 + \cos 30^O}{2}} = \sqrt{\frac{1 + \frac{\sqrt{3}}{2}}{2}} = \sqrt{\frac{2 + \sqrt{3}}{4}} = \frac{1}{2}\sqrt{2 + \sqrt{3}}$

Note that in 6-1, $\cos 15^O$ was found to be equal to $\frac{\sqrt{6} + \sqrt{2}}{4}$.

The equivalence of these two expressions is left as an exercise

for the student.

EXAMPLE: If $\cos 2\theta = \frac{-7}{25}$, $90^O < 2\theta < 180^O$, find $\cos \theta$.

SOLUTION: Formula 6-2D with $\theta = \frac{1}{2}x$ and $2\theta = x$. Substituting,

$$\cos \theta = \sqrt{\frac{1 + \cos 2\theta}{2}} = \sqrt{\frac{1 + \frac{-7}{25}}{2}} = \sqrt{\frac{18}{50}} = \sqrt{\frac{9}{25}} = \frac{3}{5} \; .$$

Note the positive radical is selected because $45^O < \theta < 90^O$.

EXAMPLE: Find $\cos \theta$ given $\cot 2\theta = \frac{5}{12}$ and $180^O < 2\theta < 270^O$.

SOLUTION: By the circular ratios, $\cot 2\theta = \frac{x}{y} = \frac{-5}{-12}$. Therefore, r = 13.
Since $180^O < 2\theta < 270^O$, Quad III, $\cos 2\theta = \frac{-5}{13}$. θ lies in quadrant II,

therefore, $\cos \theta = -\sqrt{\frac{1 + \cos 2\theta}{2}} = -\sqrt{\frac{1 - \frac{5}{13}}{2}} = -\sqrt{\frac{8}{26}} = -\sqrt{\frac{4}{13}}$

$$= \frac{-2}{\sqrt{13}} \text{ or } \frac{-2\sqrt{13}}{13} \; .$$

FORMULA 6-2E

$$\sin \frac{1}{2} x = \pm \sqrt{\frac{1 - \cos x}{2}}$$

DERIVATION

6-2B" gives $\cos 2 x = 1 - 2 \sin^2 x$.
Solving for $\sin x$ yields

$\sin x = \pm \sqrt{\frac{1 - \cos 2 x}{2}}$. The

steps are left as an exercise for

the student. Substituting $\frac{1}{2} x$ for

x and x for 2x leads to formula 6-2E.

EXAMPLE: Using $\cos 270^0 = 0$, find $\sin 135^0$.

SOLUTION: $\sin 135^0 = \sqrt{\frac{1 - \cos 270^0}{2}} = \sqrt{\frac{1 - 0}{2}} = \sqrt{\frac{1}{2}} = \frac{1}{\sqrt{2}}$

Since 135^0 is in quadrant II, the positive radical is chosen.

EXAMPLE: If $\sin 2 \theta = \frac{1}{3}$, find $\sin \theta$ if $90^0 < 2\theta < 180^0$.

SOLUTION: $\sin 2 \theta = \frac{1}{3} = \frac{y}{r}$ by circular ratios. Therefore $\cos 2 \theta = \frac{-\sqrt{8}}{3}$.

$$\sin \theta = \sqrt{\frac{1 - (\frac{-\sqrt{8}}{3})}{2}} = \sqrt{\frac{3 + \sqrt{8}}{6}} \ .$$

FORMULA 6-2F

$$\tan \frac{1}{2} x = \pm \sqrt{\frac{1 - \cos x}{1 + \cos x}}$$

DERIVATION

The result follows directly from

6-2D, 6-2E and the quotient identity.

EXAMPLE: Given $\cos 2A = \frac{-3}{5}$, $180^0 < 2A < 270^0$, find $\tan A$.

SOLUTION: Since A is in quadrant II, $\tan A = - \sqrt{\frac{1 - \cos 2A}{1 + \cos 2A}}$

$$= - \sqrt{\frac{1 - (\frac{-3}{5})}{1 + (\frac{-3}{5})}} = - \sqrt{\frac{1 + \frac{3}{5}}{1 - \frac{3}{5}}}$$

$$= - \sqrt{\frac{\frac{8}{5}}{\frac{2}{5}}} = - 2$$

FORMULA 6-2F'   <u>DERIVATION</u>

$$\tan \frac{1}{2} x = \frac{1 - \cos x}{\sin x}$$

Starting with 6-2F, the numerator and denominator of the fraction is multiplied by $1 - \cos x$:

$$\tan \frac{1}{2} x = \pm \sqrt{\frac{1 - \cos x}{1 + \cos x} \cdot \frac{1 - \cos x}{1 - \cos x}}$$

$$= \pm \sqrt{\frac{(1 - \cos x)^2}{1 - \cos^2 x}} = \pm \sqrt{\frac{(1 - \cos x)^2}{\sin^2 x}}$$

$$= \pm \frac{1 - \cos x}{\sin x} .$$

Since $1 - \cos x \geq 0$, the sign of the rational expression is determined by $\sin x$. An examination of all possible cases will show that for each x, $\tan \frac{1}{2} x = \frac{1 - \cos x}{\sin x}$ so that the $\pm$ choice will automatically be determined by $\sin x$, whatever the quadrant for $\frac{1}{2} x$.

<u>EXAMPLE:</u>   Given $\sin x = \frac{-15}{17}$, $\frac{3\pi}{2} < x < 2\pi$, find $\tan \frac{1}{2} x$.

<u>SOLUTION:</u>   If $\sin x = \frac{-15}{17}$ and x is in quadrant IV, $\cos x = \frac{8}{17}$.

Therefore by 6-2F', $\tan \frac{1}{2} x = \dfrac{1 - \frac{8}{17}}{\frac{-15}{17}} = \dfrac{\frac{9}{17}}{\frac{-15}{17}} = \dfrac{-3}{5}$

FORMULA 6-2F"   <u>DERIVATION</u>

$$\tan \frac{1}{2} x = \frac{\sin x}{1 + \cos x}$$

The student can derive this alternate formula from 6-2F' by multiplying numerator and denominator by $1 + \cos x$.

<u>EXAMPLE:</u>   Use 6-2F" to arrive at the result of the previous example.

<u>SOLUTION:</u> $\tan \frac{1}{2} x = \dfrac{\sin x}{1 + \cos x} = \dfrac{\frac{-15}{17}}{1 + \frac{8}{17}} = \dfrac{\frac{-15}{17}}{\frac{25}{17}} = \dfrac{-15}{25} = \dfrac{-3}{5} .$

EXERCISE SET 6-2.

1) Derive Formula 6-2B".

2) Derive Formula 6-2C from the quotient identity $\tan 2x = \dfrac{\sin 2x}{\cos 2x}$.

3) Derive Formula 6-2F".

4) Show that $\dfrac{1}{2} \sqrt{2 + \sqrt{3}} = \dfrac{\sqrt{6} + \sqrt{2}}{4}$.

5) Use Formula 6-2A to show that $\sin 270^0 = -1$.

6) Use Formula 6-2B' to show that $\cos 120^0 = - \cos 60^0$.

7) Use Formula 6-2C to show that $\tan 90^0$ is undefined.

Using the half-angle formulas find __exact__ values for the following:

8) $\sin 165^0$ 9) $\cos \dfrac{\pi}{8}$ 10) $\tan \dfrac{3\pi}{8}$ 11) $\cos 255^0$

12) $\sin \dfrac{5\pi}{12}$ 13) $\tan 105^0$ 14) $\sin 202.5^0$

15) $\cos \dfrac{\pi}{8} + \sin \dfrac{\pi}{8}$ 16) $\sec \dfrac{5\pi}{8} + \cot \dfrac{5\pi}{8}$

17) If $\tan x = -2$, $\dfrac{3\pi}{2} < x < 2\pi$, find $\tan \dfrac{1}{2} x$.

18) If $\sin 2t = \dfrac{-3}{5}$, $\pi < 2t < \dfrac{3\pi}{2}$, find $\sin t$.

19) If $\cos 2x = .8314$ and $\dfrac{3\pi}{2} < 2x < 2\pi$, find $\cos x$.

20) If $\tan x = -3.124$ find $\tan 2x$.

21) If $\sin x = -.4567$, $-\dfrac{\pi}{2} < x < 0$, find $\sin 2x$.

22) If $\sin 6x = .9135$, $\dfrac{\pi}{2} < 6x < \pi$, find $\cos 3x$.

23) If $\cos \theta = x$, find $\cos \dfrac{1}{2} \theta$.

24) If $\cos \theta = \sqrt{1 - x^2}$, find $\tan \dfrac{1}{2} \theta$ and $\tan 2\theta$ if $0^0 < \theta < 90^0$.

Prove the following identities:

25) $1 + \sin 2x = (\cos x + \sin x)^2$

26) $\sin 3x = 3 \sin x - 4 \sin^3 x$

27) $\sin (x + y) \sin (x - y) = \sin^2 x - \sin^2 y$

28) $\csc 2x = \dfrac{1}{2} \tan x + \dfrac{1}{2} \cot x$

29) $\cot x - \tan x = 2 \cot 2x$

30) $\dfrac{1 + \tan^2 x}{1 - \tan^2 x} = \sec 2x$

6-3. <u>CONVERSION OF SUMS TO PRODUCTS, GRAPHING THE ADDITION OF ORDINATES.</u>

Another set of formulas that are useful in advanced applications can be derived from those studied in section 6-1. To distinguish them from the earlier results there will be a change of variables.

Formula 6-3A: $\sin A + \sin B = 2 \sin \frac{A + B}{2} \cos \frac{A - B}{2}$

Formula 6-3B: $\sin A - \sin B = 2 \cos \frac{A + B}{2} \sin \frac{A - B}{2}$

Formula 6-3C: $\cos A + \cos B = 2 \cos \frac{A + B}{2} \cos \frac{A - B}{2}$

Formula 6-3D: $\cos A - \cos B = - 2 \sin \frac{A + B}{2} \sin \frac{A - B}{2}$

Formula 6-3A is derived by starting with the addition of formulas 6-1D and 6-1E:

$\sin (x + y) \qquad\qquad = \sin x \cos y + \cos x \sin y$

$\underline{\qquad\qquad + \sin (x - y) \quad = \quad \sin x \cos y - \cos x \sin y}$

$\sin (x + y) + \sin (x - y) = 2\sin x \cos y$

Now, let $A = x + y$ and $B = x - y$. Then, substituting for the left member,

$\sin (x + y) + \sin (x - y) = \sin A + \sin B$. Next, add A and B and subtract B from A.

$A + B = 2x, \qquad A - B = 2y.$

Hence, $x = \frac{A + B}{2}$ and $y = \frac{A - B}{2}$. Substituting for 2 sin x cos x yields

Formula 6-3A: $\boxed{\sin A + \sin B = 2 \sin \frac{A + B}{2} \cos \frac{A - B}{2}}$

<u>EXAMPLE</u>:   Express $\sin 30^0 + \sin 20^0$ as a product.

<u>SOLUTION</u>:   By 6-3A, $\sin 30^0 + \sin 20^0 = 2 \sin \frac{30^0 + 20^0}{2} \cos \frac{30^0 - 20^0}{2}$

$$= 2 \sin 25^0 \cos 5^0.$$

The derivations of 6-3B, 6-3C, and 6-3D are left to the student. Other useful identities can be proven by these formulas.

EXAMPLE: Prove $\dfrac{\sin A + \sin B}{\cos A + \cos B} = \tan \dfrac{A + B}{2}$

SOLUTION: Divide the left and right members of 6-3A by 6-3C respectively. Then apply the quotient identity $\dfrac{\sin \theta}{\cos \theta} = \tan \theta$.

EXAMPLE: Express $\cos 3x + \cos x$ in terms of $\cos x$.

SOLUTION: By 6-3C, $\cos 3x + \cos x = 2 \cos \dfrac{3x + x}{2} \cos \dfrac{3x - x}{2}$

$$= 2 \cos 2x \cos x$$

$$= 2(2 \cos^2 x - 1) \cos x$$

$$= 4 \cos^2 x - 2 \cos x$$

EXAMPLE: Express $2 \sin 115^0 \cos 85^0$ as the sum of sines.

SOLUTION: By 6-3A, $115^0 = \dfrac{A + B}{2}$, $85^0 = \dfrac{A - B}{2}$

Therefore, $A + B = 230^0$

$$\underline{A - B = 170^0}$$

$2A \qquad = 400^0$ \qquad Hence $A = 200^0$ and $B = 30^0$.

Substituting, $2 \sin 115^0 \cos 85^0 = \sin 200^0 + \sin 30^0$.

EXAMPLE: Express $-2 \sin \dfrac{5\pi}{6} \sin \dfrac{\pi}{8}$ as the difference of cosines.

SOLUTION: By 6-3D, $\cos A - \cos B = -2 \sin \dfrac{5\pi}{6} \sin \dfrac{\pi}{8}$ where $\dfrac{A + B}{2} = \dfrac{5\pi}{6}$ and $\dfrac{A - B}{2} = \dfrac{\pi}{8}$. Since $A + B = \dfrac{5\pi}{3}$ and $A - B = \dfrac{\pi}{4}$, adding yields

$2A = \dfrac{5\pi}{3} + \dfrac{\pi}{4} = \dfrac{23\pi}{12}$. Hence $A = \dfrac{23\pi}{24}$ and $B = \dfrac{17\pi}{24}$.

Thus $-2 \sin \dfrac{5\pi}{6} \sin \dfrac{\pi}{8} = \cos \dfrac{23\pi}{24} - \cos \dfrac{17\pi}{24}$.

SIN A + COS A

The complementary relationship between sine and cosine may be combined to derive another useful sum to product formula:

$$\sin A + \cos A = \sin A + \sin \left(\frac{\pi}{2} - A \right) = 2 \sin \frac{A + \left(\frac{\pi}{2} - A \right)}{2} \cos \frac{A - \left(\frac{\pi}{2} - A \right)}{2}$$

$$= 2 \sin \frac{\pi}{4} \cos \left(A - \frac{\pi}{4} \right) = 2 \cdot \frac{\sqrt{2}}{2} \cos \left(A - \frac{\pi}{4} \right) = \sqrt{2} \cos \left(\frac{\pi}{4} - A \right)$$

$$= \sqrt{2} \sin \left(\frac{\pi}{2} - \left(\frac{\pi}{4} - A \right) \right) = \sqrt{2} \sin \left(\frac{\pi}{4} + A \right)$$

We have derived Formula 6-3E:

$$\boxed{\sin A + \cos A = \sqrt{2} \sin \left(\frac{\pi}{4} + A \right)}$$

EXAMPLE: Express $\sin 48^0 + \cos 48^0$ as the sine of an acute angle.

SOLUTION: By 6-3E, $\sin 48^0 + \cos 48^0 = \sqrt{2} \sin (45^0 + 48^0) = \sqrt{2} \sin 93^0$

$$= \sqrt{2} \sin (180^0 - 87^0) = \sqrt{2} \sin 87^0 .$$

EXAMPLE: Find an exact value for $\sin \frac{\pi}{12} + \cos \frac{\pi}{12}$.

SOLUTION: By 6-3E, $\sin \frac{\pi}{12} + \cos \frac{\pi}{12} = \sqrt{2} \sin \left(\frac{\pi}{4} + \frac{\pi}{12} \right) = \sqrt{2} \sin \frac{\pi}{3} = \sqrt{2} \cdot \frac{\sqrt{3}}{2} = \frac{\sqrt{6}}{2}$

GRAPHING SIN X + COS X

The function $f(x) = \sin x + \cos x$ may be graphed by the technique known as "addition of ordinates ." It simply involves graphing the sine function and the cosine function separately on the same set of coordinate axes and then, adding the y-coordinates geometrically to obtain the resultant curve as shown in Figure 6-3A.

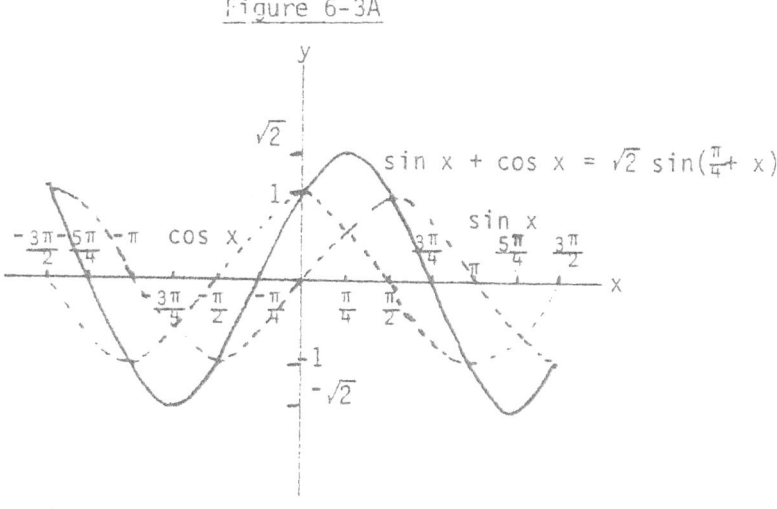

Figure 6-3A

The resultant graph illustrates convincingly the equivalence of sin x +

cos x and $\sqrt{2} \sin (\frac{\pi}{4} + x)$.

GENERALIZING A SIN X + B COS X

The foregoing is a special case of the general expression A sin x + B cos x

where A = B = 1. If the amplitudes A and B change it is still a simple task to

graph these functions by adding ordinates. Figure 6-3B shows the result of

this technique on the function f(x) = 3 sin x - 2 cos x.

Figure 6-3B

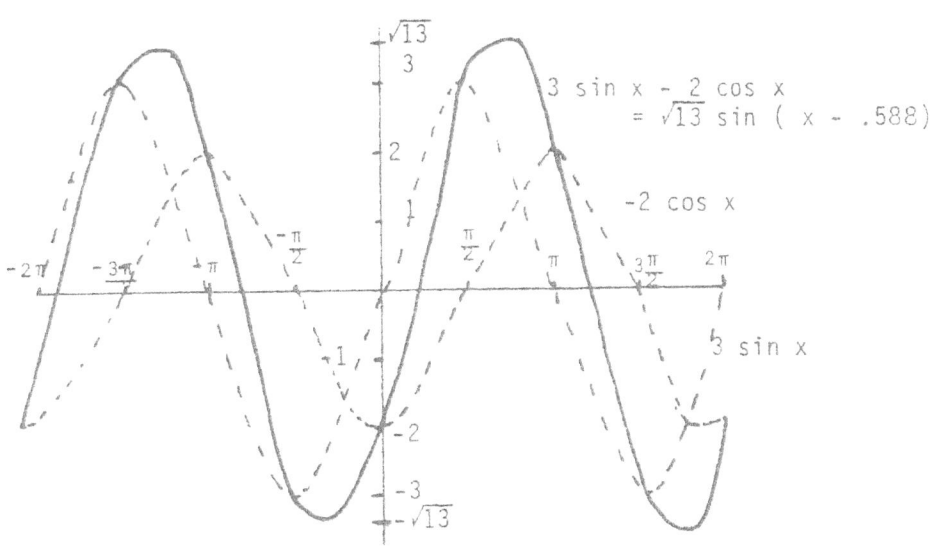

Again, it appears that $3 \sin x - 2 \cos x$ could be expressed as a sine function with a different amplitude and phase shift than that of its component functions. Assume that such is the case, $3 \sin x - 2 \cos x = C \sin (x + \theta)$ where C = amplitude and θ = phase shift of the resultant function. By the sum formula 6-1D,

$$C \sin (x + \theta) = C (\sin x \cos \theta + \cos x \sin \theta)$$
$$= (C \cos \theta) \sin x + (C \sin \theta) \cos x.$$

Let $C \cos \theta = 3$ and $C \sin \theta = -2$.

Then $\cos \theta = \frac{3}{C}$ and $\sin \theta = \frac{-2}{C}$. If $C > 0$, θ is in quadrant IV.

By the Pythagorean Identity, $C = \sqrt{13}$ and $\theta = \mathrm{Sin}^{-1} (\frac{-2}{\sqrt{13}}) = -.588$.

Therefore, $3 \sin x - 2 \cos x = C \sin (x + \theta) = \sqrt{13} \sin (x - .588)$.

To generalize the foregoing,

$$A \sin x + B \cos x = C \sin (x + \theta) \quad \text{where}$$
$$C = \sqrt{A^2 + B^2}, \quad \cos \theta = \frac{A}{\sqrt{A^2 + B^2}} \text{ and } \sin \theta = \frac{B}{\sqrt{A^2 + B^2}}.$$

EXAMPLE: Express $5 \cos x - 12 \sin x$ as a sine function.

SOLUTION: $5 \cos x - 12 \sin x = -12 \sin x + 5 \cos x.$

Let $A = -12$ and $B = 5$. Then $C = \sqrt{5^2 + (-12)^2} = 13$

Since $\cos \theta = \frac{-12}{13}$, $\sin \theta = \frac{5}{13}$, θ is in quadrant II.

Hence, $5 \cos x - 12 \sin x = 13 \sin (x + \mathrm{Cos}^{-1}(\frac{-12}{13})) \doteq 13 \sin (x + 2.75).$

EXERCISE SET 6-3.

Express the following sums as products:

1) $\sin 33^\circ + \sin 47^\circ$

2) $\cos 10^\circ - \cos 25^\circ$

3) $\sin 72^\circ - \sin 30^\circ$

4) $\cos \frac{3\pi}{8} + \cos \frac{\pi}{4}$

5) $\sin \frac{2\pi}{3} - \sin \frac{5\pi}{6}$

6) $\cos \frac{4\pi}{3} - \cos \frac{2\pi}{7}$

Express the products as sums:

7) $2 \sin 100^\circ \cos 20^\circ$

8) $-2 \sin 29^\circ \sin 4^\circ$

9) $2 \cos 8^\circ \cos 39^\circ$

10) $2 \cos \frac{7\pi}{48} \sin \frac{\pi}{48}$

11) $-2 \sin \frac{33\pi}{70} \sin \frac{23\pi}{70}$

12) $\sin \frac{\pi}{9} \cos \frac{\pi}{6}$

13) Derive 6-3B.

14) Derive 6-3C

15) Derive 6-3D.

16) Show that $\sin 35^\circ + \sin 25^\circ = \cos 5^\circ$

17) Show that $\sin \frac{\pi}{5} + \sin \frac{2\pi}{15} = \cos \frac{\pi}{30}$

18) Express $\cos 100^\circ - \cos 20^\circ$ as a function of 40°

19) Express $\cos \pi + \cos \frac{\pi}{4}$ as a function of $\frac{\pi}{8}$

20) Express $\sin x - \cos x$ as a sine function.

21) Express $\sin 10x - \sin 4x$ as a product.

22) Express $\cos x - \sin x$ as a cosine function.

23) $3 \sin x + 4 \cos x = A \sin (x + \theta)$. Find A and θ.

24) Express $25 \sin (x + \theta)$ where $\theta = \operatorname{Sin}^{-1} \frac{7}{25}$ as the sum of a sine and cosine function.

Graph the following:

25) $2 \sin x + 3 \cos x$

26) $3 \cos x - 2 \sin x$

27) $\sin x + 2 \cos x$

28) $\cos 2x + \sin x$

29) $\sin 2x - 2 \cos x$

30) $\tan x - \sin x$

Prove the following identities:

31) $\sin 3x - \sin x = 2 \sin x - 4 \sin^3 x$

32) $\dfrac{\sin x + \sin 3x}{\sin x - \sin 3x} = \dfrac{2}{\tan^2 x - 1}$

33) $\dfrac{\cos 4x - \cos 2x}{\cos 4x + \cos 2x} = - \tan 3x \tan x$

34) Use 6-3E and exercise 22 to show that:

$$\dfrac{\cos x + \sin x}{\cos x - \sin x} = \dfrac{1 + \tan x}{1 - \tan x}$$

35) $\dfrac{\cos x + \cos 2x + \cos 3x}{\sin x + \sin 2x + \sin 3x} = \cot 2x$

6-4. CONDITIONAL EQUATIONS AND THE ADVANCED FORMULAS.
USE OF THE INVERSE FUNCTIONS.

The formulas developed in the previous sections of this chapter are useful in simplifying equations.

EXAMPLE: Solve for x. $0 \le x \le 2\pi$ and $2 \sin x \cos x = 1$

SOLUTION: $2 \sin x \cos x = 1$. By 6-2A, $\sin 2x = 1$.

Therefore, $2x = \sin^{-1} 1 = \frac{\pi}{2} + 2k\pi$, $k = 0, \pm 1, \pm 2, \ldots$

and $x = \frac{\pi}{4} + k\pi$, $k = 0, \pm 1, \pm 2, \ldots$

Within the specified domain for x, $x = \frac{\pi}{4}, \frac{5\pi}{4}$

EXAMPLE: Solve for x. $1 - 2 \sin^2 x = \frac{\sqrt{2}}{2}$

SOLUTION: Two approaches are possible. Directly,

$2 \sin^2 x = 1 - \frac{\sqrt{2}}{2} = \frac{2 - \sqrt{2}}{2}$

$\sin^2 x = \frac{2 - \sqrt{2}}{4}$ and therefore, $\sin x = \frac{\pm\sqrt{2 - \sqrt{2}}}{2} = \pm.3827$

Hence, $x = \sin^{-1}(\pm.3827) = \pm.39$

Using 6-2B" exact solutions can be obtained.

$1 - 2 \sin^2 x = \cos 2x = \frac{\sqrt{2}}{2}$ where $-\frac{\pi}{2} \le x \le \frac{\pi}{2}$

Then $2x = \cos^{-1}\frac{\sqrt{2}}{2} = \pm\frac{\pi}{4}$ and $x = \pm\frac{\pi}{8}$

EXAMPLE: Solve for x. $0 \le x \le 2\pi$ and $\sin x \cos \frac{\pi}{6} + \cos x \sin \frac{\pi}{6} = \frac{1}{2}$

SOLUTION: $\sin x \cos \frac{\pi}{6} + \cos x \sin \frac{\pi}{6} = \sin(x + \frac{\pi}{6}) = \frac{1}{2}$ by 6-1D.

$x + \frac{\pi}{6} = \sin^{-1}\frac{1}{2} = \frac{\pi}{6}$ or $\frac{5\pi}{6}$

Adding $-\frac{\pi}{6}$ obtain $x = 0$ or $\frac{2\pi}{3}$

The next example requires some insight and manipulation before the simplicity of its solution can be revealed.

EXAMPLE: Solve for x. $2 \tan 2x = 3 - 3 \tan^2 2x$

SOLUTION: The equation could be treated as a quadratic in $\tan 2x$. However,

factoring $2 \tan 2x = 3(1 - \tan^2 2x)$ and dividing

$$\frac{2 \tan 2x}{1 - \tan^2 2x} = 3$$

Now by 6-2C, this reduces to $\tan 4x = 3$.

Hence $4x = \tan^{-1} 3$ and $x = \frac{1}{4} \tan^{-1} 3 = .31$

The formulas developed in section 6-3 are also useful in simplifying

conditional equations.

EXAMPLE: Solve for x. $0 \leq x \leq 2\pi$ and $\sin x + \sin 3x = 0$

SOLUTION: By 6-3A $\sin x + \sin 3x = 2 \sin 2x \cos x$.

Since $\sin 2x = 2 \sin x \cos x$ substituting yields

$\sin x + \sin 3x = 2 (2 \sin x \cos x) \cos x = 4 \sin x \cos^2 x = 0$.

Using the Pythagorean identity,

$4 \sin x (1 - \sin^2 x) = 0$

Dividing by 4 and setting each factor equal to zero,

$\sin x = 0$ or $1 - \sin^2 x = 0$.

Solving these simple equations $x = 0$ or $\frac{\pi}{2}$ or $\frac{3\pi}{2}$.

The next example demonstrates the alternative methods of solving an

equation directly or through use of the techniques developed in this chapter.

EXAMPLE: Solve for x. $0 \leq x < \pi$ and $\sin x - 3 \cos x = \sqrt{10}$

SOLUTION: DIRECT   ADVANCED

$\sin x - 3 \cos x = \sqrt{10}$ Let $\sin x - 3 \cos x = C \sin (x + \Theta)$

$\sin x = 3 \cos x + \sqrt{10}$ where $C = \sqrt{ (1)^2 + (-3)^2} = \sqrt{10}$ and

$\sin^2 x = 9 \cos^2 x + 6 \sqrt{10} \cos x + 10$ $\cos \Theta = \frac{1}{\sqrt{10}}$, $\sin \Theta = \frac{-3}{\sqrt{10}}, \Theta = \sin^{-1}(\frac{-3}{\sqrt{10}})$.

$$1 - \cos^2 x = 9 \cos^2 x + 6\sqrt{10} \cos x + 10$$

Substituting,

$$10 \cos^2 x + 6\sqrt{10} \cos x + 9 = 0$$

$$\sin x - 3 \cos x =$$

$$\cos x = \frac{-6\sqrt{10} \pm \sqrt{(6\sqrt{10})^2 - 4(10)(9)}}{20}$$

$$\sqrt{10} \sin (x + Sin^{-1}(\frac{-3}{\sqrt{10}}) = \sqrt{10}$$

$$\cos x = \frac{-6\sqrt{10}}{20} = \frac{-3\sqrt{10}}{10}$$

Therefore, $\sin (x + (-1.25)) = 1$

$$x = \cos^{-1} \frac{(-3\sqrt{10})}{10} = 2.82$$

and $x + (-1.25) = \sin^{-1} 1 = \frac{\pi}{2}$.

Hence, $x = \frac{\pi}{2} + 1.25 = 2.82$.

INVERSE FUNCTIONS AND THE ADVANCED FORMULAS.

Suppose the arguments x and y are expressed by inverse function notation. The formulas derived in this chapter may still be applicable.

EXAMPLE: Find the value of $\tan (Arccos \frac{24}{25} + Arctan \frac{4}{3})$.

SOLUTION: Let $x = Arccos \frac{24}{25}$ and $y = Arctan \frac{4}{3}$. Then,

$$\tan (Arccos \frac{24}{25} + Arctan \frac{4}{3}) = \tan (x + y) = \frac{\tan x + \tan y}{1 - \tan x \tan y} .$$

If $x = Arccos \frac{24}{25}$ then $\cos x = \frac{24}{25}$. By the Pythagorean relationship, $\tan x = \frac{7}{24}$. If $y = Arctan \frac{4}{3}$ then $\tan y = \frac{4}{3}$. Substituting,

$$\tan (x + y) = \frac{\frac{7}{24} + \frac{4}{3}}{1 - (\frac{7}{24})(\frac{4}{3})} = \frac{\frac{39}{24}}{\frac{11}{18}} = \frac{117}{44} .$$

EXAMPLE: Find the value of $\sin (2 Arccos \frac{1}{3})$

SOLUTION: Let $x = Arccos \frac{1}{3}$. Then $\sin (2 Arccos \frac{1}{3}) = \sin 2x = 2 \sin x \cos x$

Since $\cos x = \frac{1}{3}$ the Pythagorean identity will yield $\sin x = \frac{2\sqrt{2}}{3}$.

Hence, $\sin (2 Arccos \frac{1}{3}) = 2 (\frac{2\sqrt{2}}{3})(\frac{1}{3}) = \frac{4\sqrt{2}}{9}$.

EXERCISE SET 6-4.

Solve the following equations for all values of x, $0 \le x \le 2\pi$, unless the principal values are indicated. Leave answers in multiples of π whenever possible. Otherwise round off answers to hundredths.

1) $\sin 2x = \frac{1}{2}$

2) $\sqrt{2} \cos 2x = 1$

3) $\tan x = \frac{5}{2} \tan^2 x - \frac{5}{2}$

4) $\cot 2x = 3$

5) $2 \cos 2x + 3 \cos x + 1 = 0$

6) $\tan 2x = \tan x$

7) $\sin x - \cos 2x = 0$

8) $\sin^2 3x = \frac{1}{4}$

9) $\tan 3x = \sin \frac{5\pi}{8}$

10) $\tan 2x - \cot 2x = 2$

11) $\sin 2x - \sin x = 0$

12) $5 \sin x + 12 \cos x = 1$

13) $8 \cos x - 5 \sin x = 2$

14) $\sqrt{3} \sin x - \cos x = \sqrt{2}$

15) $\sin 2x \cos x - \cos 2x \sin x = 1$

16) $\cos 2x + 2 \cos^2 \frac{x}{2} = 1$

17) $2 \cos x - 2 \sin x = \sqrt{6}$

18) $3 \cos x + 4 \sin x = 3$

19) $2 \sin \frac{x}{2} \cos \frac{x}{2} = \frac{-1}{2}$

20) $\cos 3x - \cos x = 9 \cos^2 x$

21) $4 \cos 2x + 3 \cos x = 0$

22) $\sin 2x - \cos 2x = \frac{1}{2}$

23) $\sin x + \cos x = \sqrt{\frac{3}{2}}$

24) $\sqrt{1 - \cos 2x} = \sqrt{\frac{2}{3}}$

25) $\frac{\sin x}{1 + \cos x} = -2$

26) $\sqrt{3} \tan 2x = 2$

27) $e^{\cot 2x - 1} = 7$

28) $\log \sin x - \log \cos x = 0$

29) $10^{2 \sin^2 x - 1} = 4.395$

30) $\cos^2 x - \sin^2 x = -.5834$

Solve the following systems for x and y:

31) $\sin x - 3 \cos y = 2$

$2 \sin x + \cos y = 1$

32) $\tan (x + y) = \dfrac{1}{\sqrt{3}}$

$\tan (x - y) = \sqrt{3}$

Find the value of the following:

33) $\sin \left(\text{Arcsin } \dfrac{1}{2} + \text{Arccos } \dfrac{3}{4}\right)$

34) $\tan \left(\text{Sin}^{-1} \dfrac{2}{5} + \text{Sin}^{-1}\left(\dfrac{-3}{5}\right)\right)$

35) $\cos\left(\text{Tan}^{-1}\left(\dfrac{-5}{2}\right) + \text{Cos}^{-1}\left(\dfrac{3}{4}\right)\right)$

36) $\cot \left(2 \text{ Cos}^{-1}\left(\dfrac{-1}{\sqrt{2}}\right)\right)$

37) $\sin \left(\dfrac{1}{2} \text{ Arctan } \dfrac{1}{2}\right)$

38) $\cot \left(\dfrac{1}{2} \text{ Arccos } \dfrac{3}{7} \right)$

39) $\text{Arctan} \left[\dfrac{\left(\tan \dfrac{\pi}{3} + \tan \dfrac{\pi}{5} \right)}{1 - \tan \dfrac{\pi}{3} \tan \dfrac{\pi}{5}} \right]$

40) $\text{Arccos} \left[\dfrac{\left(\tan \dfrac{\pi}{3} + \tan \dfrac{\pi}{5} \right)}{1 - \tan \dfrac{\pi}{3} \tan \dfrac{\pi}{5}} \right]$

6-5. CHAPTER SUMMARY

1. The Distance Formula. The distance between two points $P_1(x_1, y_1)$ and $P_2(x_2, y_2)$ in the plane is given by:

$$d(P_1, P_2) = \sqrt{(x_1 - x_2)^2 + (y_1 - y_2)^2}$$

2. Functions for the sum and difference of two arguments:

$$\cos(x + y) = \cos x \cos y - \sin x \sin y$$

$$\cos(x - y) = \cos x \cos y + \sin x \sin y$$

$$\sin(x + y) = \sin x \cos y + \cos x \sin y$$

$$\sin(x - y) = \sin x \cos y - \cos x \sin y$$

$$\tan(x + y) = \frac{\tan x + \tan y}{1 - \tan x \tan y}$$

$$\tan(x - y) = \frac{\tan x - \tan y}{1 + \tan x \tan y}$$

3. Functions for the double argument:

$$\sin 2x = 2 \sin x \cos x \qquad \tan 2x = \frac{2 \tan x}{1 - \tan^2 x}$$

$$\cos 2x = \cos^2 x - \sin^2 x$$

$$= 2 \cos^2 x - 1$$

$$= 1 - 2 \sin^2 x$$

4. Functions for the half arguments:

$$\sin \tfrac{1}{2}x = \pm \sqrt{\frac{1 - \cos x}{2}}$$

$$\cos \tfrac{1}{2}x = \pm \sqrt{\frac{1 + \cos x}{2}}$$

$$\tan \tfrac{1}{2}x = \pm \sqrt{\frac{1 - \cos x}{1 + \cos x}}$$

$$= \frac{\sin x}{1 + \cos x}$$

$$= \frac{1 - \cos x}{\sin x}$$

5. Sum to product formulas:

$$\sin x + \sin y = 2 \sin \frac{x + y}{2} \cos \frac{x - y}{2}$$

$$\sin x - \sin y = 2 \cos \frac{x + y}{2} \sin \frac{x - y}{2}$$

$$\cos x + \cos y = 2 \cos \frac{x + y}{2} \cos \frac{x - y}{2}$$

$$\cos x - \cos y = -2 \sin \frac{x + y}{2} \sin \frac{x - y}{2}$$

6. Sums of sines and cosines:

$$A \sin x + B \cos x = C \sin (x + \Theta)$$

where $C = \sqrt{A^2 + B^2}$, $\cos \Theta = \frac{A}{C}$, $\sin \Theta = \frac{B}{C}$.

EXERCISE SET 6-5.

Find the distance between the given points:

1) $P(-3,4)$, $Q(7,10)$

2) $R(17.8,-12.9)$, $S(-4.6,7.8)$

3) $T(\cos 2, \sin 5)$, $U(\cos 4, \sin 2)$

4) $M(3\pi, -2\pi)$, $N(\frac{-\pi}{3}, \frac{\pi}{4})$

In exercises 5 - 8, $\sin x = \frac{3}{5}$ and $\cos y = \frac{-5}{13}$. If $0 < x < \frac{\pi}{2}$

and $\frac{\pi}{2} < y < \pi$, find:

5) $\cos (x + y)$

6) $\sin (x - y)$

7) $\tan (x - y)$

8) $\cot (x + y)$

In exercises 9 - 12 , $\cos x = \frac{3}{4}$. Find the value of the following:

9) $\cos 2x$

10) $\cos \frac{1}{2}x$

11) $\sin \frac{1}{2}x$

12) $\sin 2x$

Prove the following identities:

13) $(\sin x - \cos x)^2 + \sin 2x = 1$

14) $\cot x - \tan x = 2 \cot 2x$

15) $\csc^2 \frac{1}{2}x + \cot^2 \frac{1}{2}x = \frac{3 + \cos x}{1 - \cos x}$

16) $\frac{\sin 3x + \sin 7x}{\cos 3x - \cos 7x} = \cot 2x$

Simplify each of the following:

17) $\frac{2 \tan 4x}{1 - \tan^2 4x}$

18) $\cos \frac{x}{3} \cos \frac{2x}{3} - \sin \frac{x}{3} \sin \frac{2x}{3}$

19) $\sqrt{\frac{1 + \cos 8x}{2}}$

20) $\tan^2 \frac{3x}{7} - \sec^2 \frac{3x}{7}$

21) $\sin \frac{x + y}{2} \cos \frac{x - y}{2} - \cos \frac{x + y}{2} \sin \frac{x - y}{2}$

22) $\frac{\cot (90^0 - 3x) + \tan 2x}{1 - \tan 2x \cot (90^0 - 3x)}$

23) $6 \sin 6x \cos 6x$

24) $(2 \cos^2 3x - 1) - \frac{1}{\sec 6x}$

Write the following expressions as the sum or difference of two functions:

25) $2 \sin 9x \sin 5x$

26) $2 \cos 4x \sin 3x$

27) $\sin 7x \cos 2x$

28) $\cos \frac{3x}{2} \cos \frac{x}{2}$

Write each of the following expressions as the product of two functions and simplify if possible:

29) $\cos 45^0 - \cos 15^0$

30) $\sin 140^0 - \sin 60^0$

31) $\cos 110^0 + \cos 50^0$

32) $\sin 340^0 + \sin 100^0$

Solve the following equations for x , $0 \leq x < 2 \pi$:

33) $\sin 2x - \sqrt{3} \cos x = 0$

34) $2 \cos^2 2x = 5 \cos 2x + 3$

35) $\cos 2x \tan^2 2x = 3 \cos 2x$

36) $\sin x - 1 = \cos 2x$

37) $\sin 2x = \cos 2x - 1$

38) $2 \sin \frac{1}{2}x = \sin x$

39) $\tan 2x = \sin x$

40) $\sin 5x - \sin 3x = \sin x$

CHAPTER 7

COMPLEX NUMBERS
POLAR COORDINATES AND TRIGONOMETRIC FORM

7-1. TRANSFORMATION OF COORDINATES

The derivation of the formula for cos (x - y) in Section 6-1 demonstrated that the rectangular coordinates of a point P(x,y) in the plane were expressible in terms of the sine and cosine. Figure 7-1A reviews this correspondence.

Figure 7-1A

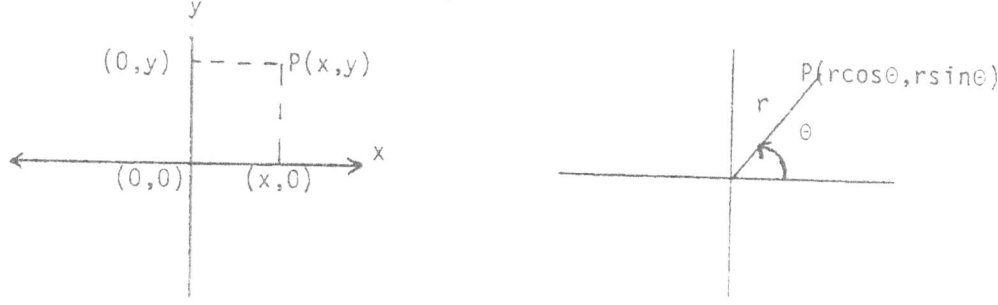

The equations of transformation are given by:

$x = r \cos \theta$ and $y = r \sin \theta$ where $r = \sqrt{x^2 + y^2}$ and $\theta = \tan^{-1} \dfrac{y}{x}$.

An ordered pair of real numbers (x,y) is referred to the rectangular coordinate system if perpendiculars drawn from P(x,y) to the coordinate axes form a rectangle with vertices at (0,0), (x,0), (x,y) and (0,y). Ordered pairs of the form (r cos θ, r sin θ) shall be referred to as <u>trigonometric coordinates</u> and their equivalence to rectangular coordinates are defined by the equations of transformation.

<u>EXAMPLE</u>: Express $(\frac{\sqrt{3}}{2}, \frac{1}{2})$ in trigonometric form.

<u>SOLUTION</u>: $x = \dfrac{\sqrt{3}}{2}$ and $y = \dfrac{1}{2}$. Therefore, $r = \sqrt{(\frac{\sqrt{3}}{2})^2 + (\frac{1}{2})^2} = 1$

$= \tan^{-1} \dfrac{\frac{1}{2}}{\frac{\sqrt{3}}{2}} = \tan^{-1} \dfrac{1}{\sqrt{3}} = 30^\circ$. It follows that $(\frac{\sqrt{3}}{2}, \frac{1}{2})$ is equivalent to $(1 \cos 30^0, 1 \sin 30^0)$ or $(\cos 30^0, \sin 30^0)$ in trigonometric form.

EXAMPLE: Express $(-2,3)$ in trigonometric form.

SOLUTION: $x = -2$ and $y = 3$. $r = \sqrt{(-2)^2 + 3^2} = \sqrt{13}$.

$\quad = \tan^{-1} (\frac{3}{-2}) \doteq 123.7^0$ or 2.16 in radians.

Therefore, $(-2,3) = (\sqrt{13} \cos (\tan^{-1}(\frac{3}{-2})), \sqrt{13} \sin (\tan^{-1}(\frac{3}{-2})))$

$\quad = (\sqrt{13} \cos 2.16, \sqrt{13} \sin 2.16)$ or

$\quad = (\sqrt{13} \cos 123.7^0, \sqrt{13} \sin 123.7^0)$

It may be noted that Θ is not unique. That is, any coterminal value of Θ is acceptable as the following example illustrates.

EXAMPLE: Express $(1,-1)$ in three different trigonometric forms.

SOLUTION: $x = 1$ and $y = -1$. $r = \sqrt{1^2 + (-1)^2} = \sqrt{2}$.

$\quad \Theta = \tan^{-1} (\frac{-1}{1}) = \frac{7\pi}{4} + 2k\pi$, $k = 0, \pm 1, \pm 2, \ldots$

Therefore, $(1,-1) = (\sqrt{2} \cos \frac{7\pi}{4}, \sqrt{2} \sin \frac{7\pi}{4}) = (\sqrt{2} \cos(\frac{-\pi}{4}), \sqrt{2} \sin(\frac{-\pi}{4}))$

$\quad = (\sqrt{2} \cos \frac{15\pi}{4}, \sqrt{2} \sin \frac{15\pi}{4})$.

Conversions from trigonometric coordinates to rectangular coordinates are easily handled by the equations of transformation.

EXAMPLE: Express $(2 \cos \frac{\pi}{6}, 2 \sin \frac{\pi}{6})$ in rectangular form.

SOLUTION: $x = 2 \cos \frac{\pi}{6} = 2 (\frac{\sqrt{3}}{2}) = \sqrt{3}$, $y = 2 \sin \frac{\pi}{6} = 2 (\frac{1}{2}) = 1$.

Therefore, $(2 \cos \frac{\pi}{6}, 2 \sin \frac{\pi}{6}) = (\sqrt{3}, 1)$.

It is common practice to abbreviate trigonometric coordinates to a form known as <u>polar coordinates</u>. In polar form, the trigonometric functions are omitted leaving an ordered pair (r,Θ). In this form, $(2 \cos \frac{\pi}{6}, 2 \sin \frac{\pi}{6})$

given in the example above becomes $(2,\frac{\pi}{6})$. Since the equations of transformation still hold in this system, the polar coordinates $(2,\frac{\pi}{6})$ are equivalent to $(\sqrt{3},1)$ in rectangular form. Here is a good example of the distinction between an equivalence and an equality. Rectangular and trigonometric coordinates are equal but rectangular and polar coordinates are equivalent but not equal. Remember that two ordered pairs (a,b) and (c,d) are equal if and only if $a = c$ and $b = d$.

Care must be exercised in using radian measurement for θ so as not to confuse rectangular form with polar form. For example, it is clear that $(3,64^{\circ})$ is in polar form. If, however, the radian approximation to 64° were given; i.e., $(3,1.117)$, it would have to be made clear that either polar coordinates or rectangular coordinates were implied.

EXAMPLE: Express the polar coordinates $(4,120^{\circ})$ in rectangular form.

SOLUTION: $r = 4$ and $\theta = 120^{\circ}$. Therefore, $x = r \cos \theta = 4 \cos 120^{\circ} = 4(\frac{-1}{2}) = -2$.

$\quad\quad y = r \sin \theta = 4 \sin 120^{\circ} = 4(\frac{\sqrt{3}}{2}) = 2\sqrt{3}$.

Hence $(4,120^{\circ}) \sim (-2,2\sqrt{3})$ where the symbol $\sim$ designates an equivalence.

The conversion from rectangular coordinates to polar coordinates also implies the application of the equations of transformation.

EXAMPLE: Express $(-3\sqrt{2},-3\sqrt{2})$ in polar form.

SOLUTION: $x = -3\sqrt{2}$ and $y = -3\sqrt{2}$.

$\quad\quad r = \sqrt{(-3\sqrt{2})^2 + (-3\sqrt{2})^2} = \sqrt{18 + 18} = \sqrt{36} = 6$

$\quad\quad \theta = \tan^{-1} \frac{(-3\sqrt{2})}{(-3\sqrt{2})} = \tan^{-1} 1$ where θ is in Quadrant III.

Therefore $\theta = \frac{5\pi}{4} + 2k\pi$, $k = 0, \pm1, \pm2, \ldots$

Hence, $(-3\sqrt{2},-3\sqrt{2}) \sim (6,\frac{5\pi}{4}) \sim (6,\frac{-3\pi}{4}) \sim (6,\frac{13\pi}{4}) \sim \ldots$

Again it can be seen that the unique rectangular coordinate pair may be expressed in an infinite number of ways in polar form. In fact, polar coordinates are not even restricted to positive values of r as in the examples above. For example, the polar coordinate pair $(-6,\frac{\pi}{4})$ is also equivalent to $(-3\sqrt{2},-3\sqrt{2})$ in rectangular form as the next example demonstrates.

EXAMPLE: Show that $(-6,\frac{\pi}{4})$ is equivalent to $(-3\sqrt{2},-3\sqrt{2})$ in rectangular form.

SOLUTION: $r = -6$ and $\Theta = \frac{\pi}{4}$.

$$x = r \cos \Theta = (-6) \cos \frac{\pi}{4} = (-6)(\frac{\sqrt{2}}{2}) = -3\sqrt{2}$$

$$y = r \sin \Theta = (-6) \sin \frac{\pi}{4} = (-6)(\frac{\sqrt{2}}{2}) = -3\sqrt{2}$$

Therefore, $(-6,\frac{\pi}{4}) \sim (-3\sqrt{2},-3\sqrt{2})$.

GRAPHING IN POLAR COORDINATES

The graphing of points in the polar coordinate plane corresponds most closely to the circle model depicted for the trigonometric ratios. In particular, all angles are generated in standard position with the initial side of each angle coincident with the polar axis as it is called. For example, the ordered pair $(3,120^{\circ})$ is graphed as the endpoint of a radius vector, a line segment drawn as an arrow that has a magnitude of 3 units and a direction of 120° as shown in Figure 7-1B.

Figure 7-1B

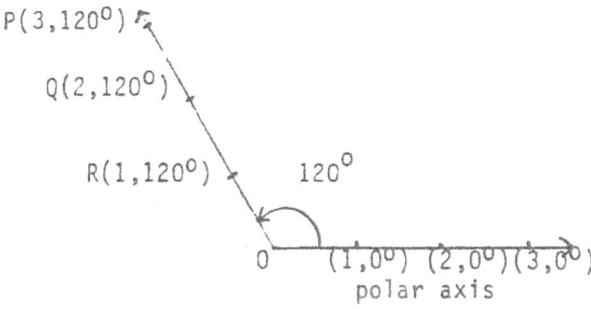

In this system the origin is called the "pole" and points along the polar axis have coordinates (a,0°) or in radians (a,0). To facilitate graphing polar coordinate paper is frequently used with concentric circles and unit radii spaced at convenient angular increments. Several points are shown graphed in Figure 7-1C.

Figure 7-1C

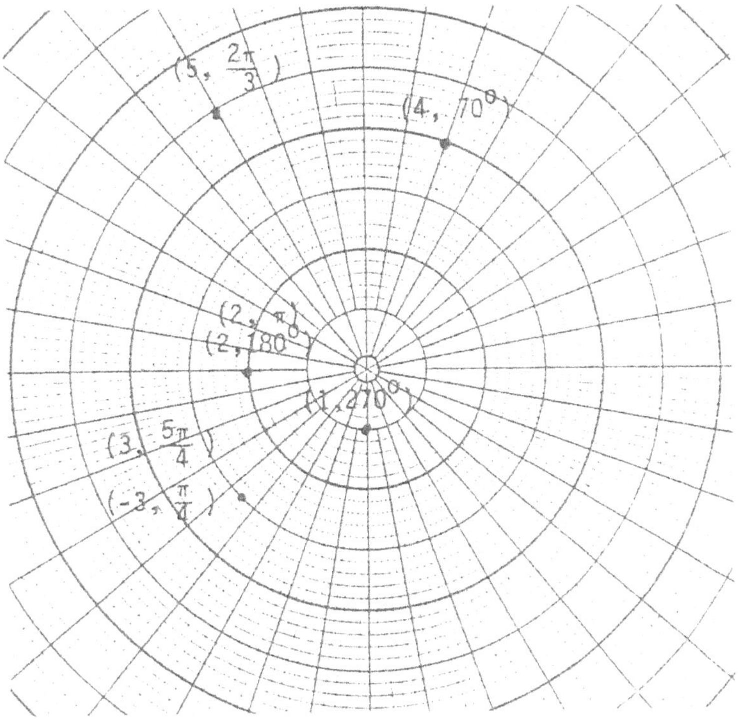

Note that the point corresponding to (-2,0) in the rectangular plane is located here as $(2,\pi)$ or (2,180°). Also note that the point $(-3,\frac{\pi}{4})$ is located at the same position as $(3,\frac{5\pi}{4})$. In other words, $(-3,\frac{\pi}{4})$ may be graphed by rotating three units of the polar axis through an angle of $\frac{\pi}{4}$ and then reflecting this segment through the pole.

EXERCISE SET 7-1.

Express the rectangular coordinates in a)trigonometric form and b)polar form.

Express θ in radians.

1) (2,2)   2) $(3\sqrt{3}, 3)$ 3) $(3\sqrt{3}, -3)$ 4) (4.-3)

5) (-12,5)   6) (-2,-3)   7) (2.8,-4.7)   8) (14,10)

Express the trigonometric coordinates in rectangular form.

9) $(2 \cos \pi , 2 \sin \pi)$ 10) $(3 \cos \frac{\pi}{3} , 3 \sin \frac{\pi}{3})$

11) $(-2\cos \frac{\pi}{3} , -2 \sin \frac{\pi}{3})$ 12) $(5 \cos \frac{7\pi}{6} , 5 \sin \frac{7\pi}{6})$

Express the polar coordinates in rectangular form.

13) $(2, \frac{\pi}{4})$ 14) $(-3, \frac{-2\pi}{3})$ 15) $(4 , 30^0)$ 16) $(7 , \frac{3\pi}{2})$

17) $(-2 , 150^0)$ 18) $(-1 , 180^0)$ 19) $(5 , 110^0)$ 20) $(4 , -75^0)$

Show that the ordered pairs of polar coordinates are equivalent.

21) $(2, \frac{\pi}{3}) , (-2, \frac{4\pi}{3})$ 22) $(4, \frac{-5\pi}{6}), (-4, \frac{\pi}{6})$

23) $(-3, \frac{-\pi}{4}) , (3, \frac{3\pi}{4})$ 24) $(1, \frac{-\pi}{2}) , (-1, \frac{5\pi}{2})$

7-2. COMPLEX NUMBERS AND VECTORS.

Historically the invention of complex numbers was a natural consequence of the inability to find real number solutions to certain equations. The simplest quadratic equation revealing this deficiency is $x^2 + 1 = 0$. For any real number substitution for x, $x^2 + 1 \geq 1$. No real number, therefore, will satisfy the equation $x^2 + 1 = 0$. A straightforward algebraic solution yields the following:

$$x^2 + 1 = 0$$

$$x^2 = -1$$

$$x = \pm\sqrt{-1}$$

Since $\sqrt{-1}$ had no application to measurement in the real world it was simply avoided by earlier civilizations. With the rapid development of the calculus in the 18th and 19th centuries, however, this number took on new significance and was designated as the "imaginary unit."[*]

The imaginary unit is now symbolized by the letter i (for imaginary). That is, $i = \sqrt{-1}$. The two solutions to $x^2 + 1 = 0$ can be written as x = i or x = -i and abbreviated as x = ±i. Other quadratic equations formerly considered unsolvable could now be expressed in terms of numerical expressions of the form $a + bi$.

EXAMPLE: Solve for x, $x^2 + 4x + 13 = 0$.

SOLUTION: By the method of completing the square trinomial,

$$x^2 + 4x = -13$$

$$x^2 + 4x + 4 = -13 + 4 = -9$$

$$(x + 2)^2 = -9 \quad \text{factoring the square trinomial}$$

$$x + 2 = \pm\sqrt{-9} \quad \text{taking square roots}$$

----

[*] By the end of the 19th century, physicists like Steinmetz realized the importance of these numbers in the development of electricity.

$$x = -2 \pm \sqrt{-9} \quad \text{Subtracting 2}$$

Since $\sqrt{-9} = \sqrt{(-1)(9)} = \sqrt{9}\sqrt{-1} = 3\sqrt{-1} = 3i$, it follows that

$$x = -2 \pm 3i .$$

The solutions $-2 + 3i$ and $-2 - 3i$ are examples of <u>complex numbers</u>. When written as $a + bi$ (or $a - bi$) where a and b are real numbers and $i = \sqrt{-1}$, complex numbers are said to be in "rectangular form ."

The quadratic formula can also be applied to solving equations with complex roots.

<u>EXAMPLE</u>: Solve for x, $2x^2 - 6x + 7 = 0$.

<u>SOLUTION</u>: $x = \dfrac{-b \pm \sqrt{b^2 - 4ac}}{2a}$ is the quadratic formula. Substituting a = 2,

b = -6 , and c = 7, $\quad x = \dfrac{-(-6) \pm \sqrt{(-6)^2 - 4(2)(7)}}{2(2)}$

Simplifying, $\quad x = \dfrac{6 \pm \sqrt{36 - 56}}{4} = \dfrac{6 \pm \sqrt{-20}}{4} = \dfrac{6 \pm \sqrt{20}\sqrt{-1}}{4}$

$$= \dfrac{6 \pm 2\sqrt{5}\, i}{4} = \dfrac{3 \pm \sqrt{5}\, i}{2} \text{ or } \dfrac{3}{2} \pm \dfrac{\sqrt{5}\, i}{2}$$

It may be noted that in the quadratic equations above, the complex roots (solutions) occurred in pairs. In general, complex numbers of the form $a + bi$ and $a - bi$ are called <u>complex conjugates</u>. For example, $-2 + 3i$ and $-2 - 3i$ are conjugates. Furthermore, in any quadratic equation with real coefficients, if the quantity $b^2 - 4ac$ under the radical sign of the formula is negative, the roots will be complex conjugates.

<u>EXAMPLE</u>: Determine whether or not the equation $3x^2 - 4x + 7 = 0$ has complex conjugate roots.

<u>SOLUTION</u>: Since $b^2 - 4ac = (-4)^2 - 4(3)(7) = 16 - 84 = -68$ is negative, the roots are complex conjugates.

Early in the 19th century, mathematicians noted a structural correspondence between the a + bi form of complex numbers and ordered pairs of real numbers.* This correspondence allowed the graphing of complex numbers as points in a complex number plane in the same manner that ordered pairs are graphed in the rectangular coordinate plane. Figure 7-2A illustrates this by the graphing of the solutions to the equations above.

Figure 7-2A

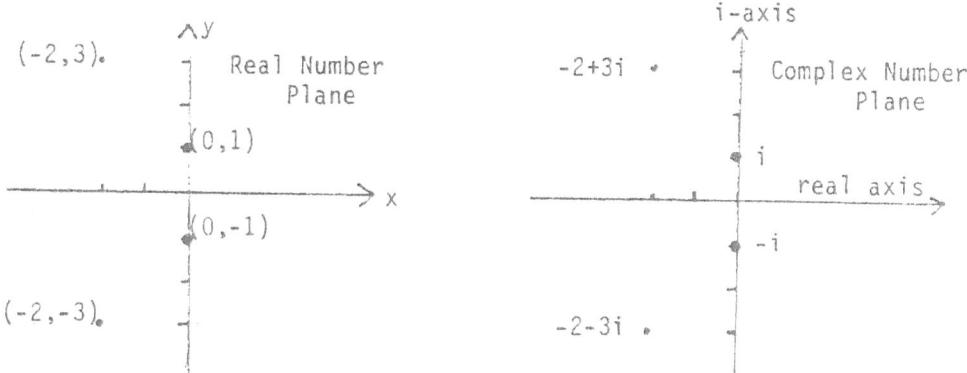

In the complex number plane, units along the real axis can be designated as ..., -2, -1, 0, 1, 2, ... or as ..., -2+0i, -1+0i, 0+0i, 1+0i, 2+0i,... and along the imaginary axis as ..., -2i, -i, 0, i, 2i, ... or as ..., 0-2i, 0-i, 0+0i, 0+i, 0+2i,... . From the structuring of complex numbers it is apparent that the real numbers form a subset of the complex numbers since every real number can be written as a+0i . Numbers of the form bi or 0+bi are called "pure imaginaries."

----

* A structural correspondence of this kind is known as an "isomorphism". A rigorous development of the complex numbers as ordered pairs of real numbers is studied in advanced mathematics.

OPERATIONS WITH PURE IMAGINARY NUMBERS.

Returning to the definition $i = \sqrt{-1}$, it follows that $i^2 = -1$. To preserve this result, the multiplication of square roots of negative reals must be handled differently than multiplication of the square roots of positive reals. For example it is permissible to simplify $\sqrt{4}\ \sqrt{9}$ as follows:

$$\sqrt{4}\ \sqrt{9}\ =\ \sqrt{(4)(9)}\ =\ \sqrt{36}\ =\ 6.$$

If however, the product is $\sqrt{-4}\ \sqrt{-9}$, the following would be incorrect:

$$\sqrt{-4}\ \sqrt{-9}\ =\ \sqrt{(-4)(-9)}\ =\ \sqrt{36}\ =\ 6.$$

To avoid this error, the following steps are recommended:

$$\sqrt{-4}\ \sqrt{-9}\ =\ \sqrt{(4)(-1)}\ \sqrt{(9)(-1)}\ =\ (\ \sqrt{4}\ \sqrt{-1}\ )\ (\sqrt{9}\ \sqrt{-1}\ )\ =\ (2i)(3i)\ =\ (2)(3)\ i^2$$
$$=\ 6\ i^2\ =\ 6\ (-1)\ =\ -6.$$

Additional examples will emphasize the procedure of converting to i-form first.

EXAMPLE: Simplify $\sqrt{-75}$.

SOLUTION: $\sqrt{-75}\ =\ \sqrt{75}\ \sqrt{-1}\ =\ \sqrt{25}\ \sqrt{3}\ i\ =\ 5\sqrt{3}\ i$.

Note: Some authors prefer the answer to be expressed as $5i\sqrt{3}$ so that it is clear that i is not included beneath the radical sign.

EXAMPLE: Simplify the product $\sqrt{-32}\ \sqrt{-8}$.

SOLUTION: $\sqrt{-32}\ \sqrt{-8}\ =\ \sqrt{32}\ i\ \sqrt{8}\ i\ =\ \sqrt{32}\ \sqrt{8}\ i^2\ =\ \sqrt{256}\ i^2\ =\ 16\ i^2\ =\ (16)(-1)\ =\ -16$.

POWERS OF i.

Since $i = \sqrt{-1}$ and $i^2 = -1$, higher powers of i can be obtained by applying the law of exponents.

$i^3 = i^2 (i) = (-1)\ i = -i$

$i^4 = (i^2)(i^2) = (-1)(-1) = 1$

$i^5 = i^4 (i) = (1)\ i = i$

$i^6 = (i^4)(i^2) = (1)(-1) = -1$

$i^7 = i^4(i^3) = (1)(-i) = -i$

$i^8 = (i^4)(i^4) = (1)(1) = 1$

$i^9 = i^8(i) = (1)i = i\ ...$

Study of the foregoing results reveals that if m is a positive integer, i^m is one of the following: i, -1, -i, or 1, and this sequence repeats itself for every four successive powers of i . It is only necessary, therefore, to know that $i^1 = i$, $i^2 = -1$, $i^3 = -i$, and $i^4 = 1$. For completeness and consistency we define $i^0 = 1$. Then for any other value of m, i^m can be determined by dividing m by 4 and considering the remainder only. That is, $i^m = i^r$ where $r = m - 4q$ and q is the partial quotient obtained by dividing m by 4.

EXAMPLE: Find the value of i^{75} .

SOLUTION: $75 \div 4 = 18 \frac{3}{4}$. Since the remainder r = 3, $i^{75} = i^3 = -i$.

      Note that if $r = m - 4q$, substituting gives r = 75 - 4(18) = 75 - 72 = 3.

If the remainder is zero then $i^m = i^0 = 1$.

EXAMPLE: Find the value of i^{216} .

SOLUTION: $216 \div 4 = 54$ and r = 0 . Therefore $i^{216} = i^0 = 1$.

SUMS AND DIFFERENCES.

A complex number in the form a + bi has two parts. The number a is the "real" part and bi is the "imaginary" part. In adding two complex numbers in this form the rule to remember is "add real parts and imaginary parts separately." Formally, (a + bi) + (c + di) = (a + c) + (b + d)i .

EXAMPLE: Find the sum of 3 - 2i and 4 + 5i .

SOLUTION: (3 - 2i) + (4 + 5i) = (3 + 4) + (-2 + 5)i = 7 + 3i .

The same rule applies to subtraction.

EXAMPLE: Find the difference of 24 + 5i and -3 - 17i .

SOLUTION: (24 + 5i) - (-3 - 17i) = (24 - (-3)) + (5 - (-17))i = 27 + 22i .

If the result has a negative imaginary coefficient, simplify as follows:

EXAMPLE: Find the difference, (-2 + 3i) - (4 + 7i) .

SOLUTION: (-2 + 3i) - (4 + 7i) = (-2 - 4) + (3 - 7)i

$$= -6 + (-4)i = -6 - 4i .$$

VECTORS.

Geometric interpretations of addition and subtraction of complex numbers are examples of a mathematical concept that has broad practical applications. It is known as vector addition (or subtraction) and employs the so called parallelogram method. Consider the first example above: (3 - 2i) + (4 + 5i) = 7 + 3i . Figure 7-2B demonstrates the method.

Figure 7-2B

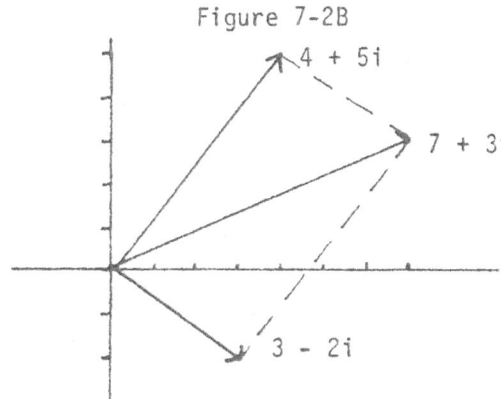

The arrows drawn from the origin to the points in the complex number plane are called vectors. If a parallelogram is constructed with the addend vectors as adjacent sides, the diagonal drawn from the origin will be the sum vector. In vector terminology, the addend vectors are called "components" and the sum vector is called the "resultant".

Subtraction is accomplished by remembering that it is the inverse operation of addition. The resultant vector behaves as the minuend in the subtraction format: minuend (resultant) - subtrahend (component) = difference (component).

EXAMPLE: Show the result of the subtraction $(-2 + 3i) - (4 + 7i)$ geometrically.

SOLUTION: $(-2 + 3i) - (4 + 7i) = -6 - 4i$ as shown above.

First draw the vectors $-2 + 3i$ and $4 + 7i$ and connect the arrow heads with a vector as shown in Figure 7-2C.

Figure 7-2C

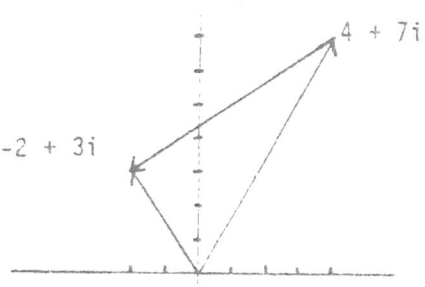

Since $-2 + 3i$ is the resultant diagonal vector, the connecting vector must be translated to the origin to form the opposite side of the parallelogram as in Figure 7-2D.

Figure 7-2D

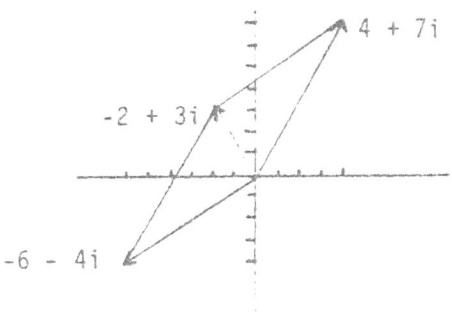

The remaining side is drawn to complete the parallelogram. By the arithmetic of complex numbers, $(-2 + 3i) - (4 + 7i) = -6 - 4i$ and $(-6 - 4i) + (4 + 7i) = -2 + 3i$ as confirmed by the parallelogram of vectors.

As a special sum, consider $(2 + 5i) + (2 - 5i) = 4 + 0i = 4$.

Figure 7-2E

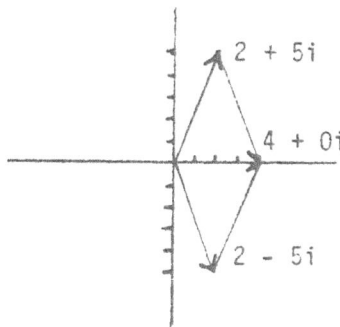

The vector sum shows that the sum of complex conjugates is a real number. Figure 7-2E shows this resultant sum and illustrates that conjugate vectors are symmetric with respect to the real axis.

EXAMPLE: Prove that the difference of complex conjugates is a pure imaginary.

SOLUTION: Let a + bi and a - bi be complex conjugates. Then,

$$(a + bi) - (a - bi) = (a - a) + (b - (-b))i = 0 + 2bi = 2bi .$$

THE LENGTH (MAGNITUDE) OF A VECTOR IN THE COMPLEX PLANE.

In the real number plane the distance of a point from the origin is determined by application of the distance formula.

EXAMPLE: Find the distance of P(-3,4) from the origin (0,0).

SOLUTION: $d(OP) = \sqrt{(-3 - 0)^2 + (4 - 0)^2} = \sqrt{(-3)^2 + (4)^2} = \sqrt{9 + 16} = \sqrt{25} = 5$.

Now consider the point -3 + 4i in the complex number plane. The length of the vector from the origin to the point corresponding to -3 + 4i is determined in exactly the same manner. In general, if a + bi is a complex number, the length or magnitude of the vector from the origin to the point corresponding to a + bi is given by $\sqrt{a^2 + b^2}$. For complex numbers of the form a + bi, the length or magnitude of its vector is symbolized by $|a + bi|$.

EXAMPLE: Find $|-2 - 5i|$.

SOLUTION: $|-2 - 5i| = \sqrt{(-2)^2 + (-5)^2} = \sqrt{4 + 25} = \sqrt{29}$.

Note that in finding the length of the vector corresponding to the point a + bi, only the real coefficients a and b enter into the computations. By definition, the imaginary unit i does not enter.

EXERCISE SET 7-2.

Solve the following equations:

1) $x^2 + 2x + 5 = 0$ 2) $x^2 - 4x + 13 = 0$ 3) $9x^2 - 2x + 3 = 0$

4) $4x^2 - 3x + 10 = 0$ 5) $4x^2 - 2x + 7 = 0$ 6) $9x^2 + 8 = 0$

7) $5x^2 - 4x + 5 = 0$ 8) $3x^2 + 2x + 1 = 0$

Simplify the following:

9) $\sqrt{-8}\ \sqrt{-2}$ 10) $\sqrt{-25}\ \sqrt{-4}$ 11) $\sqrt{49}\ \sqrt{-16}$ 12) $\sqrt{-32}$

13) $\sqrt{-50} + \sqrt{-2}$ 14) $\sqrt{-72} - \sqrt{-98}$ 15) $\sqrt{-1} + \sqrt{-4}$

16) $5\ \sqrt{-3}\ (\sqrt{-3} - \sqrt{-12}\ )$ 17) i^{67} 18) i^{142} 19) i^{59} 20) $i^{47}\ i^{35}$

Express the following in a + bi form:

21) $(2 + 5i) + (-7 - 13i)$ 22) $(-4 - i) + (17 + 6i)$ 23) $(6 + 5i) + 2i$

24) $(11 - 19i) - (-5 - 2i)$ 25) $(3 - 2i) + (3 + 2i)$ 26) $(3i - 5) + (2 + 6i)$

27) $(\frac{2}{3} + \frac{4i}{3}) - (\frac{-2}{5} + \frac{7i}{6})$ 28) $\frac{6 + 7i}{3} + \frac{-2 - 5i}{4}$

Find the magnitudes of the given complex numbers:

29) $-3 + 5i$ 30) $12 - 4i$ 31) $11 + 7i$ 32) $-9 - 3i$

Express the following in a + bi form:

33) $|-2 + 6i|$ 34) $|2 - 5i| \cdot |6 - 3i|$ 35) $|3 + i|(2) + 4i$

36) $|-i|$ 37) $|3.4 + 5i|$ 38) $|(-3) + (7 + 2i)|$

Find the following sums and differences by the parallelogram method for vectors.

Draw the appropriate diagram of vectors for each exercise:

39) $(-3 + 2i) + (-1 - i)$ 40) $(2 - 5i) + (6 + 2i)$

41) $(5 + i) - (2 - 6i)$ 42) $(1 + i) - (3 - i)$

43) $(4 + i) + (4 - i)$ 44) $(-6 + 3i) - (-6 - 3i)$

The parallelograms of vectors below represent vector sums and differences.

Find the magnitude of the vector v in each exercise:

45)   46)

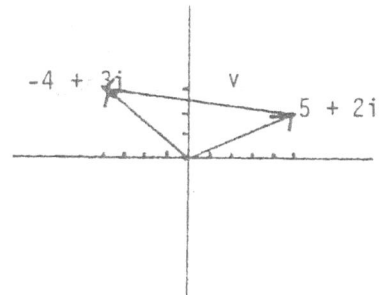   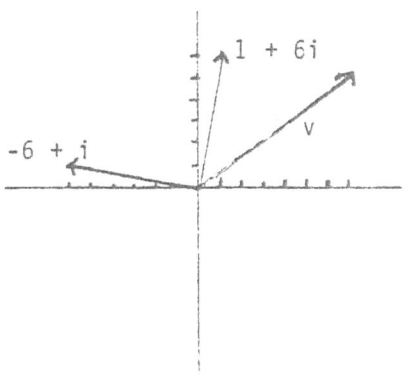

47)   48)

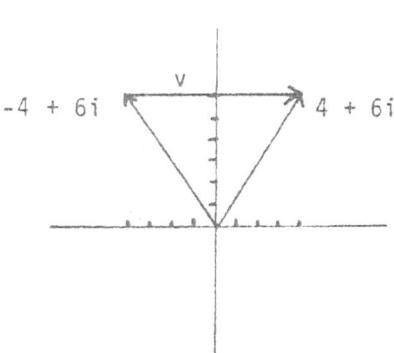   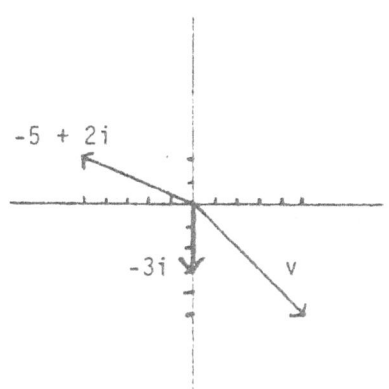

7-3. RECTANGULAR AND TRIGONOMETRIC FORMS OF COMPLEX NUMBERS. POLAR FORM.
MULTIPLYING AND DIVIDING COMPLEX NUMBERS.

SCALAR MULTIPLICATION

The multiplication of complex numbers is a direct application of the distributive law. If t is a real number, $t(a + bi) = at + bti$.

EXAMPLE: Find the product of 2 and $-1 + 3i$ and show the geometric results in the complex plane.

SOLUTION: $2(-1 + 3i) = -2 + 6i$.

Figure 7-3A

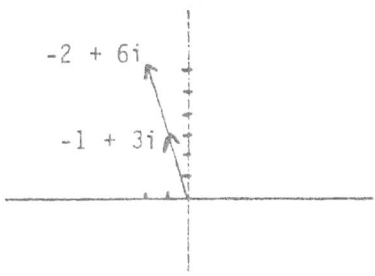

Multiplying by 2 has the effect of stretching the vector for $-1 + 3i$ to twice its original length. That is, $|-2 + 6i| = \sqrt{(-2)^2 + (6)^2} = \sqrt{4 + 36}$

$$= \sqrt{4(1 + 9)} = 2\sqrt{(-1)^2 + (3)^2}$$

$$= 2|-1 + 3i|$$

This is an example of what is called "scalar multiplication" in the terminology of vectors. If the real number multiplier, called a "scalar", is between 0 and 1, the original vector will be contracted in the same direction. If the scalar is negative, the product vector will be reflected through the origin in the opposite direction.

EXAMPLE: Find the product $\left(\frac{-2}{3}\right)(6 - 3i)$ and show the scalar product geometrically.

SOLUTION: $\left(\frac{-2}{3}\right)(6 - 3i) = -4 + 2i$.

Figure 7-3B

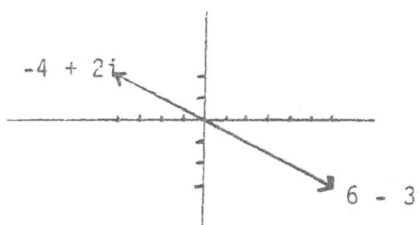

The scalar product -4 + 2i is represented by a vector that is two-thirds the length of the vector representation for 6 - 3i and is pointed in the opposite direction.

MULTIPLICATION BY i .

Given the complex number a + bi, the product (a + bi)i is found as follows: $(a + bi)i = ai + bi^2 = ai + b(-1) = ai - b = -b + ai$.

EXAMPLE: Express the product (3 + 4i)i in a + bi form and plot the result.

SOLUTION: $(3 + 4i)i = 3i + 4i^2 = 3i + 4(-1) = 3i - 4 = -4 + 3i$.

Figure 7-3C

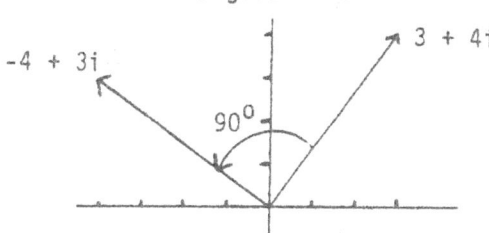

The vector result is significant in that it shows that multiplying a complex number by i will rotate the original vector through an angle of 90°. The length of the product vector is identical to the original;i.e.,

$$|-4 + 3i| = |3 + 4i| = \sqrt{(-4)^2 + 3^2} = \sqrt{25} = 5 .$$

THE PRODUCT (a + bi)(c + di) .

Recall from algebra the method of multiplying binomials.

EXAMPLE: Find the product (3x - 5)(2x + 7).

SOLUTION: The so-called FOIL method is applied where

F is the product of the <u>First</u> terms, $(3x)(2x) = 6x^2$,

O is the product of the <u>Outer</u> terms, $(3x)(7) = 21x$,

I is the product of the <u>Inner</u> terms, $(-5)(2x) = -10x$,

L is the product of the <u>Last</u> terms, $(-5)(7) = -35$.

Combining the <u>Outer</u> and <u>Inner</u> products yields $21x - 10x = 11x$.

Arranging the product as a polynomial in standard form,

$$(3x - 5)(2x + 7) = 6x^2 + 11x - 35.$$

The same method applies to the multiplication of two complex numbers.

EXAMPLE: Find the product $(3 - 5i)(2 + 7i)$ in $a + bi$ form.

SOLUTION: $F = (3)(2) = 6$

$O = (3)(7i) = 21i$

$I = (-5i)(2) = -10i$

$L = (-5i)(7i) = -35i^2 = (-35)(-1) = 35$

The difference between this product and the product of binomials described above is that F and L are combined to form a and O and I are combined to form bi. Therefore, $(3 - 5i)(2 + 7i) = 41 + 11i$.

In general, $(a + bi)(c + di) = (ac - bd) + (ad + bc)i$.

Another example will demonstrate the effect on the length of the product vector.

EXAMPLE: Graph the vector product of $(1 + 2i)(3 - i)$.

SOLUTION: $(1 + 2i)(3 - i) = 3 - i + 6i + 2 = 5 + 5i$.

Figure 7-3D

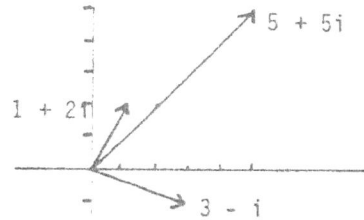

The respective vector magnitudes are:

$$|1 + 2i| = \sqrt{1^2 + 2^2} = \sqrt{5}$$

$$|3 - i| = \sqrt{3^2 + (-1)^2} = \sqrt{10}$$

$$|5 + 5i| = \sqrt{5^2 + 5^2} = \sqrt{50} = 5\sqrt{2}$$

It follows that, in general, the product of the lengths of two vectors in the complex plane is equal to the length of the product vector. This statement is formalized as follows: Let $z_1 = a + bi$ and $z_2 = c + di$, then,

$$|z_1||z_2| = |z_1 z_2|.$$

THE DIRECTION OF A VECTOR.

A vector has two unique properties: length and direction. The direction is defined by an angle in standard position. In the complex number plane this angle is measured with its initial side coincident with the positive direction of the real number axis and its terminal side coincident with the vector as in Figure 7-3E.

Figure 7-3E

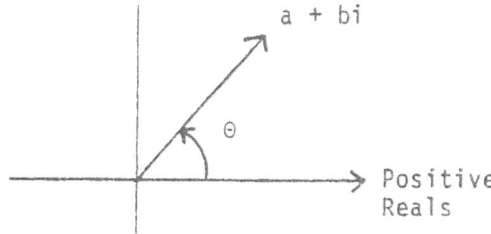

The angle Θ may be detemined in the same manner as the polar coordinate Θ in the real number plane:

$$\Theta = \tan^{-1} \frac{b}{a}$$

In fact, there is a direct correspondence between the polar coordinates of a point in the real number plane (r, Θ) and the length and direction of a vector in the complex number plane ($|a + bi|$, Θ). The parallel is drawn below:

Polar Coordinates	Vector in the Complex Number Plane
Given P(x,y),	Given a + bi,
$r = \sqrt{x^2 + y^2}$	$\|a + bi\| = \sqrt{a^2 + b^2}$
$\Theta = \tan^{-1} \dfrac{y}{x}$	$\Theta = \tan^{-1} \dfrac{b}{a}$

Figure 7-3F

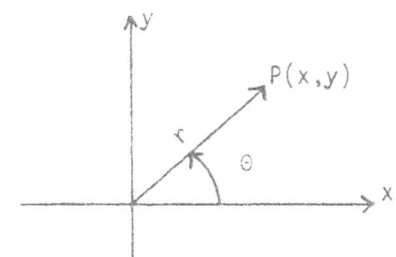 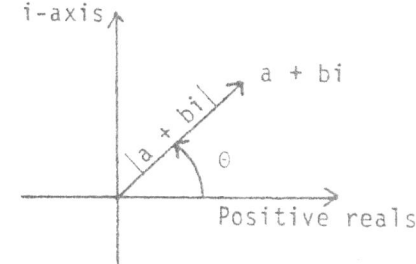

POLAR AND TRIGONOMETRIC FORMS OF A COMPLEX NUMBER.

The equations of transformation between rectangular coordinates and polar coordinates are identical to the equations giving the length and direction of vectors. It is because of this identity that polar coordinates (r, Θ) may be identified as the polar form of a complex number. Furthermore, if $r = |a + bi|$ then $a = r \cos \Theta$ and $b = r \sin \Theta$. From these equations comes the trigonometric form of a complex number:

$$a + bi = r \cos \Theta + (r \sin \Theta)i = r(\cos \Theta + i \sin \Theta).$$

EXAMPLE: Express 1 + 2i in polar and trigonometric forms.

SOLUTION: a = 1 and b = 2. $r = \sqrt{1^2 + 2^2} = \sqrt{5}$. $\Theta = \tan^{-1} \dfrac{2}{1} = 63^\circ 26'$.

Therefore, $1 + 2i \backsim (\sqrt{5}, 63^\circ 26') \backsim \sqrt{5}(\cos 63^\circ 26' + i \sin 63^\circ 26')$.

EXAMPLE: Express $2(\cos 120^\circ + i \sin 120^\circ)$ in a + bi form.

SOLUTION: $2(\cos 120^\circ + i \sin 120^\circ) = 2((\dfrac{-1}{2}) + i(\dfrac{\sqrt{3}}{2})) = -1 + i\sqrt{3}$ or $-1 + \sqrt{3} i$.

EXAMPLE: Express $(5\sqrt{2}, 45^\circ)$ in a + bi form.

SOLUTION: $(5\sqrt{2},\ 45^\circ) \sim 5\sqrt{2}(\cos 45^\circ + i \sin 45^\circ) = 5\sqrt{2}(\frac{1}{\sqrt{2}} + \frac{1}{\sqrt{2}} i) = 5 + 5i.$

THE PRODUCT OF TWO COMPLEX NUMBERS IN TRIGONOMETRIC FORM.

Referring to Figure 7-3D above, $5 + 5i$ was found to be the product of $(1 + 2i)$ and $(3 - i)$. Now, $1 + 2i = \sqrt{5}(\cos 63^\circ 26' + i \sin 63^\circ 26')$ and

$$3 - i = \sqrt{10}(\cos(-18^\circ 26') + i \sin(-18^\circ 26')).$$

The product $5 + 5i = 5\sqrt{2}(\cos 45^\circ + i \sin 45^\circ).$

It has already been shown that the length of the product vector $5 + 5i$ is equal to the <u>product</u> of the lengths of the factors $1 + 2i$ and $3 - i$. That is, $5\sqrt{2} = \sqrt{5}\ \sqrt{10}$. It is equally important to see that the direction angle of the product vector is equal to the <u>sum</u> of the direction angles of the factor vectors. That is, $45^\circ = 63^\circ 26' + (-18^\circ 26')$. This is an example of the following generalization: Given $a + bi = r_1(\cos \Theta_1 + i \sin \Theta_1)$ and

$$c + di = r_2(\cos \Theta_2 + i \sin \Theta_2),$$

$$(a + bi)(c + di) = r_1 r_2(\cos (\Theta_1 + \Theta_2) + i \sin(\Theta_1 + \Theta_2)).$$

EXAMPLE: Find the product of $3(\cos 30^\circ + i \sin 30^\circ)$ and $2(\cos 60^\circ + i \sin 60^\circ)$ in $a + bi$ form.

SOLUTION: $3(\cos 30^\circ + i \sin 30^\circ) \cdot 2(\cos 60^\circ + i \sin 60^\circ)$

$= (3)(2)(\cos(30^\circ + 60^\circ) + i \sin(30^\circ + 60^\circ))$

$= 6(\cos 90^\circ + i \sin 90^\circ) = 6(0 + i(1)) = 6(0 + i) = 0 + 6i$ or $6i.$

As a check, convert the factors to $a + bi$ form first and find the product by the FOIL method.

$3(\cos 30^\circ + i \sin 30^\circ) = 3(\frac{\sqrt{3}}{2} + \frac{1}{2} i) = \frac{3\sqrt{3}}{2} + \frac{3}{2} i$

$2(\cos 60^\circ + i \sin 60^\circ) = 2(\frac{1}{2} + \frac{\sqrt{3}}{2} i) = 1 + \sqrt{3}\ i$

Hence, $(\frac{3\sqrt{3}}{2} + \frac{3}{2} i)(1 + \sqrt{3}\ i) = \frac{3\sqrt{3}}{2} + \frac{9}{2} i + \frac{3}{2} i - \frac{3\sqrt{3}}{2} = (\frac{9}{2} + \frac{3}{2})i = 6i$

The equivalence of trigonometric form and polar form permits an abbreviation in the notation involved in finding products.

EXAMPLE: Find the product of $(-5)(\cos 72^0 + i \sin 72^0)$ and

$3(\cos (-30^0) + i \sin (-30^0))$.

SOLUTION: $(-5)(\cos 72^0 + i \sin 72^0) \sim (-5,72^0)$ and $3(\cos (-30^0) + i \sin (-30^0))$

$\sim (3,-30^0)$. Therefore, the product may be written as $(-5,72^0)(3,-30^0)$

$= (-15,42^0) = -15(\cos 42^0 + i \sin 42^0)$ in trigonometric form.

EXAMPLE: Express the product $(4,\frac{\pi}{6})(5, \frac{2\pi}{3})$ in trigonometric form.

SOLUTION: $(4, \frac{\pi}{6})(5, \frac{2\pi}{3}) = (20, \frac{5\pi}{6}) \sim 20(\cos \frac{5\pi}{6} + i \sin \frac{5\pi}{6})$.

r cis θ form.

Another abbreviation frequently employed for trigonometric form is

$$r(\cos \theta + i \sin \theta) = r \text{ cis } \theta$$

where the "cis" part telescopes that portion of the trigonometric form located within the parenthesis. Thus, $4 \text{ cis } \frac{\pi}{3}$ is the same as $4(\cos \frac{\pi}{3} + i \sin \frac{\pi}{3})$.

EXAMPLE: Find the product of $3 \text{ cis } \frac{\pi}{9}$ and $2 \text{ cis } \frac{5\pi}{4}$ in a + bi form.

SOLUTION: $(3 \text{ cis } \frac{\pi}{9})(2 \text{ cis } \frac{5\pi}{4}) = 6 \text{ cis } (\frac{\pi}{9} + \frac{5\pi}{4}) = 6 \text{ cis } \frac{49\pi}{36}$

$= 6(\cos \frac{49\pi}{36} + i \sin \frac{49\pi}{36}) = 6(-.4226 + i(-.9063)) = -2.536 - 5.438 \text{ i }$.

EXAMPLE: Express $3\sqrt{3} - 3i$ in r cis θ form where θ is in degrees.

SOLUTION: $a = 3\sqrt{3}$, $b = -3$. $r = \sqrt{(3\sqrt{3})^2 + (-3)^2} = \sqrt{27 + 9} = \sqrt{36} = 6$.

$\theta = \tan^{-1}(\frac{-1}{\sqrt{3}}) = 330^0$. Hence, $3\sqrt{3} - 3i = 6 \text{ cis } 330^0$.

THE QUOTIENT OF TWO COMPLEX NUMBERS.

If z_1 and z_2 represent two complex numbers and $z_2 \neq 0$, then z_1/z_2 is also a complex number. If z_1 and z_2 are in $a + bi$ form, finding their quotient makes use of an important property of conjugates. It is easy to show that

$$(a + bi)(a - bi) = a^2 + b^2$$

That is, the product of complex conjugates <u>is a real number</u> and that number is the sum of the squares of the real coefficients. The following example demonstrates the use of this fact in finding a quotient.

<u>EXAMPLE</u>: Express $\dfrac{2 + 3i}{-1 - 4i}$ in a + bi form.

<u>SOLUTION</u>: The conjugate of the divisor is $-1 + 4i$. Multiplying numerator and denominator by the conjugate is the same as multiplying by 1. It does not change the value of the fraction:

$$\frac{2 + 3i}{-1 - 4i} \cdot \frac{-1 + 4i}{-1 + 4i} = \frac{(2 + 3i)(-1 + 4i)}{(-1 - 4i)(-1 + 4i)} = \frac{-14 + 5i}{17} = \frac{-14}{17} + \frac{5}{17}i$$

With practice the student will be able to write down the denominator without hesitation since the product of the conjugates is equal to the sum of the squares of a and b. The numerator is obtained as before using the FOIL method.

<u>EXAMPLE</u>: Express $\dfrac{2 + 5i}{2 - 5i}$ in a + bi form.

<u>SOLUTION</u>: $\dfrac{2 + 5i}{2 - 5i} \cdot \dfrac{2 + 5i}{2 + 5i}$ ← These must $= \dfrac{(2 + 5i)(2 + 5i)}{(2 - 5i)(2 + 5i)} = \dfrac{-21 + 20i}{29}$
← be the same

These are conjugates

$= \dfrac{-21}{29} + \dfrac{20}{29}i$.

If the divisor is a pure imaginary, the simplest method is to multiply numerator and denominator by i .

<u>EXAMPLE</u>: Express $\dfrac{2 - i}{3i}$ in a + bi form.

SOLUTION: $\dfrac{2 - i}{3i} \cdot \dfrac{i}{i} = \dfrac{(2 - i)(i)}{(3i)(i)} = \dfrac{2i - i^2}{3i^2} = \dfrac{2i + 1}{-3} = \dfrac{1 + 2i}{-3} = \dfrac{-1}{3} - \dfrac{2}{3} i$

THE QUOTIENT OF TWO COMPLEX NUMBERS IN TRIGONOMETRIC FORM.

Since multiplication and division are inverse operations, this will be reflected in the process of finding the quotient of two complex numbers in trigonometric form.

Let $z_1 = r_1(\cos \Theta_1 + i \sin \Theta_1)$ and $z_2 = r_2(\cos \Theta_2 + i \sin \Theta_2)$ where $r_2 \neq 0$.
Then $z_1/z_2 = (r_1/r_2)(\cos(\Theta_1 - \Theta_2) + i \sin(\Theta_1 - \Theta_2))$.

EXAMPLE: Let $z_1 = 6(\cos 72^0 + i \sin 72^0)$ and $z_2 = 2(\cos 42^0 + i \sin 42^0)$.

Find z_1/z_2 in trigonometric form.

SOLUTION: $z_1/z_2 = \dfrac{6(\cos 72^0 + i \sin 72^0)}{2(\cos 42^0 + i \sin 42^0)} = \dfrac{6}{2} (\cos (72^0 - 42^0) + i \sin(72^0 - 42^0))$

$$= 3 (\cos 30^0 + i \sin 30^0).$$

The result can be checked by multiplication since the product of the quotient and the divisor is equal to the dividend.

CHECK: $3(\cos 30^0 + i \sin 30^0) \cdot 2(\cos 42^0 + i \sin 42^0)$

$= (3)(2)(\cos(30^0 + 42^0) + i \sin(30^0 + 42^0)) = 6(\cos 72^0 + i \sin 72^0)$.

The next example shows the use of the abbreviated "cis" notation in division.

EXAMPLE: Show that $\dfrac{8 \text{ cis } 270^0}{2 \text{ cis } 120^0}$ yields equivalent results in trigonometric form

and rectangular form.

SOLUTION: $\dfrac{8 \text{ cis } 270^0}{2 \text{ cis } 120^0} = 4 \text{ cis}(270^0 - 120^0) = 4 \text{ cis } 150^0$

$$= 4(\cos 150^0 + i \sin 150^0) = 4(\dfrac{-\sqrt{3}}{2} + \dfrac{1}{2} i) = -2\sqrt{3} + 2i.$$

Now, 8 cis 270^0 = 8(cos 270^0 + i sin 270^0)= 8(0 + i(-1)) = -8i

and 2 cis 120^0 = 2(cos 120^0 + i sin 120^0) = $2(\frac{-1}{2} + i(\frac{\sqrt{3}}{2}))$ = -1 + $\sqrt{3}$ i.

Therefore, $\frac{8 \text{ cis } 270^0}{2 \text{ cis } 120^0}$ = $\frac{-8i}{-1 + \sqrt{3}i}$ = $\frac{-8i}{-1 + \sqrt{3}i}$ · $\frac{-1 - \sqrt{3}i}{-1 - \sqrt{3}i}$

$$= \frac{(-8i)(-1 - \sqrt{3}i)}{(-1 + \sqrt{3}i)(-1 - \sqrt{3}i)} = \frac{(-8i)(-1 - \sqrt{3}i)}{4}$$

$$= (-2i)(-1 - \sqrt{3}i) = 2i - 2\sqrt{3} = -2\sqrt{3} + 2i$$

and this checks with the result obtained above.

EXERCISE SET 7-3.

Find the products in a + bi form:

1) 6(-4 + 5i)   2) (-7)(-3 - 6i)   3) (2i)(-7 + 6i)   4) $(\frac{3}{5}i)$(-10 + 15i)

5) (-2i)(5i + 7)   6) (-2 - 5i)(1 + 2i) 7) $(\sqrt{3}$ - 2i)(2 - i) 8) $(5 - \sqrt{2}$ i$)^2$

9) $(-\sqrt{2} + \sqrt{2}$ i)(-2 - $\sqrt{2}$ i)   10) $(\frac{\sqrt{3}}{2} - \frac{1}{2}$ i$)(\frac{\sqrt{3}}{2} + \frac{1}{2}i)$ 11) $(\frac{\sqrt{2}}{2} + \frac{\sqrt{2}}{2}$ i$)^2$

12) $(2 + i)^3$ 13) (4 + 3i)(2 + 5i) 14) (-7 + i)(3 - 2i) 15) $(1 - i)^4$

16) $(\sqrt{3} + i)^3$

Find the magnitudes and direction angles of the vector products in the odd problems above for exercises 17-24.

Express the following in a + bi form:

25) 3(cos 60^0 + i sin 60^0)   26) (-2)(cos $\frac{5\pi}{6}$ + i sin $\frac{5\pi}{6}$)

27) 4(cos $\frac{3\pi}{2}$ + i sin $\frac{3\pi}{2}$)   28) 5(cos 120^0 + i sin 120^0)

29) $(2,45^0)$ 30) $(4,330^0)$ 31) $(-3,150^0)$ 32) $(-3,-180^0)$

33) $(7,\frac{-2\pi}{3})$ 34) $(6,\frac{-7\pi}{6})$ 35) 4(cos 28^0 + i sin 28^0)

36) (-2)(cos 115^0 + i sin 115^0)

Express the following in a) trigonometric form and b) polar form:

37) $3 - 3i$ 38) $\sqrt{3} + i$ 39) $-2 + 2\sqrt{3}\ i$ 40) $-6\sqrt{2} + 6\sqrt{2}\ i$

41) $2 + \sqrt{3}\ i$ 42) $5 - 6i$ 43) $1 + 2i$ 44) $-3 - 4i$

Find the products and express them in trigonometric form:

45) $5(\cos 72^{0} + i \sin 72^{0}) \cdot 3(\cos 46^{0} + i \sin 46^{0})$

46) $(-3)(\cos 254^{0} + i \sin 254^{0}) \cdot 4(\cos (-142^{0}) + i \sin (-142^{0}))$

47) $4(\cos \frac{5\pi}{6} + i \sin \frac{5\pi}{6}) \cdot 2(\cos \frac{6\pi}{5} + i \sin \frac{6\pi}{5})$

48) $(-5)(\cos 3.84 + i \sin 3.84) \cdot (-7)(\cos 4.63 + i \sin 4.63)$

49) $(2,78^{0}) \cdot (-4,53^{0})$ 50) $(2,\frac{\pi}{8}) \cdot (5,\frac{4\pi}{5})$ 51) $3 \text{ cis } \frac{\pi}{9} \cdot 7 \text{ cis } \frac{2\pi}{3}$

52) $(-6)\text{cis } 85^{0} \cdot 2 \text{ cis } (-37^{0})$

Find the quotients in a + bi form:

53) $\dfrac{-2 + 5i}{1 - 2i}$ 54) $\dfrac{2i}{3 - 5i}$ 55) $\dfrac{1 + 3i}{2 - i}$ 56) $\dfrac{2 - \sqrt{3}\ i}{\sqrt{3} + 2i}$

Find the quotients and express them in trigonometric form:

57) $\dfrac{36(\cos 192^{0} + i \sin 192^{0})}{4(\cos 48^{0} + i \sin 48^{0})}$ 58) $\dfrac{15(\cos 330^{0} + i \sin 330^{0})}{5(\cos 90^{0} + i \sin 90^{0})}$

59) $\dfrac{27(\cos \frac{4\pi}{3} + i \sin \frac{4\pi}{3})}{6(\cos \frac{\pi}{4} + i \sin \frac{\pi}{4})}$ 60) $\dfrac{324(\cos \frac{5\pi}{8} + i \sin \frac{5\pi}{8})}{64(\cos \frac{\pi}{2} + i \sin \frac{\pi}{2})}$

61) $\dfrac{(-18,420^{0})}{(9, 240^{0})}$ 62) $\dfrac{(7.46,75.8^{0})}{(2.47, 9.6^{0})}$ 63) $\dfrac{24 \text{ cis } \frac{3\pi}{14}}{6 \text{ cis } \frac{\pi}{7}}$ 64) $\dfrac{278 \text{ cis } \frac{2\pi}{9}}{48 \text{ cis } \frac{5\pi}{3}}$

Show that the quotients below are equivalent in a + bi form and trigonometric form:

65) $\dfrac{12 \text{ cis } 240^{0}}{3 \text{ cis } 90^{0}}$ 66) $\dfrac{(21,\frac{5\pi}{6})}{(-3,\frac{4\pi}{3})}$ 67) $\dfrac{-2\sqrt{3} - 2i}{\sqrt{3} + i}$ 68) $\dfrac{16 - 16i}{-1 - i}$

7-4. POWERS AND ROOTS OF COMPLEX NUMBERS. DE MOIVRE'S THEOREM.

The simple rule for multiplication of complex numbers in trigonometric form leads to an important generalization for raising such numbers to any natural number power. We can leap to this generalization by induction:

Let $z = r(\cos \theta + i \sin \theta)$ be a complex number.

Then $z^2 = (r(\cos \theta + i \sin \theta))^2 = r(\cos \theta + i \sin \theta)\cdot(r)(\cos \theta + i \sin \theta)$

$\quad\quad = r^2(\cos(\theta + \theta) + i \sin(\theta + \theta)) = r^2(\cos 2\theta + i \sin 2\theta)$

and $\quad z^3 = z^2 z = r^2(\cos 2\theta + i \sin 2\theta)\cdot r(\cos \theta + i \sin \theta)$

$\quad\quad = r^3(\cos(2\theta + \theta) + i \sin(2\theta + \theta)) = r^3(\cos 3\theta + i \sin 3\theta).$

Continuing in this manner it can be seen that if n is a natural number,

$$z^n = r^n(\cos n\theta + i \sin n\theta).$$

This theorem was known as early as the beginning of the 18th century and is attributed to the French mathematician, Abraham De Moivre.

EXAMPLE: Express $(3(\cos 25^{\circ} + i \sin 25^{\circ}))^4$ in trigonometric form.

SOLUTION: By De Moivre's theorem, $(3(\cos 25^{\circ} + i \sin 25^{\circ}))^4$

$$= 3^4(\cos (4)(25^{\circ}) + i \sin (4)(25^{\circ}))$$

$$= 81(\cos 100^{\circ} + i \sin 100^{\circ}).$$

EXAMPLE: Express $(1 + i)^5$ in a + bi form.

SOLUTION: Without De Moivre's theorem, expanding $(1 + i)^5$ is a lengthy exercise.

Instead, since $(1 + i) = \sqrt{2}(\cos 45^{\circ} + i \sin 45^{\circ})$,

$$(1 + i)^5 = (\sqrt{2})^5(\cos (5)(45^{\circ}) + i \sin (5)(45^{\circ}))$$

$$= 4\sqrt{2}(\cos 225^{\circ} + i \sin 225^{\circ})$$

$$= 4\sqrt{2}(\frac{-1}{\sqrt{2}} + i(\frac{-1}{\sqrt{2}})) = -4 + (-4)i$$

$$= -4 - 4i$$

ROOTS OF COMPLEX NUMBERS.

Finding square roots, cube roots, fourth roots, etc. of complex numbers can be simplified by framing the search for such roots in the form of equations. For example, in algebra it is learned that if x is a square root of 4, then $x^2 = 4$. The solution is obtained by taking square roots, so that if

$$x^2 = 4$$

$$\sqrt{x^2} = \sqrt{4}$$

$$|x| = 2$$

Hence, x = 2 or x = -2

Suppose $x^2 = -4$. Then, following the solution given above, x = 2i or x = -2i. A simple check will confirm that $(2i)^2 = 4i^2 = -4$ and $(-2i)^2 = 4i^2 = -4$.

Now consider finding the square roots of 4i. Solving in the same manner,

$$x^2 = 4i$$

$$\sqrt{x^2} = \sqrt{4i}$$

$$|x| = 2\sqrt{i}$$

Hence, x = $2\sqrt{i}$ or x = $-2\sqrt{i}$.

Unfortunately, this solution is unsatisfactory as it does not express the square roots of 4i in a + bi form. To do so another tactic is employed resorting to trigonometric form. If $x^2 = 4i$ then $x^2 = 4(\cos 90^0 + i \sin 90^0)$ and the two square roots must satisfy De Moivre's theorem. That is, if $r(\cos \theta + i \sin \theta)$ is a square root of $4(\cos 90^0 + i \sin 90^0)$ then

$$r^2(\cos 2\theta + i \sin 2\theta) = 4(\cos 90^0 + i \sin 90^0)$$

and by equating corresponding parts, $r^2 = 4$ and $2\theta = 90^0$. Hence r = ±2 and $\theta = 45^0$. The square roots of 4i are $2(\cos 45^0 + i \sin 45^0) = \sqrt{2} + \sqrt{2} i$ and $-2(\cos 45^0 + i \sin 45^0) = -\sqrt{2} - \sqrt{2} i$.

Another example will help to develop a generalized method.

EXAMPLE: Find the cube roots of 8i.

SOLUTION: Let x be a cube root of 8i. Then $x^3 = 8i = 8(\cos 90^\circ + i \sin 90^\circ)$.

If $x = r(\cos \theta + i \sin \theta)$ is a cube root of $8(\cos 90^\circ + i \sin 90^\circ)$,

then $r^3(\cos 3\theta + i \sin 3\theta) = 8(\cos 90^\circ + i \sin 90^\circ)$.

Since $r^3 = 8$, $r = 2$ and since $3\theta = 90^\circ$, $\theta = 30^\circ$.

It follows that $2(\cos 30^\circ + i \sin 30^\circ)$ is a cube root of 8i.

In a + bi form, $2(\cos 30^\circ + i \sin 30^\circ) = 2(\frac{\sqrt{3}}{2} + i(\frac{1}{2})) = \sqrt{3} + i$.

The procedure followed above has revealed only one of the cube roots of 8i and is therefore incomplete. For example, it is easily shown that -2i is also a cube root of 8i since $(-2i)^3 = (-2i)(-2i)(-2i) = (-2)^3(i)^3 = -8(-i) = 8i$. It is necessary then to expand the procedure so that all possible cube roots are found. First, note that 8i is expressible in an infinity of trigonometric forms owing to the periodicity of the sine and cosine functions. Adding 360° successively to the direction angle:

$8i = 8(\cos 90^\circ + i \sin 90^\circ) = 8(\cos 450^\circ + i \sin 450^\circ) = 8(\cos 810^\circ + i \sin 810^\circ)$

$= 8(\cos 1170^\circ + i \sin 1170^\circ) = \ldots$, or in general,

$8i = 8(\cos(90^\circ + k \cdot 360^\circ) + i \sin(90^\circ + k \cdot 360^\circ))$ for $k = 0, \pm1, \pm2, \ldots$.

The cube roots corresponding to these are as follows:

$(8i)^{1/3} = (8(\cos 90^\circ + i \sin 90^\circ))^{1/3} = 2(\cos 30^\circ + i \sin 30^\circ) = \sqrt{3} + i$

$(8i)^{1/3} = (8(\cos 450^\circ + i \sin 450^\circ))^{1/3} = 2(\cos 150^\circ + i \sin 150^\circ) = -\sqrt{3} + i$

$(8i)^{1/3} = (8(\cos 810^\circ + i \sin 810^\circ))^{1/3} = 2(\cos 270^\circ + i \sin 270^\circ) = -2i$

$(8i)^{1/3} = (8(\cos 1170^\circ + i \sin 1170^\circ))^{1/3} = 2(\cos 390^\circ + i \sin 390^\circ)$

$= 2(\cos 30^\circ + i \sin 30^\circ) = \sqrt{3} + i$

Continuing this process further it will become obvious that no additional cube roots other than $\sqrt{3} + i$, $-\sqrt{3} + i$, and $-2i$ will be obtained. This is a general result that can be stated as follows: Every complex number has at most n distinct nth roots. To formalize the process, if $z = r(\cos \theta + i \sin \theta)$ then

$$z^{1/n} = r^{1/n}\left(\cos \frac{\theta+k(360^0)}{n} + i \sin \frac{\theta+k(360^0)}{n}\right)$$

or $z^{1/n} = r^{1/n}\left(\cos \frac{\theta+2\pi k}{n} + i \sin \frac{\theta+2\pi k}{n}\right)$ if θ is in radians, where

$k = 0, 1, 2, \ldots , (n - 1)$.

EXAMPLE: Use the formal definition above to find the four fourth roots of 16.

SOLUTION: Let $z = 16 = 16(\cos 0^0 + i \sin 0^0)$.

If $k = 0$ then $z^{1/4} = 16^{1/4}\left(\cos \frac{0^0}{4} + i \sin \frac{0^0}{4}\right) = 2(\cos 0^0 + i \sin 0^0)$

$= 2(1 + 0i) = 2.$

If $k = 1$ then $z^{1/4} = 16^{1/4}\left(\cos \frac{0^0+360^0}{4} + i \sin \frac{0^0+360^0}{4}\right)$

$= 2(\cos 90^0 + i \sin 90^0) = 2(0 + i) = 2i.$

If $k = 2$ then $z^{1/4} = 16^{1/4}\left(\cos \frac{0^0+720^0}{4} + i \sin \frac{0^0+720^0}{4}\right)$

$= 2(\cos 180^0 + i \sin 180^0) = 2(-1 + 0i) = -2.$

If $k = 3$ then $z^{1/4} = 16^{1/4}\left(\cos \frac{0^0+1080^0}{4} + i \sin \frac{0^0+1080^0}{4}\right)$

$= 2(\cos 270^0 + i \sin 270^0) = 2(0 - i) = -2i.$

Therefore, the four fourth roots of 16 are 2, 2i, -2, -2i .

SIMPLIFYING THE SEARCH FOR COMPLEX ROOTS.

A direct approach to finding roots of complex numbers is made possible by analyzing the distribution of roots geometrically. The graphs of the square roots of 4i, the cube roots of 8i, and the fourth roots of 16 as found above are shown below.

Figure 7-4A

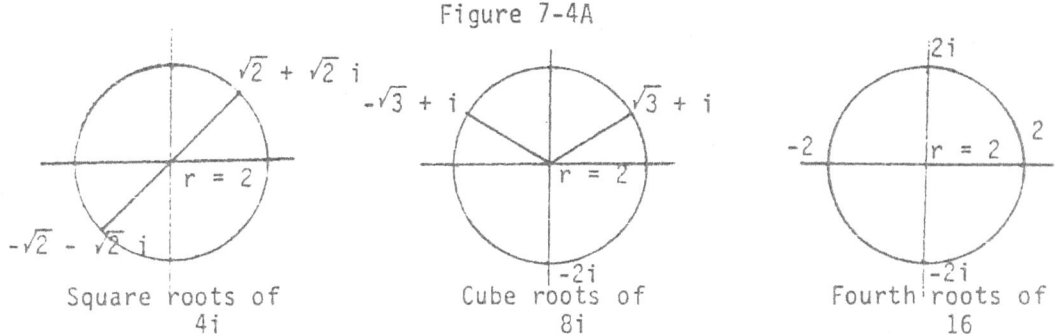

| Square roots of 4i | Cube roots of 8i | Fourth roots of 16 |

Figure 7-4A shows that in each example, the n distinct nth roots of a complex number $z = r(\cos \theta + i \sin \theta)$ divide a circle of radius $r^{1/n}$ into n equal parts where the roots are separated by an angle, $\frac{360^o}{n}$. Thus the cube roots of 8i are located on a circle with radius $8^{1/3} = 2$ and are separated by angles, $\frac{360^o}{3}$

$= 120^o$. The next example shows how this approach is used with abbreviated polar coordinate notation.

EXAMPLE: Find the five fifth roots of -243.

SOLUTION: From the formal definition, if $z = -243 = (243, 180^o)$ then

$$z^{1/5} = (243, 180^o)^{1/5} = (243^{1/5}, \frac{180^o}{5}) = (3, 36^o) \text{ when } k = 0.$$

The five fifth roots of -243 divide a circle of radius 3 into five equal parts where the roots are separated by angles, $\frac{360^o}{5} = 72^o$.

Adding 72° successively to each of the angles beginning with 36°, the five roots are: $(3, 36^o)$, $(3, 108^o)$, $(3, 180^o)$, $(3, 252^o)$, $(3, 324^o)$. Figure 7-4B shows the distribution of these roots corresponding to points on a circle of radius 3 units.

Figure 7-4B

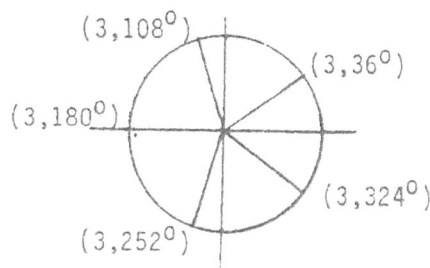

Conversion to a + bi form yields:

$(3,36^0) \sim 3(\cos 36^0 + i \sin 36^0) = 3(.8090 + .5878i) = 2.427 + 1.763i$

$(3,108^0) \sim 3(\cos 108^0 + i \sin 108^0)= 3(-.3090 + .9511i) = -.9271 + 2.853i$

$(3,180^0) \sim 3(\cos 180^0 + i \sin 180^0) = 3(-1 + 0i) = -3$

$(3,252^0) \sim 3(\cos 252^0 + i \sin 252^0) = 3(-.3090 - .9511i) = -.9271 - 2.853i$

$(3,324^0) \sim 3(\cos 324^0 + i \sin 324^0) = 3(.8090 - .5878i) = 2.427 - 1.763i$

Solutions of equations that involve complex numbers are made possible by the method described above.

EXAMPLE: Solve $x^3 = -2 + 5i$ for all complex values of x .

SOLUTION: $-2 + 5i \sim (\sqrt{29}, 111.8^0)$ in polar form. Since the solution to the equation is analogous to finding the three cube roots of $-2 + 5i$, let x_1, x_2, and x_3 represent the three possible complex solutions.

$x_1 = (\sqrt{29}, 111.8^0)^{1/3} = (29^{1/6}, \frac{111.8^0}{3}) \doteq (1.75, 37.3^0) \sim 1.39 + 1.06i$

The remaining solutions are obtained by adding 120^0 to 37.3^0 twice:

$x_2 = (1.75, 157.3^0) \sim -1.61 + 0.68i$

$x_3 = (1.75, 277.3^0) \sim .22 - 1.74i$

The solutions correspond to points on a circle of radius 1.75 units that divide the circumference into three equal parts:

Figure 7-4C

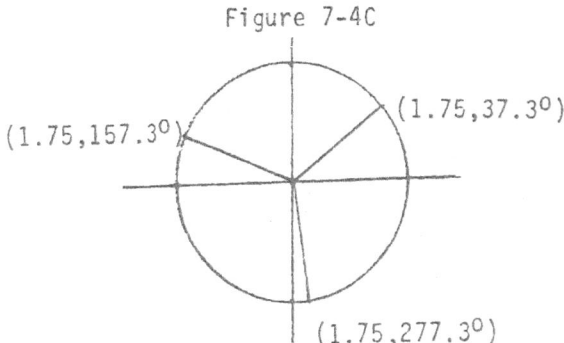

(1.75,157.3°) (1.75,37.3°)

(1.75,277.3°)

    If the direction angle is expressed in radian measure, successive nth roots are found by adding the appropriate nth fractional part of 2π: $\frac{2\pi}{2}$, $\frac{2\pi}{3}$, $\frac{2\pi}{4}$, ...

<u>EXAMPLE</u>: Find the complex solutions to the equation $x^4 + 2 - \sqrt{3}\,i = 0$.

<u>SOLUTION</u>: If $x^4 + 2 - \sqrt{3}\,i = 0$ then $x^4 = -2 + \sqrt{3}\,i \sim (\sqrt{7}, 2.43)$ where 2.43 is a radian measure. Let x_1, x_2, x_3, x_4 be the four fourth roots of $(\sqrt{7}, 2.43)$. Then $x_1 = (\sqrt{7}, 2.43)^{1/4} = (1.28, .61)$ where $1.28 = \sqrt{7}^{1/4}$ and $.61 = \frac{2.43}{4}$ rounded to hundredths. The remaining roots are obtained by adding $\frac{2\pi}{4} = \frac{\pi}{2} \doteq 1.57$ successively: $x_1 = (1.28, .61) \sim 1.05 + .73i$

$x_2 = (1.28, .61 + 1.57) = (1.28, 2.18) \sim -.73 + 1.05i$

$x_3 = (1.28, 2.18 + 1.57) = (1.28, 3.75) \sim -1.05 - .73i$

$x_4 = (1.28, 3.75 + 1.57) = (1.28, 5.32) \sim .73 - 1.05i$

Figure 7-4D

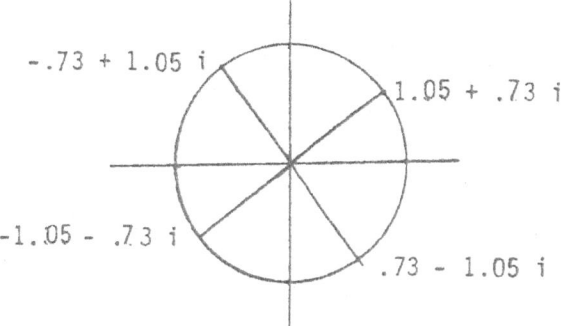

-.73 + 1.05 i   1.05 + .73 i

-1.05 - .73 i

.73 - 1.05 i

EXERCISE SET 7-4.

Find the indicated power of the given complex number by De Moivre's Theorem. Express answers in trigonometric form.

1) $(2(\cos 48^o + i \sin 48^o))^3$

2) $(3(\cos 12^o + i \sin 12^o))^6$

3) $(5(\cos (-130^o) + i \sin (-130^o)))^2$

4) $((-2)(\cos \frac{\pi}{12} + i \sin \frac{\pi}{12}))^4$

5) $(2 \text{ cis } 147^o)^3$

6) $(6 \text{ cis } \frac{\pi}{8})^3$

7) $(\frac{1}{2} \text{ cis } \frac{\pi}{3})^4$

8) $((-\sqrt{3}) \text{ cis } \frac{3\pi}{4})^4$

9) $(2,36^o)^5$

10) $(\sqrt{3},45^o)^{10}$

11) $(1 + \sqrt{3} \, i)^6$

12) $(-2 - 2\sqrt{3} \, i)^4$

13) $(2 - 2i)^5$

14) $(3\sqrt{2} + 3\sqrt{2} \, i)^2$

15) $(2 - i)^4$

16) $(-4 - 3i)^3$

Find all of the indicated roots of the given complex numbers. Express answers in a) trigonometric form and b) a + bi form.

17) The square roots of -4.

18) The square roots of -9i.

19) The cube roots of 1.

20) The cube roots of -27i.

21) The fourth roots of 16i.

22) The fourth roots of -81.

23) The square roots of -2 + 2i.

24) The cube roots of 27 cis 330°.

25) The square roots of 5 + 2i.

26) The cube roots of $4\sqrt{2} - 4\sqrt{2} \, i$.

27) The cube roots of $4(\cos 147^o + i \sin 147^o)$.

28) The fourth roots of $72(\cos \frac{8\pi}{5} + i \sin \frac{8\pi}{5})$.

Solve the following equations using De Moivre's Theorem.

29) $x^3 + 64i = 0$

30) $x^5 - 243 = 0$

31) $x^2 = 5 - 3i$

32) $x^5 - 1024i = 0$

7-5. CHAPTER SUMMARY

1. TRANSFORMATION OF COORDINATES

A. To transform a given rectangular coordinate pair (x,y) to either trigonometric coordinates (r cos θ, r sin θ) or polar coordinates (r,θ), the equations of transformation are:

$$r = \sqrt{x^2 + y^2} \quad \text{and} \quad \theta = \tan^{-1} \frac{y}{x}$$

B. To transform a given trigonometric ordered pair (r cos θ, r sin θ) or a given polar coordinate ordered pair (r,θ) to rectangular coordinates the equations of transformation are:

$$x = r \cos \theta \quad \text{and} \quad y = r \sin \theta$$

C. Rectangular coordinates and trigonometric coordinates are unique. Polar coordinates are not. That is, if (r_1,θ_1) and (r_2,θ_2) are mapped into the same point in the polar coordinate plane, it does not necessarily follow that $r_1 = r_2$ and/or $\theta_1 = \theta_2$. For example, $(2,\frac{\pi}{4})$, $(2,\frac{9\pi}{4})$, and $(-2,\frac{5\pi}{4})$ all map into the same point in the polar plane.

2. COMPLEX NUMBERS

A. The imaginary unit is defined as the positive square root of -1. $i = \sqrt{-1}$, $i^2 = -1$, $i^3 = -\sqrt{-1} = -i$, $i^4 = i^0 = 1$. If m is a natural number then $i^m \in \{i, -1, -i, 1\}$ as determined by the remainder when m is divided by 4. For example, $i^{34} = i^2 = -1$. If p and q are negative real numbers, $\sqrt{p}\sqrt{q} = -\sqrt{|p||q|}$. For example, $\sqrt{-3}\sqrt{-5} = -\sqrt{(3)(5)} = -\sqrt{15}$.

B. Complex numbers in rectangular form have a real part a and an imaginary part bi expressed as the sum a + bi. The real numbers can be placed in one to one correspondence with a subset of the complex numbers and expressed in the form a + 0i. Complex numbers of the form 0 + bi are called pure imaginary numbers.

C. If $z_1 = a + bi$ and $z_2 = c + di$ then

$$z_1 + z_2 = (a + c) + (b + d)i$$

$$z_1 - z_2 = (a - c) + (b - d)i$$

$$z_1 \cdot z_2 = (ac - bd) + (ad + bc)i$$

$$z_1 / z_2 = \frac{ac + bd}{c^2 + d^2} + \frac{bc - ad}{c^2 + d^2} i \,, \ z_2 \neq 0$$

D. Two numbers of the form $a + bi$ and $a - bi$ are called complex conjugates.

3. VECTORS

Complex numbers may be identified with directed line segments in a complex number plane. These segments are called vectors. They are added and subtracted by the parallelogram method (see section 7-2). If $z = a + bi$, the length or magnitude of z is defined as $|z| = |a + bi| = \sqrt{a^2 + b^2}$. The direction of z is given by $\theta = \tan^{-1} \frac{b}{a}$, an angle in standard position. If t is a real number then $tz = t(a + bi) = at + bti$. This product is called a scalar multiple. A vector can be stretched, contracted, and/or reversed in direction by being multiplied by a scalar.

If $z = a + bi$ then $zi = -b + ai$ and the vector corresponding to zi is rotated 90° counterclockwise from the vector corresponding to z .

4. TRIGONOMETRIC (POLAR) FORM OF A COMPLEX NUMBER

A. A complex number in $a + bi$ form can be transformed into $r(\cos \theta + i \sin \theta)$ or (r,θ) form by the equations of transformation:

$$r = \sqrt{a^2 + b^2} \qquad \theta = \tan^{-1} \frac{b}{a}$$

A complex number in $r(\cos \theta + i \sin \theta)$ or (r,θ) form can be transformed into $a + bi$ form by the equations of transformation:

$$a = r \cos \theta \qquad b = r \sin \theta$$

B. If $z_1 = r_1(\cos \Theta_1 + i \sin \Theta_1)$ and $z_2 = r_2(\cos \Theta_2 + i \sin \Theta_2)$ then

$$z_1 \cdot z_2 = r_1 \cdot r_2 (\cos (\Theta_1 + \Theta_2) + i \sin (\Theta_1 + \Theta_2)) \text{ and}$$

$$z_1 / z_2 = (r_1 / r_2)(\cos (\Theta_1 - \Theta_2) + i \sin (\Theta_1 - \Theta_2)) , \; z_2 \neq 0$$

C. DE MOIVRE'S THEOREM

If $z = r(\cos \Theta + i \sin \Theta)$ and n is a natural number then

$$z^n = r^n (\cos n\Theta + i \sin n\Theta)$$

$$z^{1/n} = r^{1/n} (\cos \frac{\Theta + k(360^0)}{n} + i \sin \frac{\Theta + k(360^0)}{n}), \; \Theta \text{ in degrees,}$$

or $z^{1/n} = r^{1/n} (\cos \frac{\Theta + 2\pi k}{n} + i \sin \frac{\Theta + 2\pi k}{n}), \; \Theta$ in radians,

where $k = 0, 1, 2, \ldots, (n-1)$.

EXERCISE SET 7-5.

Express the given ordered pairs in (r, Θ) form where $r > 0$ and $0 \leq \Theta < 360^0$.

1) $(-1,1)$ 2) $(2,-2)$ 3) $(\sqrt{3},-1)$ 4) $(5,-5\sqrt{3})$

5) $(3,7)$ 6) $(-4,5)$ 7) $(-11,-2)$ 8) $(1,-8)$

9) $(3 \cos 20^0, 3 \sin 20^0)$ 10) $(-4 \cos \frac{\pi}{3}, -4 \sin \frac{\pi}{3})$

Express the given ordered pairs in $a + bi$ form (exact values):

11) $(2,30^0)$ 12) $(-4,15^0)$ 13) $(3,75^0)$

14) $(5,150^0)$ 15) $(-4, \pi)$ 16) $(3,\frac{5\pi}{6})$

Solve the quadratic equations:

17) $5x^2 + 2x + 1 = 0$ 18) $7x^2 + 4x + 2 = 0$ 19) $2x^2 - 3x + 4 = 0$

20) $4x^2 + 3x - 1 = 0$

Perform the indicated operations and express answers in $a + bi$ form:

21) $(-5 + 2i) + (2 - 6i)$ 22) $(3 - 5i) + (-17 + i)$

23) $(2 - 7i) - (-11 - 3i)$ 24) $(-9 + 5i) - (-3 + 7i)$

25) $(3 + i)(3 - i)$ 26) $(\sqrt{2} + i)^2$ 27) $(9 - 5i)(-2 - 3i)$

28) $(-6 - i)(6 + i)$ 29) i^{45} 30) i^{319} 31) $(7 + i) / (4 - i)$

32) $(-3 + 4i) / 2i$ 33) $(3\sqrt{-2})(\sqrt{-50})$ 34) $(\sqrt{-75}) / (\sqrt{-3})$

35) $(\sqrt{-20}) / (\sqrt{-45})$ 36) $\sqrt{-32} + \sqrt{-50}$

Find the magnitudes of the following:

37) $-3 + 2i$ 38) $4 + 7i$ 39) $\sqrt{3} - \sqrt{2}\, i$ 40) $11\sqrt{3} + 2\sqrt{5}\, i$

Find the resultant vector sums and differences by the parallelogram method:

41) $(-2 - 2i) + (3 - 4i)$ 42) $(5 - 6i) + (-4 + 3i)$

43) $(2 + i) - (-3 - 4i)$ 44) $(-7 - 2i) - (2 + 5i)$

Express the following in $r(\cos \Theta + i \sin \Theta)$ form:

45) $-2 + 2i$ 46) $(1 + i)^6$ 47) $(-\sqrt{3} + i)^5$ 48) $(-1 - \sqrt{3}\, i)^4$

Express the following in $a + bi$ form:

49) $(2(\cos 120^0 + i \sin 120^0))^4$ 50) $(3(\cos \frac{11\pi}{6} + i \sin \frac{11\pi}{6}))^3$

51) $(3(\cos 45^0 + i \sin 45^0))(6(\cos 150^0 + i \sin 150^0))$

52) $(15(\cos 330^0 + i \sin 330^0)) / (3(\cos 90^0 + i \sin 90^0))$

53) $4 \text{ cis } 58^0 \cdot 7 \text{ cis } 149^0$ 54) $12 \text{ cis } \frac{7\pi}{5} / 4 \text{ cis } \frac{3\pi}{4}$

Find the following:

55) The square roots of $-\sqrt{3} - i$ 56) The cube roots of 8

57) The fourth roots of $-2\sqrt{2} + 2\sqrt{2}\, i$ 58) The fifth roots of $32i$

Solve the equations:

59) $x^6 - 64 = 0$ 60) $x^4 + 16i = 0$

CHAPTER 8

APPLICATIONS.   THE LAWS OF TRIANGLES

8-1. RIGHT TRIANGLE TRIGONOMETRY.

The word "trigonometry" derives from the Greeks and means the measurement of three-sided figures or triangles. Throughout this text such measurements have been implied, particularly in the applications made of the Pythagorean theorem and the definition of the trigonometric ratios.

Recall that if P(x,y) is a point on a circle of radius r with center at the origin then a right triangle △OPT is formed if a perpendicular is dropped from P to the X-axis as in Figure 8-1A.

Figure 8-1A

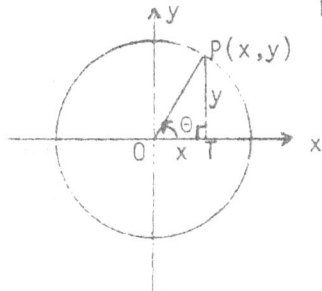 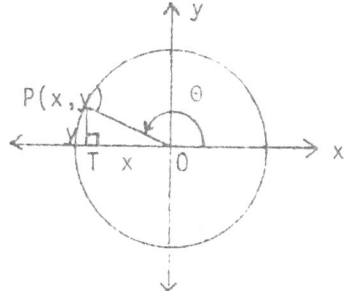

If θ is any angle in standard position:

$$\sin \theta = \frac{y}{r} \ , \ \cos \theta = \frac{x}{r} \ , \ \tan \theta = \frac{y}{x} \ , \ \cot \theta = \frac{x}{y} \ , \ \sec \theta = \frac{r}{x} \ , \ \csc \theta = \frac{r}{y}$$

If △OPT is abstracted from the coordinate system and the vertices relabeled as A, B, C, then it is customary to label the sides opposite the vertices by corresponding lower case letters as in Figure 8-1B.

FIGURE 8-1B

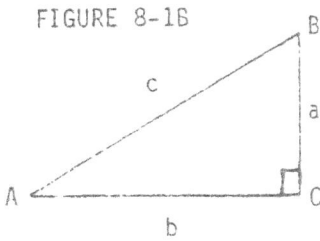

If ABC is a right triangle, vertex C is traditionally located at the right angle. With respect to the angle at A, a is called the opposite side, b is called the adjacent side, and side c opposite the right angle is called the hypotenuse. The trigonometric ratios of angle A are defined as follows:

$$\sin A = \frac{\text{opposite side}}{\text{hypotenuse}} = \frac{a}{c}, \qquad \csc A = \frac{\text{hypotenuse}}{\text{opposite side}} = \frac{c}{a}$$

$$\cos A = \frac{\text{adjacent side}}{\text{hypotenuse}} = \frac{b}{c}, \qquad \sec A = \frac{\text{hypotenuse}}{\text{adjacent side}} = \frac{c}{b}$$

$$\tan A = \frac{\text{opposite side}}{\text{adjacent side}} = \frac{a}{b}, \qquad \cot A = \frac{\text{adjacent side}}{\text{opposite side}} = \frac{b}{a}$$

With respect to angle B, the complement of A, the following trigonometric ratios hold:

$$\sin B = \frac{\text{opposite side}}{\text{hypotenuse}} = \frac{b}{c}, \qquad \csc B = \frac{\text{hypotenuse}}{\text{opposite side}} = \frac{c}{b}$$

$$\cos B = \frac{\text{adjacent side}}{\text{hypotenuse}} = \frac{a}{c}, \qquad \sec B = \frac{\text{hypotenuse}}{\text{adjacent side}} = \frac{c}{a}$$

$$\tan B = \frac{\text{opposite side}}{\text{adjacent side}} = \frac{b}{a}, \qquad \cot B = \frac{\text{adjacent side}}{\text{opposite side}} = \frac{a}{b}$$

The complementary relationships are readily apparent since $B = 90^{\circ} - A$:

$$\sin (90^{\circ} - A) = \sin B = \frac{b}{c} = \cos A, \quad \cos (90^{\circ} - A) = \cos B = \frac{a}{c} = \sin A$$

$$\tan (90^{\circ} - A) = \tan B = \frac{b}{a} = \cot A, \quad \cot (90^{\circ} - A) = \cot B = \frac{a}{b} = \tan A$$

$$\sec (90^{\circ} - A) = \sec B = \frac{c}{a} = \csc A, \quad \csc (90^{\circ} - A) = \csc B = \frac{c}{b} = \sec A$$

It is also possible to show the fundamental identities hold with the definitions above.

EXAMPLE: Prove that $\tan^2 A + 1 = \sec^2 A$.

SOLUTION: $\tan^2 A + 1 = (\frac{a}{b})^2 + 1 = \frac{a^2}{b^2} + \frac{b^2}{b^2} = \frac{a^2 + b^2}{b^2} = \frac{c^2}{b^2} = (\frac{c}{b})^2$

$$= \sec^2 A.$$

Expressions involving the advanced formulas may also be derived in terms of the sides of the triangle.

EXAMPLE: Express sin 2B in terms of the sides of right triangle ABC.

SOLUTION: $\sin 2B = 2 \sin B \cos B = 2 \cdot \frac{b}{c} \cdot \frac{a}{c} = \frac{2ab}{c^2}$.

If two sides of a right triangle are known the remaining side may be determined by the Pythagorean theorem. The trigonometric ratios for the acute angles are obtained from the definitions above.

EXAMPLE: In right triangle ABC, a = 7.3, b = 4.9 and C = 90°. Find the remaining side and the trigonometric ratios for angle A.

SOLUTION: The unknown side is the hypotenuse c. $c = \sqrt{(7.3)^2 + (4.9)^2} = 8.8$

$$\sin A = \frac{a}{c} = \frac{7.3}{8.8} = .8295 \qquad \cos A = \frac{b}{c} = \frac{4.9}{8.8} = .5568$$

$$\tan A = \frac{a}{b} = \frac{7.3}{4.9} = 1.490 \qquad \cot A = \frac{b}{a} = \frac{4.9}{7.3} = .6712$$

$$\sec A = \frac{c}{b} = \frac{8.8}{4.9} = 1.796 \qquad \csc A = \frac{c}{a} = \frac{8.8}{7.3} = 1.205$$

EXAMPLE: In triangle ABC the hypotenuse is 1 unit. Express the six trigonometric ratios for angle B in terms of side b.

SOLUTION: Figure 8-1C shows the arrangement of sides and angles. Note that side a = $\sqrt{1 - b^2}$ by the Pythagorean theorem.

FIGURE 8-1C

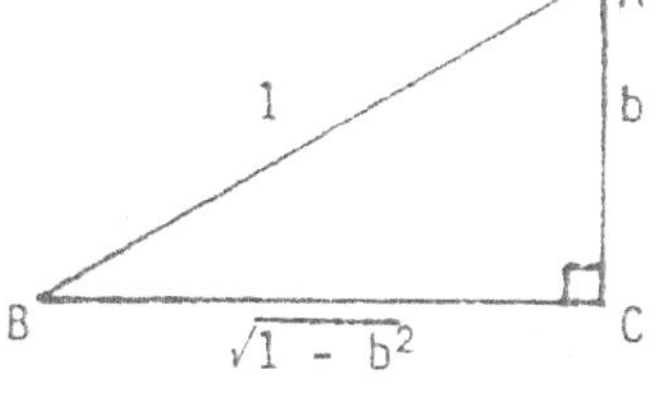

$$\sin b = \frac{b}{1} = b, \quad \cos B = \frac{\sqrt{1 - b^2}}{1} = \sqrt{1 - b^2}, \quad \tan B = \frac{b}{\sqrt{1 - b^2}}$$

$$\cot B = \frac{\sqrt{1 - b^2}}{b}, \quad \sec B = \frac{1}{\sqrt{1 - b^2}}, \quad \csc B = \frac{1}{b}$$

RIGHT TRIANGLE MEASUREMENT

A large number of practical problems involving indirect measurement can be solved through right triangle trigonometry. These are explored through examples.

EXAMPLE: A bridge is to be built to span a river. At the point chosen, two sightings 100 feet apart are made on a tree located on the opposite bank as in Figure 8-1D.

Figure 8-1D

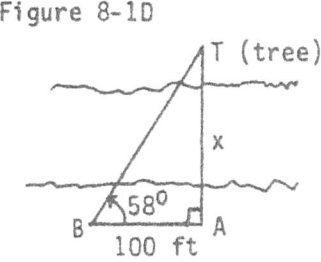

If angle ABT = 58°, how far is it from point A to the tree?

SOLUTION: In right triangle ABT, $\tan 58° = \frac{x}{100 \text{ ft}}$.

Therefore, $x = 100 \text{ ft } (\tan 58°) = 160 \text{ ft.}$

To measure the height of an object from ground level it is often advantageous to determine an "angle of elevation" at a known distance from the base of the object to be measured.

EXAMPLE: From a point on the ground 250 feet away from the base of a building, the angle of elevation is 33°40'. Find the height of the building to the nearest foot.

SOLUTION: Figure 8-1E

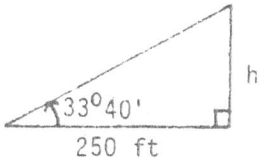

250 ft

Let h be the height of the building. Then tan $33°40' = \dfrac{h}{250\ ft}$

and h = 250ft(tan $33°40'$) = 166.52 ft = 167 ft.

An "angle of depression" is a sighting made to an object that lies below
an observer.

EXAMPLE: From the top of a vertical cliff 430 feet high, a boat is sighted
at an angle of depression of $14°27'$. How far is the boat from
the foot of the cliff (to the nearest ten feet)?

SOLUTION: Figure 8-1F illustrates the situation.

Figure 8-1F

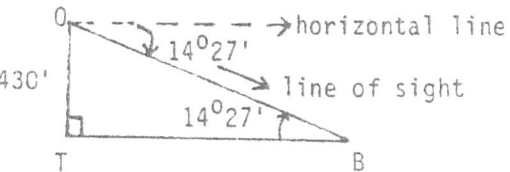

The angle of depression is measured from a horizontal line from
the observer to a line of sight connecting the observer to the
boat. By alternate interior angles between parallel lines,
angle OBT is equal to the angle of depression $14°27'$.

Therefore, tan $14°27' = \dfrac{430\ ft}{x}$ or x = $\dfrac{430\ ft}{\tan 14°27'}$ = 1670 ft

to the nearest ten feet.

EXAMPLE: Find the lengths of the sides of a rectangle if the diagonal is
25 cm and the angle between the diagonal and the longer side is
$34°$.

315

SOLUTION: Figure 8-1G

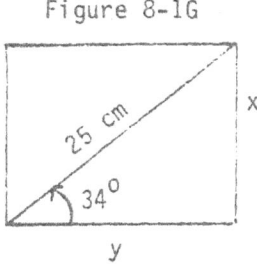

Let x be the length of the shorter side and y be the length of

the longer side. From Figure 8-1G we have sin 34^0 = $\frac{x}{25cm}$

and cos 34^0 = $\frac{y}{25cm}$. Hence, x = 25 cm (sin 34^0) = 13.98 cm

and y = 25 cm (cos 34^0) = 20.73 cm.

EXAMPLE: A road 570 feet long rises to an overpass that is 45 feet high.

Find the angle of elevation between the road and the horizontal

to the nearest ten minutes.

Figure 8-1H

SOLUTION: Let θ be the angle of elevation. sin θ = $\frac{45}{570}$.

θ = $\sin^{-1}\frac{45}{570}$ = 4.528^0 = 4^0 30' to nearest ten minutes.

DOUBLE SIGHTINGS

When an object to be measured is inaccessible it is possible to take

its measure by obtaining two sightings that are a known distance apart.

EXAMPLE: A knight and his band of followers plan to scale a wall of a

castle that is surrounded by a moat. At two points directly in line with the wall that are 20 feet apart they take sightings of the angles of elevation of the wall. The angle measures are 30° and 40°. Find the height of the wall to the nearest foot.

Figure 8-1 I

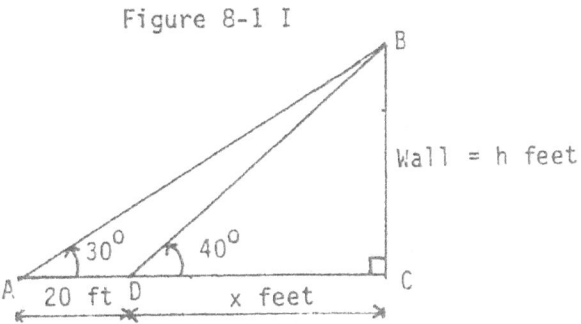

Wall = h feet

SOLUTION: $\tan 40^O = \dfrac{h}{x}$, $h = x \tan 40^O$ and $\tan 30^O = \dfrac{h}{20+x}$, $h = (x+20)\tan 30^O$

Since both measures are equal to h, they are equal to each other.

$x \tan 40^O = (x+20)\tan 30^O = x \tan 30^O + 20 \tan 30^O$

Then $x \tan 40^O - x \tan 30^O = 20 \tan 30^O$. Factoring,

$x (\tan 40^O - \tan 30^O) = 20 \tan 30^O$ and solving for x,

$x = \dfrac{20 \tan 30^O}{\tan 40^O - \tan 30^O}$. Substituting for h,

$h = x \tan 40^O = \dfrac{20 \tan 30^O \tan 40^O}{\tan 40^O - \tan 30^O} = 37$ feet.

EXERCISE SET 8-1.

In exercises 1 - 10 right triangle ABC is given with C = 90⁰ . Find the six trigonometric ratios from the given data. Exact answers.

1) a = 5 , b = 12 , c = 13 .   2) a = 7 , b = 24 , c = 25 .

3) a = 3 , b = 4 .   4) a = 15 , c = 17 .

5) $\tan A = \frac{3}{4}$, c = 15 .   6) $\sec B = \frac{7}{6}$, b = 18 .

7) $\cos B = \frac{8}{17}$, c = 68 .   8) $\csc A = \sqrt{5}$, a = 3 .

9) $\cot B = \frac{1}{\sqrt{3}}$, a = 6 .   10) $\sin A = \frac{3}{4}$, c = 0.12 .

In exercises 11 - 16 right triangle EFG is given with G = 90⁰ . Find the number of degrees in E from the given information.

11) sin E = cos E   12) csc 4E = sec 5E

13) $\tan \frac{1}{3} E = \cot \frac{5}{3} E$ 14) sin (2E + 40⁰) = cos (3E - 20⁰)

15) sec (F - 10⁰) = csc (2F +40⁰)

16) $\tan \frac{3}{2} (F + 20^0) = \cot \frac{1}{2} (F - 30^0)$

In exercises 17 - 24 a right triangle ABC is given with C = 90⁰ . Find the unknown side or angle from the given data. Round off answers to the same precision as the given data.

17) A = 38⁰ , c = 24.6 . Find a .

18) B = 71.6⁰ , a = 19.8 . Find b .

19) a = 212 , b = 317 . Find A .

20) b = 32.416 ,   c = 58.723 . Find B.

21) A = 13⁰40' , a = 11.63 . Find c .

22) B = 29⁰15' , b = 33.7 . Find a .

23) $c = 117$, $a = 91$. Find B .

24) $a = 95.6$, $b = 51.8$. Find B .

25) In right triangle ABC , $C = 90^0$ and $c = 1$. Express tan A in terms of side b .

26) In right triangle ABC , $C = 90^0$ and $a = 3$. Express sin B in terms of side c .

27) Express cos 2A in terms of the sides of right triangle ABC with $C = 90^0$.

28) Express tan 2A in terms of the sides of right triangle ABC , $C = 90^0$.

29) Express sin ½ B in terms of the sides of right triangle ABC, $C = 90^0$.

30) Using the sides of right triangle ABC , $C = 90^0$, and the formula for cos $(x + y)$, show that cos $(A + B) = 0$.

31) With the same information as given in exercise 30, show that sin $(A + B) = 1$.

32) With the same information as given in exercise 30, show that
$$\cos (A - B) = \frac{2ab}{a^2 + b^2}$$

33) A kite string is staked to the ground. The kite and string make an angle of 72^0 with the ground. If 178 feet of string has been played out, how high is the kite?

34) A tree casts a shadow 28 feet long. If the sun's elevation is $57^0 20'$, find the height of the tree to the nearest foot.

35) A ladder is leaning against a wall. The foot of the ladder is 4 feet from the wall and the angle formed by the top of the ladder and the wall is 53^0 . How long is the ladder?

36) On the top of a peak that is 12350 feet high a sighting is taken of the top of Mt. McKinley in Alaska. If the angle of elevation is $16^\circ 48'$ find the height of Mt. McKinley to the nearest ten feet if the sighting is 5 miles away. (1 mile = 5280 feet).

37) The tallest building in the United States is the Sears Tower of Chicago, Illinois. An observer on top of the Sears Tower measures an angle of depression of 72.7° to the roof edge of a building 240 feet high that is located 378 feet from the base of the Sears Tower. Determine the height of the Sears Tower to the nearest foot.

38) A 20 foot guy wire is attached to a TV mast at a point 6 feet from the top of the mast. If the angle between the guy wire and the mast is $49^\circ 16'$, how high is the TV mast?

39) A commando team plans to scale a vertical cliff that borders the opposite bank of a river. Two angles of elevation of 32° and 25° are taken 100 feet apart directly in line with face of the cliff. If the first reading is at the edge of the river, find the height of the cliff and the width of the river.

40) An observer walks 40 feet down a slope from the base of a tree and measures an angle of elevation of 47° to the top of the tree.

(Remember that angles of elevation are measured from the horizontal.)

If the slope is inclined at an angle of 10° to the horizontal, find the height of the tree.

8-2. <u>SOLVING OBLIQUE TRIANGLES. THE LAW OF SINES.</u>

If a triangle in the plane does not contain a right angle it is called an oblique triangle. Oblique triangles fall into two categories: either one of the angles is obtuse (greater than 90^0) or all of the angles are acute (less than 90^0). The two cases are shown below.

Figure 8-2A

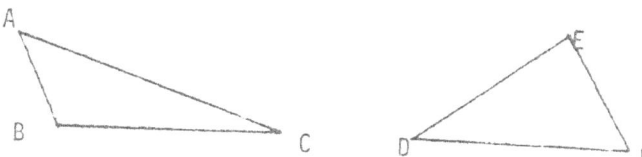

To "solve" a triangle means to use whatever information is given to determine the measures of all of its interior angles and sides. In this section an important relationship called the Law of Sines will be derived and the appropriate situations for its application will be investigated.

<u>DERIVATION OF THE LAW OF SINES.</u>

Consider an oblique triangle with all of its angles acute.

Figure 8-2B

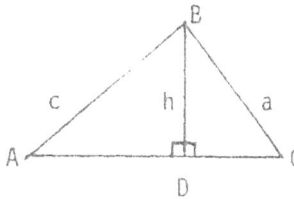

From vertex B a perpendicular to side AC is drawn terminating at point D as shown in Figure 8-2B above. The two right triangles formed give the following trigonometric relationships:

$$\sin A = \frac{h}{c} \qquad\qquad\qquad \sin C = \frac{h}{a}$$

Since c sin A = h and a sin C = h, it follows that c sin A = a sin C.

Dividing by ac, $\dfrac{c \sin A}{ac} = \dfrac{a \sin C}{ac}$

and simplifying, $\dfrac{\sin A}{a} = \dfrac{\sin C}{c}$

If a perpendicular is drawn to side BC, the same derivation would give

$$\frac{\sin B}{b} = \frac{\sin C}{c}$$

Similarly, if the perpendicular is drawn to side AB, the derivation would give $\dfrac{\sin A}{a} = \dfrac{\sin B}{b}$

The generalization that can be drawn from these results may be stated as follows: In any oblique triangle the ratio of the sine of any angle to the side opposite is a constant. Symbolically,

$$\frac{\sin A}{a} = \frac{\sin B}{b} = \frac{\sin C}{c}$$

and this is called The Law of Sines. The derivation of The Law of Sines for oblique triangles in which one of the angles is obtuse is left as an exercise for the student.

APPLICATIONS OF THE LAW OF SINES.

Since the sum of the angle measures of a plane triangle is equal to 180°, this fact may be used in conjunction with The Law of Sines provided that at least one side is given.

EXAMPLE: Given A = 24°, C = 49°, and side b = 15, solve for the remaining
         sides (to the nearest hundredths) and angle (to the nearest degree).

SOLUTION: Draw a triangle showing the given information.

Figure 8-2C

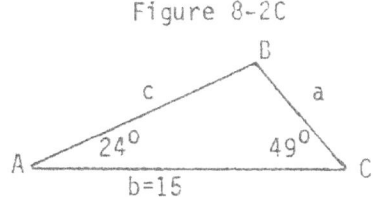

$B = 180^O - (24^O + 49^O) = 180^O - 73^O = 107^O$

By The Law of Sines, $\dfrac{\sin 24^O}{a} = \dfrac{\sin 107^O}{15} = \dfrac{\sin 49^O}{c}$

Therefore, $a = \dfrac{15 \sin 24^O}{\sin 107^O} = 6.38$ from the first equality,

and $c = \dfrac{15 \sin 49^O}{\sin 107^O} = 11.84$ from the second equality.

The example above reveals that a solution by The Law of Sines is possible if two angles and a side are given. Now consider the situation where one angle and two sides are given. Figure 8-2D shows the possible arrangements. The given parts are circled.

Figure 8-2D

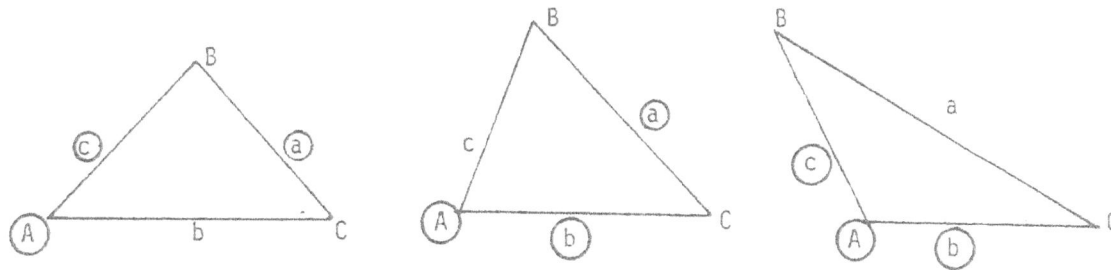

In arrangements 1 and 2, $\dfrac{\sin A}{a}$ is known, so it is possible to use The Law of Sines in these situations. However the drawings could be misleading so these arrangements will need further discussion. In arrangement 3 $\dfrac{\sin A}{a}$ cannot be determined since side a isn't given. The law covering this arrangement is known as The Law of Cosines and will be derived in Section 8-3.

To return to arrangement 1 for the moment, the given information allows setting up the equation from The Law of Sines:

$$\frac{\sin Ⓐ}{ⓐ} = \frac{\sin C}{ⓒ}$$

and solving for sin C, $\sin C = \frac{ⓒ \sin Ⓐ}{ⓐ}$. However, since sin C = sin ($180°$ - C), there is no certainty as to whether the quantity $\frac{ⓒ \sin Ⓐ}{ⓐ}$ represents sin C or the sine of the supplement of C. Both cases are possible as the following figure demonstrates.

Figure 8-2E

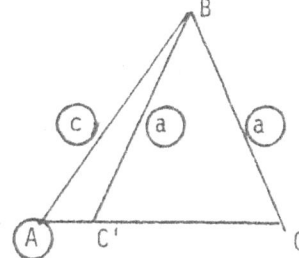

In triangle ABC', C' = $180°$ - C

and sin C' = sin C = $\frac{ⓒ \sin Ⓐ}{ⓐ}$.

As a result of this ambiguity (which also applies to arrangement 2), further analysis is required to solve triangles where <u>two sides and the angle opposite one of them</u> is given.

<u>THE AMBIGUOUS CASE</u>.

Given sides a, c and angle A, the following table summarizes all of the possible (or impossible) triangles that may be constructed:

<u>AMBIGUOUS CASE TABLE</u>

	A > 90°	A = 90°	A < 90°
a > c	one triangle	one right triangle	one triangle
a = c	no triangle	no triangle	one isosceles triangle
a < c	no triangle	no triangle	a) no triangle if a < c sin A b) one right triangle if a = c sin A c) two triangles if a> c sin A

The table is to be read as a grid where the conditions given at the top and side intersect. Reading down column A > 90°, if a > c then a triangle as shown below may be constructed:

Figure 8-2F

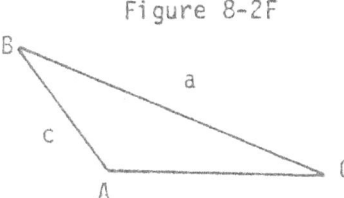

If a = c and A > 90°, the next figure shows that the triangle is impossible since A + C > 180°. A similar figure would result if a < c and A > 90°.

Figure 8-2G

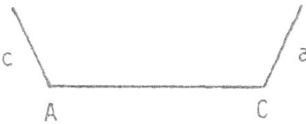

In the next column of the table A = 90°. If a > c, a right triangle is possible with side a as the hypotenuse as shown below.

Figure 8-2H

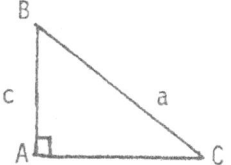

If a = c then it must follow that C = A = 90° an impossible triangle.

If a < c then it must follow that C > A = 90°, also an impossible triangle.

Figure 8-2I

 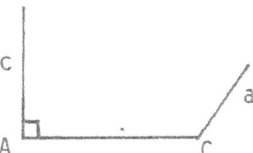

In the last column A is acute. If a > c then it must follow that C < A. If a = c then △ABC is an isosceles triangle and A = C. The two situations are shown below.

Figure 8-2J

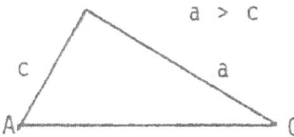 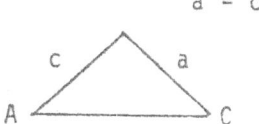

Finally, if A < 90° and a < c then there are three possibilities. First, recall that by The Law of Sines, $\sin C = \dfrac{c \sin A}{a}$. If a < c sin A then sin C > 1, an impossibility for the sine function. The quantity c sin A is equal to the length of the altitude drawn from vertex B to side AC.

Figure 8-2K

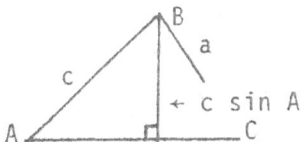

Therefore no triangle is constructible with the given data. Next, if $a = c \sin A$ then $\sin C = 1$ and $C = 90^\circ$, so a right triangle may be constructed as shown.

Figure 8-2L

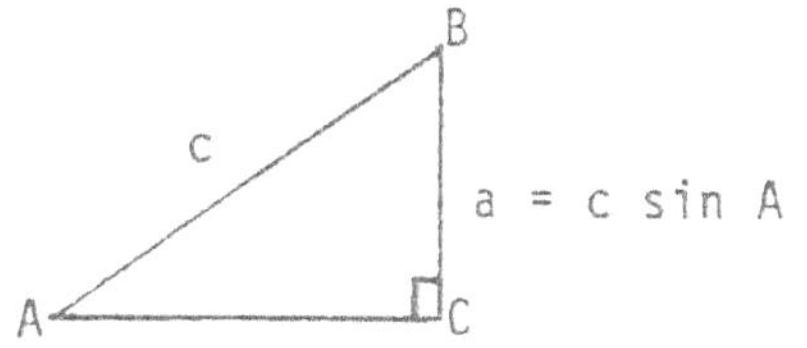

$$a = c \sin A$$

In the third and last possible subcase, if $a > c \sin A$ then $\sin C < 1$. Figure 8-2M shows that two triangles, $\triangle ABC$ and $\triangle ABC'$ may be constructed. Note that $C' = 180^\circ - C$ and that there will be two sets of values for B and side b corresponding to the two triangles.

Figure 8-2M

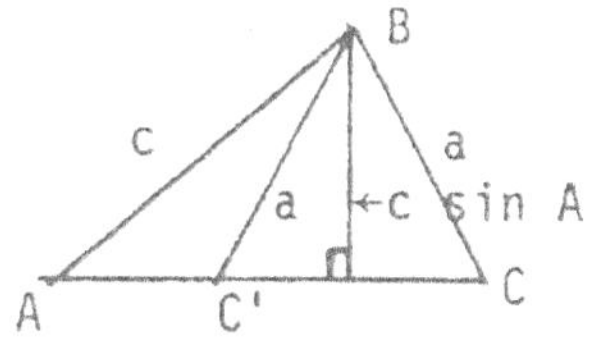

The next series of examples will focus on the use of the table given above for the ambiguous case. Determine the number of triangles that can be constructed with the given data.

EXAMPLE: $A = 105^\circ$, $a = 14$, $c = 23$.

SOLUTION: Since $A > 90^\circ$ and $a < c$, the intersection of column 1 and row 3 shows that no triangle is possible.

EXAMPLE: $A = 38^\circ$, $a = 15.6$, $c = 21.4$.

SOLUTION: Since $A < 90^\circ$ and $a < c$, the intersection of column 3 and row 3 applies. To determine the subcases that is operative, $c \sin A$ must be determined. $c \sin A = 21.4 \sin 38^\circ = 13.2$ rounded. Since $a = 15.6 > c \sin A = 13.2$, there are two possible triangles.

In the next example the data has been labeled differently but the same table relationships hold.

EXAMPLE: B = 17°, b = 12.4, a = 13.8.

SOLUTION: Here B corresponds to A in the table, side b with side a in the table, and side a with side c in the table. Since B < 90° and b < a, the intersection of column 3 and row 3 applies.

a sin B = 13.8 sin 17° = 4.03. Since b > a sin B there are two possible solutions.

EXAMPLE: Complete the solutions for the triangles in the example above.

SOLUTION: B = 17°, b = 12.4, a = 13.8

$$\sin A = \frac{a \sin B}{b} = \frac{13.8 \sin 17°}{12.4} = .3254.$$ Hence A = 19° or 161°.

If A = 19°, C = 180° - (17° + 19°) = 144°.

If A = 161°, C = 180° - (17° + 161°) = 2°.

The two possible values for side c are found by The Law of Sines as follows: $\frac{\sin 144°}{c} = \frac{\sin 17°}{12.4}$, $c = \frac{12.4 \sin 144°}{\sin 17°} = 24.9.$

or $\frac{\sin 2°}{c} = \frac{\sin 17°}{12.4}$, $c = \frac{12.4 \sin 2°}{\sin 17°} = 1.5.$

The solution triangles are shown below:

Figure 8-2N

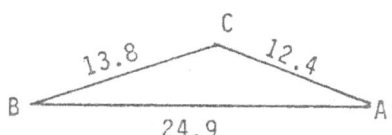
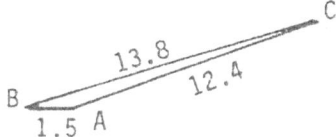

EXAMPLE: In the final demonstration example of Section 8-1, a band of
knights were planning to scale a wall surrounded by a moat. The
solution of that problem was completed by right triangle trigonometry. Solve the problem again using the Law of Sines.

SOLUTION: Figure 8-20 recasts the problem for the Law of Sines

Figure 8-2 0

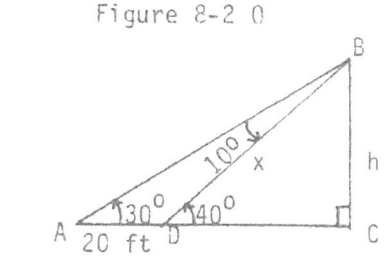

In $\triangle$ ABD, $\dfrac{x}{\sin 30^{\circ}} = \dfrac{20 \text{ ft}}{\sin 10^{\circ}}$, so $x = \dfrac{20 \sin 30^{\circ}}{\sin 10^{\circ}}$.

In $\triangle$ BDC $h = x \sin 40^{\circ}$ and substituting for x,

$$h = \frac{20 \sin 30^{\circ} \sin 40^{\circ}}{\sin 10^{\circ}} = 37 \text{ feet.}$$

EXERCISE SET 8-2 .

In △ ABC two angles and a side are given. Find the remaining parts of the triangle.

1) A = 38° , B = 45° , c = 38.3 .

2) B = 24° , C = 95° , b = 10 .

3) A = 102° , B = 11° , a = 93 .

4) A = 48° , C = 73° , b = 24.8 .

5) C = 20° 35' , B = 55° 17' , c = 13.65 .

In △ EFG two sides and the angle opposite one of them is given. How many triangles are determined by the given data?

6) e = 17 , f = 9 , E = 73° .   7) e = 24 , g = 18 , G = 32° .

8) f = 16.3 , g = 5.4 , F = 107° . 9) e = 3 , f = 4 , E = 93° .

10) g = 15.6 , f = 6.8 , F = 57° .

In the following triangles solve for the remaining parts:

11) △ ABC , A = 35° , a = 14 , b = 11 .

12) △ EFG , F = 103° , f = 78.6 , g = 23. 8 .

13) △ RST , T = 37.8° , t = 24 , r = 30 .

14) △ UVW , V = 57.4° , v = 22.7 , w = 25.6 .

15) △ XYZ , X = 84.3° , x = 12.3 , y = 9.8 .

16) The deepest gorge in the United States is at Hells Canyon on the Snake River, Idaho. Its greatest depth is 7900 feet. Two observers at the same altitude are on opposite sides of the gorge. They measure angles of depression to the bottom of the gorge of 79° and 84° respectively. How far apart are the observers? Solve by the Law of Sines.

17) A television mast 30 feet high is located on top of a building. From

a point on the ground the angles of elevation of the top and bottom of the mast are 57° and 49° respectively. Find the height of the building to the nearest foot.

18) A tower is located on the bank of a river. From a point on the opposite bank the angle of elevation is 35.8° to the top of the tower. At a second point in line with the base of the tower and the first observation point, the angle of elevation is 27.6°. If the observation points are 148 feet apart, how wide is the river?

19) From the top of a hill angles of depression of 54° and 28° are measured to two markers that are directly in line on the horizontal plane below. How high is the hill if the markers are 150 feet apart?

20) A tree is growing vertically on a slope that is inclined 18° to the horizontal. At a point 100 feet downslope from the base of the tree the angle of elevation of the tree is 38° $20'$. Find the height of the tree to the nearest foot.

8-3. <u>SOLVING OBLIQUE TRIANGLES. THE LAW OF COSINES. THE LAW OF TANGENTS.</u>

In section 8-2 (Figure 8-2D) it was shown that The Law of Sines could not be applied directly to solving an oblique triangle where the given data consisted solely of two sides and the angle included between them. Another situation where the Law of Sines is inapplicable occurs when the given data consists only of three given sides and none of the angles of the oblique triangle. To handle these situations the following relationship is commonly applied:

<u>LAW OF COSINES</u>.

Given oblique triangle ABC with opposite sides a,b,c respectively, if the given data consists of sides a and b and angle C, then

$$c^2 = a^2 + b^2 - 2ab \cos C.$$

<u>DERIVATION OF THE LAW OF COSINES</u>.

Let triangle ABC be positioned on the rectangular coordinate system so that the given angle C is in standard position.

Figure 8-3A

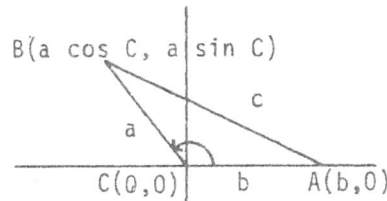

The trigonometric coordinates of vertex B are found by looking at point B as if it were on a circle with radius a. Since c is the distance between points A and B, the distance formula gives:

$$c = \sqrt{(a \cos C - b)^2 + (a \sin C - 0)^2}$$

Squaring and expanding the binomials under the radical sign:

$$c^2 = (a^2 \cos^2 C - 2ab \cos C + b^2) + a^2 \sin^2 C$$

But $a^2 \cos^2 C + a^2 \sin^2 C = a^2 (\cos^2 C + \sin^2 C) = a^2 (1) = a^2$.

Therefore, $c^2 = a^2 + b^2 - 2ab \cos C$ and the derivation is complete.

The Law of Cosines is a generalization of the Pythagorean Theorem to all plane triangles. For, if $C = 90^\circ$, $\cos C = \cos 90^\circ = 0$ and the formula reduces to $c^2 = a^2 + b^2$ for a right triangle.

EXAMPLE: In triangle ABC, a = 20, b = 24 and C = 54°. Find side c to the nearest tenth.

SOLUTION: $c^2 = (20)^2 + (24)^2 - 2(20)(24) \cos 54^\circ$

$c^2 = 400 + 576 - 960 \cos 54^\circ = 411.7$

Therefore, $c = 20.3$

The pattern of The Law of Cosines is adaptable to the other sides and angles:

$$a^2 = b^2 + c^2 - 2bc \cos A$$
$$b^2 = a^2 + c^2 - 2ac \cos B$$

EXAMPLE: In triangle ABC, b = 14, c = 28.4 and A = 95°. Find a to the nearest tenth.

SOLUTION: $a^2 = (14)^2 + (28.4)^2 - 2(14)(28.4) \cos 95^\circ$

$a^2 = 196 + 806.56 - (-69.31)$

$a^2 = 1071.87$

Therefore, $a = 32.7$

If three sides of the triangle are given any of the angles may be determined by The Law of Cosines. Since $c^2 = a^2 + b^2 - 2ab \cos C$, solving for $\cos C$ yields:

$$\cos C = \frac{a^2 + b^2 - c^2}{2ab} \quad \text{and} \quad C = \cos^{-1}\left(\frac{a^2 + b^2 - c^2}{2ab}\right)$$

Similarly, $\cos A = \dfrac{b^2 + c^2 - a^2}{2bc}$ and $A = \cos^{-1}\left(\dfrac{b^2 + c^2 - a^2}{2bc}\right)$

Also, $\cos B = \dfrac{a^2 + c^2 - b^2}{2\ ac}$ and $B = \cos^{-1}\left(\dfrac{a^2 + c^2 - b^2}{2\ ac}\right)$

EXAMPLE: $a = 2$, $b = 3$, $c = 4$. Find angle C to tenths of a degree.

SOLUTION: $C = \cos^{-1}\left(\dfrac{a^2 + b^2 - c^2}{2\ ab}\right) = \cos^{-1}\left(\dfrac{2^2 + 3^2 - 4^2}{2(2)(3)}\right)$

$C = \cos^{-1}\left(\dfrac{-3}{12}\right) = \cos^{-1}(-.25) = 104.5^0$ rounded.

The remaining angles may be found either by application of The Law of Sines or the Law of Cosines and the supplementary relationship.

EXAMPLE: Find angles A and B in the example above.

SOLUTION: BY LAW OF SINES

$\dfrac{\sin A}{2} = \dfrac{\sin 104.5^0}{4}$

$\sin A = \dfrac{2 \sin 104.5^0}{4}$

$A = \sin^{-1} \dfrac{2 \sin 104.5^0}{4}$

$A = 29^0$

BY LAW OF COSINES

$A = \cos^{-1}\left(\dfrac{b^2 + c^2 - a^2}{2\ bc}\right)$

$A = \cos^{-1}\left(\dfrac{3^2 + 4^2 - 2^2}{2(3)(4)}\right)$

$A = \cos^{-1}(.875)$

$A = 29^0$

Therefore, $B = 180^0 - (104.5^0 + 29^0) = 46.5^0$.

Combinations of the Law of Sines and the Law of Cosines are frequently applied in problems of physics, engineering, navigation, surveying and many other practical fields. Of particular importance are those problems involving vector quantities.

EXAMPLE: A body is acted upon by two forces of 158 pounds and 235 pounds acting at an angle of 73^0. Find the angle at which the resultant force is inclined to the smaller force to the nearest degree.

SOLUTION: The vector parallelogram of forces is shown in Figure 8-3B.

Figure 8-3B

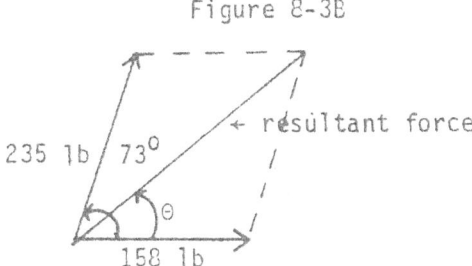

235 lb 73° ← resultant force

Θ

158 lb

From this diagram a triangle can be abstracted showing the known

quantities and suggesting the appropriate method for solution.

Figure 8-3C

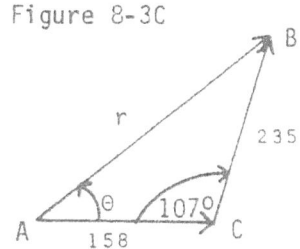

Since $C = 180^O - 73^O = 107^O$, the Law of Cosines gives the

resultant, $r^2 = (158)^2 + (235)^2 - 2(158)(235) \cos 107^O$

$r = 319.$

By the Law of Sines, $\dfrac{\sin \Theta}{235} = \dfrac{\sin 107^O}{319}$. Solving for Θ,

$\Theta = \sin^{-1} \dfrac{235 \sin 107^O}{319} = 44'.75^O.$

THE LAW OF TANGENTS.

The problem above can be solved without requiring that the resultant

be found first through use of a relationship known as the Law of Tangents.

If two sides, a and b, and the included angle, C, are given for a triangle

then, since $180^O - C = A + B$,

$$\frac{\tan \frac{1}{2}(A - B)}{\tan \frac{1}{2}(A + B)} = \frac{a - b}{a + b}$$

EXAMPLE: Solve the example above using the Law of Tangents.

SOLUTION: If a = 235, b = 158, and C = 107° then

$$\frac{\tan \frac{1}{2}(A - B)}{\tan \frac{1}{2}(73°)} = \frac{235 - 158}{235 + 158} = \frac{77}{393}$$

$$\tan \frac{1}{2}(A - B) = \frac{77 \tan 36.5°}{393} = .1450$$

$\frac{1}{2}(A - B) = \tan^{-1}(.1450) = 8.25°$, A - B = 16.5°.

Since A + B = 73° and A - B = 16.5°, adding gives 2A = 89.5°.

Therefore, A = 44.75° the angle between the resultant force and

the smaller force as found above.

DERVIATION OF THE LAW OF TANGENTS.

The derivation of this law is interesting in that it makes use of the

sum to product formulas studied in Chapter 6. Beginning with the Law of

Sines: $\frac{\sin A}{a} = \frac{\sin B}{b}$ and rearranging the terms of the proportion,

$\frac{\sin A}{\sin B} = \frac{a}{b}$. Then subtracting equivalents of 1 from both sides,

$\frac{\sin A}{\sin B} - \frac{\sin B}{\sin B} = \frac{a}{b} - \frac{b}{b}$ or $\frac{\sin A - \sin B}{\sin B} = \frac{a - b}{b}$.

Similarly, adding equivalents of 1 , $\frac{\sin A + \sin B}{\sin B} = \frac{a + b}{b}$.

Then $\dfrac{\dfrac{\sin A - \sin B}{\sin B}}{\dfrac{\sin A + \sin B}{\sin B}} = \dfrac{\dfrac{a - b}{b}}{\dfrac{a + b}{b}}$ or $\dfrac{\sin A - \sin B}{\sin A + \sin B} = \dfrac{a - b}{a + b}$.

Now, sin A - sin B = 2 sin $\frac{1}{2}$(A - B) cos $\frac{1}{2}$(A + B) by 6-3B

and sin A + sin B = 2 cos $\frac{1}{2}$(A - B) sin $\frac{1}{2}$(A + B) by 6-3A.

Substituting, $\dfrac{\sin A - \sin B}{\sin A + \sin B} = \dfrac{2 \sin \frac{1}{2}(A - B) \cos \frac{1}{2}(A + B)}{2 \cos \frac{1}{2}(A - B) \sin \frac{1}{2}(A + B)}$

$= \tan \frac{1}{2}(A - B) \cot \frac{1}{2}(A + B) = \dfrac{\tan \frac{1}{2}(A - B)}{\tan \frac{1}{2}(A + B)} = \dfrac{a - b}{a + b}$.

The last equality is the Law of Tangents. As seen in the example above, the Law of Tangents is used as an alternative to the Law of Cosines when two sides and the included angle of a triangle are given. Prior to the advent of electronic calculators the Law of Tangents lent itself more readily to logarithmic computations. Another example will illustrate this practice.

EXAMPLE: Find the distance from point A to point B across a lake if the distances from A and B to a third point C on the lake shore are 785 feet and 468 feet respectively, and the angle subtended by segment AB at point C is 58.14° .

SOLUTION: Figure 8-3D

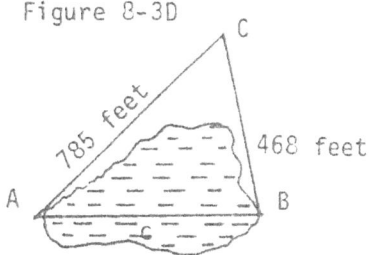

In this problem, the Law of Tangents takes the form:

$$\frac{\tan \frac{1}{2}(B - A)}{\tan \frac{1}{2}(B + A)} = \frac{b - a}{b + a}$$ with $\frac{1}{2}(B + A) = \frac{1}{2}(180^{\circ} - 58.14^{\circ}) = 60.93^{\circ}$,

$$b - a = (785 - 468) \text{feet} = 317 \text{ feet},$$

and $b + a = (785 + 468) \text{feet} = 1253 \text{ feet}.$

Substituting and solving for $\tan \frac{1}{2}(B - A)$,

$$\tan \frac{1}{2}(B - A) = \frac{317 \tan 60.93^{\circ}}{1253} .$$

$$\log \tan \frac{1}{2}(B - A) = \log 317 + \log \tan 60.93^{\circ} - \log 1253.$$

$$= 2.5011 + 0.2550 - 3.0980 = 0.6581 - 1 .$$

Using tables or calculators, $\frac{1}{2}(B - A) = 24.47^{\circ}$.

Since $B - A = 48.94^{\circ}$ and $B + A = 121.86^{\circ}$, $B = 85.40^{\circ}$.

The distance, c , across the lake is found by the Law of Sines:

$$c = \frac{b \sin C}{\sin B} = \frac{785 \sin 58.14^{\circ}}{\sin 85.40^{\circ}} \text{ and taking logarithms again,}$$

$$\log c = \log 785 + \log \sin 58.14^{\circ} - \log \sin 85.40^{\circ}$$

$$= 2.8949 + (0.9291 - 1) - (0.9986 - 1) = 2.8254.$$

Hence c = 669 feet.

Checking by the Law of Cosines,

$$c^2 = a^2 + b^2 - 2ab \cos C = (468)^2 + (785)^2 - 2(468)(785) \cos 58.14^{\circ}$$

$$c^2 = 447,409.23 \text{ and therefore, } c = 669 \text{ feet.}$$

EXERCISE SET 8-3.

In exercises 1 - 4 use the Law of Cosines to find the remaining side of the triangle:

1) a = 13 , b = 16 , C = 49° . 2) c = 12.8 , b = 31.4 , A = 34.6° .

3) a = 264 , c = 275 , B = 72.8° .

4) a = 159 , b = 314 , C = 102° .

In exercises 5 - 8 , use the Law of Tangents to find the unknown angles of $\triangle$ ABC :

5) a = 33.6 , b = 17.8 , C = 92.8° .

6) b = 145 , c = 285 , A = 33° 45' .

7) a = 1.95 , c = 3.91 , B = 67° .

8) b = 74.33 , c = 67.92 , A = 114.3° .

9) Two forces of 73 pounds and 84 pounds are acting on a body at an angle of 38° . Find the resultant to the nearest pound.

10) The diagonals of a parallelogram are 40 centimeters and 74 centimeters respectively. They intersect at an angle of 54° . Find the length

of the longer side of the parallelogram.

11) The diagonals of a parallelogram measure 46 and 68 inches respectively and the shorter side is 15 inches. Find the acute angle formed by the diagonals to the nearest tenth of a degree.

12) Two sides of a triangular plot are 135 feet and 275 feet respectively and the angle included between them is 47^0. Find the perimeter.

13) A body is acted upon by two forces of 28 pounds and 57 pounds. If the resultant is 73 pounds, find correct to the nearest degree, the angle between the resultant and the smaller force.

14) A boat is located 700 feet from a raft on a lake. Another raft is located 500 feet from the boat. If the angle subtended at the boat by lines of sight to the two rafts is 128^0, find the distance between the rafts to the nearest foot.

15) An air race is flown on a triangular course. If the distance for the first leg is 450 miles and the distance of the second leg is 600 miles find the length of the course if the angle between the first and second legs is 74^0.

Solve the following triangles completely. Round off to the nearest degree for angles and the nearest integer for lengths.

16) $A = 63^0$, $b = 76$, $c = 43$ 17) $C = 55^0$, $a = 164$, $b = 280$

18) $a = 95$, $b = 52$, $c = 64$ 19) $a = 117$, $b = 243$, $c = 184$

20) $a = 1438$, $b = 2854$, $c = 3452$

8-4. APPLICATIONS: AREAS OF TRIANGLES.

The basic formula for the area of a triangle that is learned in geometry is Area = ½ bh , where b is the base and h is the altitude that is drawn to the given base. This formula assumes that a side of a triangle is given and the length of the perpendicular from the opposite vertex to the given side is also known. Figure 8-4A shows the typical cases.

Figure 8-4A

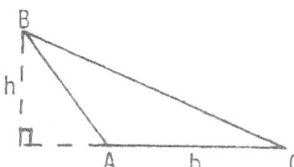

EXAMPLE: Find the area of a triangle with base 6 cm and altitude 9 cm.

SOLUTION: Let K = the area of the triangle.

K = ½ (6 cm)(9 cm) = 27 cm^2 . Note that the area is expressed in square units.

If the altitude isn't given it may possibly be determined by the given data. In Figure 8-4B, sides a and b and the included angle C are given.

Figure 8-4B

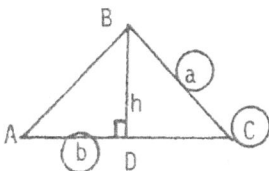

If an altitude is dropped from vertex B to D on side AC, then in right triangle CDB, sin C = $\dfrac{h}{a}$ or h = a sin C. Hence, K = ½ (a sin C)b = ½ ab sin C. The same pattern may be used if any two sides and the included angle are given: K = ½ ac sin B = ½ bc sin A.

EXAMPLE: Find the area of a triangle if a = 7, c = 10, and B = 94^0.

SOLUTION: K = ½ ac sin B = ½(7)(10)sin 94^0 = 35 sin 94^0 = 34.9 square

units rounded.

AREA: ASA

If two angles and the included side are given:

Figure 8-4C

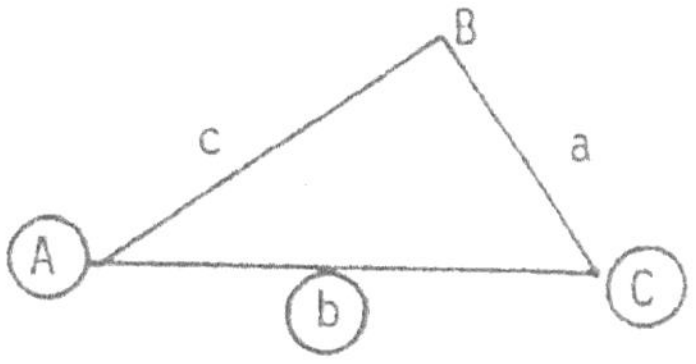

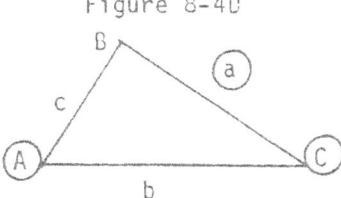

then since K = ½ ab sin C, side a must be determined. B is known since

B = 180^0 - (A + C). Hence by the Law of Sines, a = $\frac{b \sin A}{\sin B}$.

Substituting, K = ½ ab sin C = ½ $\frac{(b \sin A)}{\sin B}$ b sin C = ½ b^2 $\frac{\sin A \sin C}{\sin B}$.

EXAMPLE: Given B = 73^0, C = 44^0, and a = 13 inches, find the area of

triangle ABC.

SOLUTION: K = ½ a^2 $\frac{\sin B \sin C}{\sin A}$ = ½ (13)^2 $\frac{\sin 73^0 \sin 44^0}{\sin 63^0}$ = 63 in^2 .

AREA: AAS

If two angles and a side opposite one of them is given:

Figure 8-4D

then since B is obtainable by B = 180^0 - (A + C) and K = ½ ab sin C,

only b has to be determined by the Law of Sines. b = $\frac{a \sin B}{\sin A}$. Hence,

K = ½ ab sin C = ½ a($\frac{a \sin B}{\sin A}$)sin C = ½ a^2 $\frac{\sin B \sin C}{\sin A}$, a result that is identical to that obtained above in the ASA case.

AREA: SSA

If two sides and an angle opposite one of them is given the conditions for the ambiguous case are operative. Following the analysis of possible triangles the area(s) may be determined as follows:

Figure 8-4E

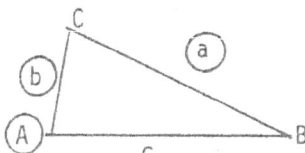

Since $K = \frac{1}{2} ab \sin C$, it is essential to express $\sin C$ in terms of the known quantities. $C = 180^0 - (A + B)$ and $\sin C = \sin(180^0 - (A + B))$. However, the reduction formula $\sin(180^0 - \theta) = \sin \theta$ yields $\sin C = \sin(A + B)$. With the given information B can be determined by the Law of Sines: $\dfrac{\sin B}{b} = \dfrac{\sin A}{a}$ or $\sin B = \dfrac{b \sin A}{a}$. Hence $B = \sin^{-1}\left(\dfrac{b \sin A}{a}\right)$.

Substituting, $\sin C = \sin(A + B) = \sin\left(A + \sin^{-1}\left(\dfrac{b \sin A}{a}\right)\right)$. Then the area

$K = \frac{1}{2} ab \sin\left(A + \sin^{-1}\left(\dfrac{b \sin A}{a}\right)\right)$. The formula looks complicated but is

easily programmable on a calculator. If the data results in two possible triangles, then either $B = \sin^{-1}\left(\dfrac{b \sin A}{a}\right)$ or $B = 180^0 - \sin^{-1}\left(\dfrac{b \sin A}{a}\right)$

and two different areas will be determined.

EXAMPLE: Given $a = 8$ cm, $b = 13$ cm, and $A = 35^0$, find the area of $\triangle ABC$.

SOLUTION: Draw a figure.

Figure 8-4F

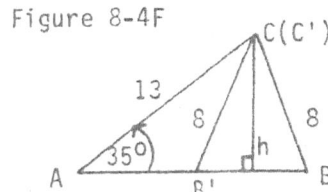

Since a < b and a > h = 13 sin 35° = 7.46, two triangles are possible as shown in Figure 8-4F. Let K_1 = area of $\triangle$ ABC and K_2 = area of $\triangle$ AB'C'. From the foregoing discussion,

$$K_1 = \tfrac{1}{2} \, ab \, \sin \left(A + \sin^{-1}\left(\frac{b \, \sin \, A}{a}\right)\right)$$

$$= \tfrac{1}{2} \, (8)(13) \, \sin \left(35^\circ + \sin^{-1}\left(\frac{(13) \, \sin \, 35^\circ}{(8)}\right)\right) = 50.5 \text{ cm}^2$$

$$K_2 = \tfrac{1}{2} \, ab \, \sin (A + B') = \tfrac{1}{2} \, ab \, \sin (A + (180^\circ - B))$$

$$= \tfrac{1}{2} \, ab \, \sin \left(A + \left(180^\circ - \sin^{-1}\left(\frac{b \, \sin \, A}{a}\right)\right)\right)$$

$$= \tfrac{1}{2} \, (8)(13) \, \sin \left(35^\circ + \left(180^\circ - \sin^{-1}\left(\frac{(13) \, \sin \, 35^\circ}{(8)}\right)\right)\right) = 28.9 \text{ cm}^2.$$

AREA: SSS

If three sides of a triangle are given its area may be determined with the aid of the Law of Cosines.

Figure 8-4G

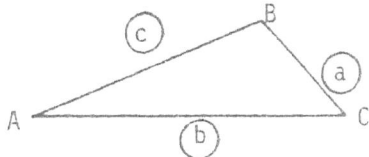

Starting with the basic formula, $K = \tfrac{1}{2} \, ab \, \sin C$, C may be found by

$$\cos C = \frac{a^2 + b^2 - c^2}{2 \, ab} \quad \text{or} \quad C = \text{Cos}^{-1}\left(\frac{a^2 + b^2 - c^2}{2 \, ab}\right).$$

Therefore, $K = \tfrac{1}{2} \, ab \, \sin \left(\text{Cos}^{-1}\left(\frac{a^2 + b^2 - c^2}{2 \, ab}\right)\right).$

EXAMPLE: Find the area of a triangle with sides 8, 12, 15 respectively.

SOLUTION: Let a = 8, b = 12, and c = 15. Then

$$K = \tfrac{1}{2} \, (8)(12) \, \sin \left(\text{Cos}^{-1}\left(\frac{8^2 + 12^2 - 15^2}{2 \, (8)(12)}\right)\right) = 47.8 \text{ square units}.$$

HERON'S FORMULA. SSS.

If the three sides of a triangle are given there is a simpler formula attributed to the Greek mathematician, Heron of Alexandria (ca.100 A.D.) that is more convenient to use:

$$K = \sqrt{s\,(s-a)(s-b)(s-c)} \text{ , where } s = \tfrac{1}{2}(a+b+c).$$

As defined, the letter s is the semiperimeter of the triangle.

DERIVATION OF HERON'S FORMULA.

The derivation of Heron's formula begins with the basic formula $K = \tfrac{1}{2} ab \sin C$. Since a and b are known, an expression for sin C in terms of the given sides is sought. Beginning with the Pythagorean identity,

$\sin C = \sqrt{1 - \cos^2 C}$ and substituting by the Law of Cosines,

$$\sin C = \sqrt{1 - \frac{(a^2 + b^2 - c^2)^2}{(2\,ab)^2}} = \sqrt{\left(1 + \frac{a^2 + b^2 - c^2}{2\,ab}\right)\left(1 - \frac{a^2 + b^2 - c^2}{2\,ab}\right)}$$

$$= \sqrt{\frac{\left((a+b)^2 - c^2\right)}{2\,ab} \cdot \frac{\left(c^2 - (a-b)^2\right)}{2\,ab}}$$

$$= \frac{1}{2ab}\sqrt{(a+b+c)(a+b-c)(c+a-b)(c-a+b)}$$

Substituting in $K = \tfrac{1}{2} ab \sin C$,

$$K = \frac{1}{4}\sqrt{(a+b+c)(a+b-c)(a+c-b)(b+c-a)}$$

Since $2s = a+b+c$, $2s - 2c = a+b-c$, $2s - 2b = a+c-b$, and $2s - 2a = b+c-a$, substituting these quantities yields

$$K = \frac{1}{4}\sqrt{2s\,(2s-2c)(2s-2b)(2s-2a)}$$

$$= \frac{1}{4}\sqrt{16\,s\,(s-a)(s-b)(s-c)} = \sqrt{s\,(s-a)(s-b)(s-c)}\,.$$

EXAMPLE: Find the area of a triangle with sides 8, 12, 15 by Heron's

formula.

SOLUTION: $s = \frac{1}{2}(a + b + c) = \frac{1}{2}(8 + 12 + 15) = 17.5$.

$s - a = 17.5 - 8 = 9.5$, $s - b = 17.5 - 12 = 5.5$,

$s - c = 17.5 - 15 = 2.5$

$K = \sqrt{s(s - a)(s - b)(s - c)} = \sqrt{(17.5)(9.5)(5.5)(2.5)}$

$K = 47.8$ square units.

The results are identical to the solution obtained in the

previous example.

EXERCISE SET 8-4 .

In exercises 1 - 10 find the area of $\triangle$ ABC from the given data.

1) $a = 7$, $b = 12$, $C = 59^{O}$ 2) $b = 15.86$, $c = 13.95$, $A = 112^{O}$

3) $a = 754$, $c = 343$, $C = 28^{O}$ 4) $a = 1243$, $b = 768$, $B = 55^{O}$

5) $a = 19.79$, $b = 74.3$, $B = 95^{O}$ 6) $b = 57$, $c = 94$, $C = 115^{O}$

7) $a = 13.8$, $b = 7.95$, $c = 13.42$ 8) $a = 94$, $b = 117$, $c = 63$

9) $A = 57^{O}$, $B = 93^{O}$, $c = 48.6$ 10) $B = 33^{O}$, $C = 48^{O}$, $a = 98.7$

11) In triangle ABC , $c = 8$, $b = 14$, and $\tan C = 2/3$. Find the area
     of the triangle.

12) The area of triangle EFG is 748 square centimeters. If $e = 35$ centi-
     meters and $G = 78^{O}$, find f .

13) If the vertex angle of an isosceles triangle is 38° and each leg is
     11 , find the area of the triangle.

14) Find the area of an equilateral triangle that measures 17 units for
     each side.

Use Heron's formula to find the areas of the triangles with the given sides:

15) a = 14 , b = 17 , c = 19   16) a = 24 , b = 18 , c = 38

17) The sides of a pentagon measure 15 . Find its area .

18) Two sides of a parallelogram are 43 and 57 respectively. The included angle measures 48^0 . Find the area of the parallelogram.

19) A triangular plot measures 78 feet, 94 feet, and 125 feet. Find its area.

20) Prove that the area of any quadrilateral is equal to one half the product of its two diagonals multiplied by the sine of the included angle.

8-5. CHAPTER SUMMARY

Given $\triangle$ ABC with sides a, b, c and opposite angles A, B, C respectively:

1. If $C = 90^0$ the trigonometric ratios for angle A are given as follows:

$$\sin A = \frac{\text{opposite side}}{\text{hypotenuse}} = \frac{a}{c} \qquad \csc A = \frac{\text{hypotenuse}}{\text{opposite side}} = \frac{c}{a}$$

$$\cos A = \frac{\text{adjacent side}}{\text{hypotenuse}} = \frac{b}{c} \qquad \sec A = \frac{\text{hypotenuse}}{\text{adjacent side}} = \frac{c}{b}$$

$$\tan A = \frac{\text{opposite side}}{\text{adjacent side}} = \frac{a}{b} \qquad \cot A = \frac{\text{adjacent side}}{\text{opposite side}} = \frac{b}{a}$$

2. LAW OF SINES: $\dfrac{a}{\sin A} = \dfrac{b}{\sin B} = \dfrac{c}{\sin C}$

3. THE AMBIGUOUS CASE: SSA. If two sides and an angle opposite one of them are given, then either one triangle, two triangles, or no triangle can be formed from the given data. See section 8-2 for a complete analysis.

4. LAW OF COSINES: SAS. If two sides, a and b, and the included angle C are given, then $c^2 = a^2 + b^2 - 2ab \cos C$.

5. LAW OF COSINES: SSS. If sides a, b, and c are given then

$$\cos C = \frac{a^2 + b^2 - c^2}{2ab}$$

6. LAW OF TANGENTS: SAS. If two sides, a and b, and the included angle C are given, then $\dfrac{\tan \frac{1}{2}(A - B)}{\tan \frac{1}{2}(A + B)} = \dfrac{a - b}{a + b}$, with $C = 180^0 - (A + B)$.

7. AREAS OF TRIANGLES.

Let K = area of $\triangle$ ABC

a) If a side is given and the altitude to that side is known, then $K = \frac{1}{2}$ (side)(altitude). The given side is usually called the base and denoted by the letter b. The altitude to the base is lettered as h. Then the formula is $K = \frac{1}{2}bh$.

b) Given two sides, a and b, and the included angle C,

$$K = \tfrac{1}{2} \, ab \sin C$$

c) Given two angles and a side,

$$K = \tfrac{1}{2} \, a^2 \, \frac{\sin B \sin C}{\sin A} = \tfrac{1}{2} \, b^2 \, \frac{\sin A \sin C}{\sin B} = \tfrac{1}{2} \, c^2 \, \frac{\sin A \sin B}{\sin C}$$

d) Given two sides, a and b, and the angle opposite one of them, say A,

then either $K = \tfrac{1}{2} \, ab \sin \left(A + \sin^{-1}\left(\frac{b \sin A}{a}\right)\right)$ or

$$K = \tfrac{1}{2} \, ab \sin \left(A + \left(180^0 - \sin^{-1}\left(\frac{b \sin A}{a}\right)\right)\right).$$

The choice of K above depends upon the outcome of the analysis for

the ambiguous case.

e) Given the three sides a, b, and c then

$$K = \tfrac{1}{2} \, ab \sin \left(Cos^{-1}\left(\frac{a^2 + b^2 - c^2}{2 \, ab}\right)\right) \text{ or}$$

$$K = \sqrt{s \, (s - a)(s - b)(s - c)} \text{ where } s = \tfrac{1}{2} \, (a + b + c).$$

The last formula is known as Heron's formula.

EXERCISE SET 8-5.

In exercises 1 - 10 solve the triangles completely from the given data:

1) $\triangle$ ABC , $C = 90^0$, $a = 7$, $b = 3$.

2) $\triangle$ EFG , $G = 90^0$, $\sin F = 1/3$, $g = 16$.

3) $\triangle$ HIK , $h = 37$, $I = 74^0$, $K = 39^0$.

4) $\triangle$ LMN , $m = 943$, $M = 58^0$, $N = 7^0$.

5) $\triangle$ OPQ , $p = 17.8$, $q = 11.6$, $O = 103^0$.

6) $\triangle$ RST , $r = 95.7$, $s = 84.8$, $t = 39.5$.

7) $\triangle$ DEF , $\sin D = .6843$, $\cos E = .9437$, $f = 67$.

8) $\triangle$ XYZ , $\tan X = -1.432$, $x = 133$, $y = 96$.

9) $\triangle$ UVW , u = 78.6 , v = 117.6 , area = 3838 square units.

10) $\triangle$ JKL , k/j = 3/4 , k = 54.6 , cos L = -.8456 .

Find the areas of the triangles from the given data:

11) $\triangle$ ABC , a = 7, b = 9 , c = 14 .

12) $\triangle$ DEF , d = 67 , e = 95 , F = 78° .

13) $\triangle$ GHI , g = 94.6 , h = 117.6 , G = 57° .

14) $\triangle$ JKL , j = 17 , k = 47 , L = 90° .

15) $\triangle$ MNO , m = 9 , N = 43° , O = 102° .

16) $\triangle$ PQR , p = 65 , q = 44 , P = 74° .

SOLVE THE FOLLOWING PROBLEMS:

17) From a point 115 feet from the foot of a flagpole the angle of elevation to the top is 37° . Find the height of the flagpole to the
nearest foot.

18) A balloon is hovering directly over a marker on the ground. The
balloonist sights a second marker at an angle of depression of 39° 15'
located 7480 yards from the first marker. What is the height of the
balloon to the nearest yard?

19) A TV antennae mast is braced by a guy wire 12 feet from its top. If
the guy wire is 79 feet long and makes an angle of 55° with the mast
how high is the mast?

20) A climber is on top of a cliff that descends to a river bed below.
The angle of depression of the face of the cliff measures 83° . If
the cliff is 342 feet high what is the minimum amount of rope needed
to descend to the bottom if the rope is fastened at the top of the
cliff? Assume the rope will touch the bottom of the cliff.

21) From a point 115 feet from the base of a building with a sign located on its roof an observer measures two angles of elevation. If the angle to the top of the sign is 57° and the angle to its bottom is 38°, how tall is the sign? Assume the angles are measured from ground level.

22) At the top of a hill an observer measures angles of depression of 16° and 39° to the top and bottom respectively of a tree located on the slope of the hill below him. If the tree is 165 feet high, how far is the observer from the base of the tree?

23) Two forces of 116 pounds and 148 pounds act at an angle of 47° on a body. Find the magnitude of the resultant force and the angle between the resultant and the smaller force.

24) A balloon hovers 2840 feet directly over a docking point on the ground. A second balloon is located at an angle of depression of 38° from the first balloon. If the angle of depression from the second balloon to the docking point is 41°, find the distance between the two balloons.

SOLUTIONS TO ODD NUMBERED EXERCISES

CHAPTER 1

Exercise Set 1-1.

1) a) Perimeter of square = $8\sqrt{2}$ $\doteq$ 11.31 units.

   b) Perimeter of octagon = $16\sqrt{2 - \sqrt{2}}$ $\doteq$ 12.25 units

      Note that the circumference of the circle = $4\pi \doteq$ 12.57 units.

3) a) 23.24   b) 2.83   c) .5   d) .01

5) .05%   7) 12.08 meters

9) a) $\frac{17\pi}{12}$ = 4.45   b) $\frac{5\pi}{12}$ = 1.31   c) $\frac{9\pi}{2}$ = 14.14   d) $\frac{2\pi}{3}$ = 2.09

11)

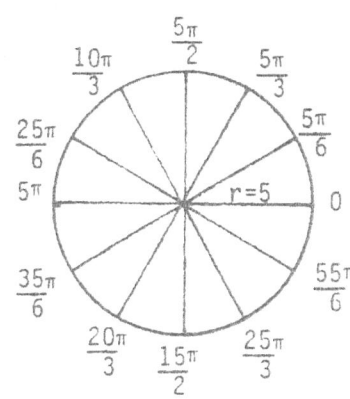

13) Each division number is divided by 5 .

15) 66.6 π 17) 126 revolutions

19) 42 revolutions in 20 seconds or 2.1 revolutions per second.

Exercise Set 1-2.

1) IV   3) II   5) III   7) IV   9) III

11) III   13) III   15) I   17) none ($\frac{3\pi}{2}$ divides quadrants

                                                III and IV)

Exercise Set 1-2.

19) I   21)   147.25   23)   -1802.70

25) $\frac{\pi}{4} < 1 < \frac{\pi}{3}$ $\frac{\pi}{2} < 2 < \frac{2\pi}{3}$ $\frac{5\pi}{6} < 3 < \pi$

Exercise Set 1-3.

1) 120^{0} 3) 252^{0} 5) -67.5^{0} 7) -243.51^{0} 9) 280^{0}

11) 63^{0} 13) 60^{0} 15) 193.09^{0} 17) 157.5^{0} 19) -40.58^{0}

Exercise Set 1-4.

1) $\frac{\pi}{4}$ 3)   2.58 or $\frac{37\pi}{45}$ 5) $-\frac{11\pi}{6}$ 7)   4.43   9)   1.35

11) 150^{0} 13) -240^{0} 15) 143.24^{0} 17) -4.01^{0} 19) 19.48^{0}

21)   1.6   23) $\frac{2\pi}{3}$ 25)   3.2   27) $\frac{4\pi}{3}$ 29)   -4.36

31) -48^{0} 21' 00"   33) 245^{0} 49' 48"   35) 337^{0} 39' 00"

37) 45.26^{0} 39) -243.91^{0} 41)   III   43)   III

Review Exercises.   Exercise Set 1-5.

1)   14.45   3) $6\pi^{2}$ 5)   3.14   7) $\frac{2\pi - 1}{2\pi}$ or 0.84   9)   II

11)   IV   13) $\frac{7\pi}{6} < 3.8 < \frac{8\pi}{6}$ 15) $\frac{4\pi}{6} < 2.1 < \frac{5\pi}{6}$ 17) $\frac{7\pi}{6}$

19) $\frac{7\pi}{4}$ 21)   -1.93   23)   2.52   25) 144^{0} 27) -280^{0}

29) 158.14^{0} 31) -275.02^{0} 33) $\frac{7\pi}{32}$ 35)   .51   37)   1.61

39) $\frac{24\pi}{5}$ or 4.8π 41) 45^{0}, 135^{0}, 225^{0}, 315^{0}

43) 90^{0}, 180^{0}, 270^{0}, 360^{0} 45) $\frac{\pi}{6}$, $\frac{\pi}{4}$, $\frac{\pi}{3}$, $\frac{\pi}{2}$

47) $\frac{5\pi}{3}$, $\frac{11\pi}{6}$, $\frac{7\pi}{4}$, $\frac{4\pi}{3}$

CHAPTER 2

Exercise Set 2-1.

1) $x^2 + y^2 = 4$ 3) $x^2 + y^2 = \pi^2$ 5) $x^2 + y^2 = 7$

7) $x^2 + y^2 = 0.1369$ 9) $\sqrt{15}$ 11) .2   13) 2.41

15) Yes   17) No   19) No   21) IV   23) I

25) II   27) III

Exercise Set 2-2.

1) $\cos \theta = \dfrac{-3}{\sqrt{13}}$, $\sin \theta = \dfrac{2}{\sqrt{13}}$ 3) $\cos \theta = \dfrac{-1}{\sqrt{10}}$, $\sin \theta = \dfrac{-3}{\sqrt{10}}$

5) $\cos \theta = \dfrac{-15}{17}$, $\sin \theta = \dfrac{8}{17}$ 7) $\cos \theta = \dfrac{-3}{5}$, $\sin \theta = \dfrac{-4}{5}$

9) $\cos \theta = \dfrac{12}{13}$, $\sin \theta = \dfrac{-5}{13}$

11) II, III   13) III   15) None   17) III, IV   19) IV

21) $\cos \theta = \dfrac{x}{r}$, $\sin \theta = \dfrac{y}{r}$ definition

      $x^2 + y^2 = r^2$ definition

      $\dfrac{x^2}{r^2} + \dfrac{y^2}{r^2} = 1$ dividing by r^2

      $\left(\dfrac{x}{r}\right)^2 + \left(\dfrac{y}{r}\right)^2 = 1$ $\dfrac{a^2}{b^2} = \left(\dfrac{a}{b}\right)^2$

      $(\cos \theta)^2 + (\sin \theta)^2 = 1$ substitution

23) 90°, 270° 25) $|\sin \theta| \le 1$ 27) $\cos 270^\circ = 0 = \sin 0^\circ$

29) $\cos \dfrac{\pi}{2} + \sin \dfrac{\pi}{2} = 0 + 1 = 1$

Exercise Set 2-3.

1) $\dfrac{-1}{\sqrt{2}}$ 3) $\dfrac{1}{2}$ 5) $\dfrac{1}{2}$ 7) $\dfrac{3}{4}$ 9) $\dfrac{\sqrt{3}}{2}$ 11) 0

13) $\sqrt{3}$ 15) $\dfrac{\sqrt{3}+1}{2}$ 17) $\dfrac{1}{2}$ 19) $\sqrt{3}$ 21) $\dfrac{3\pi}{4}$, $\dfrac{5\pi}{4}$

23) $\frac{7\pi}{6}$, $\frac{11\pi}{6}$ 25) $\frac{\pi}{2}$ 27) None   29) 0, π, 2π

31) All θ, $0 \le \theta \le 2\pi$ 33) $\frac{\pi}{4}$, $\frac{5\pi}{4}$ 35) $\frac{\pi}{2}$

37) $\frac{7\pi}{18}$, $\frac{11\pi}{18}$, $\frac{19\pi}{18}$, $\frac{23\pi}{18}$, $\frac{31\pi}{18}$, $\frac{35\pi}{18}$ 39) $\frac{2\pi}{9}$, $\frac{14\pi}{9}$

Exercise Set 2-4.

1) 115^0 3) 156^0 5) 298^0 7) 284^0 9) $\frac{\pi}{2}$

11) 4.52   13) $\frac{15\pi}{8}$ 15) 2.67   17) 0   19) 0.5

21) 0   23) $-\sqrt{3}$ 25) $\frac{\sqrt{3}}{2}$ 27) 1   29) $\frac{-1}{\sqrt{2}}$

31) $\sin 45^0$ 33) $-\cos 60^0$ 35) $-\cos \frac{\pi}{4}$ 37) $-\sin 45^0$

39) $\sin 30^0$ 41) False   43) True   45) True

47) False   49) True

Exercise Set 2-5.

1) .2193   3) .7790   5) -.9967   7) -.7234   9) -.0785

11) .9511   13) -.9135   15) .7986   17) .2952   19) .9998

21) 35^0 30'   23) 56^0 40'   25) 49^0 50'   27) 0^0 40'

29) ..-4^0 50'   31) 69^0 40'   33) 150^0 30'   35) 139^0

37)

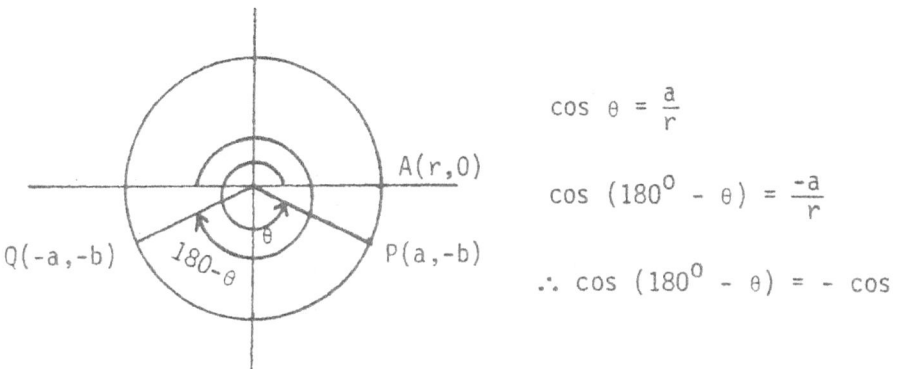

$\cos \theta = \frac{a}{r}$

$\cos (180^0 - \theta) = \frac{-a}{r}$

$\therefore \cos (180^0 - \theta) = -\cos \theta$

39)

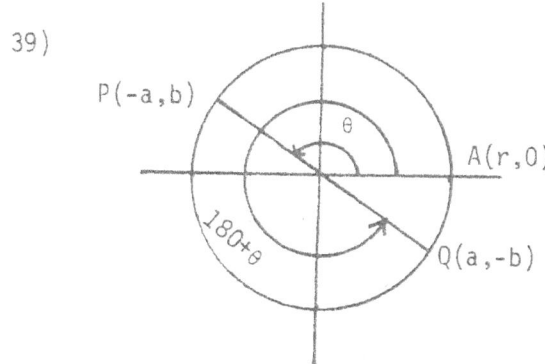

$$\sin \theta = \frac{b}{r}$$

$$\sin (180^0 + \theta) = \frac{-b}{r}$$

$$\therefore \sin (180^0 + \theta) = - \sin \theta$$

Exercise Set 2-6.

1) .6749   3) .9940   5) .2280   7) 1.01   9) 1.23

11) 1.00   13) .89   15) .2867   17) .9891   19) .9449

21) .9998   23) .3827   25) .8090   27) .4161

29) .5403   31) .9093   33) 5.62, 3.80   35) .98, 5.30

37) 1.03, 2.11   39) .68, 5.60

Exercise Set 2-7. Chapter Review.

1) $x^2 + y^2 = 25$ 3) $x^2 + y^2 = 81.09$ 5) $\frac{4}{5}$ 7) .4997

9) $-.3420$ 11) .1825   13) 340^0 15) 10.52^0 17) 0

19) $\frac{1}{\sqrt{2}}$ 21) $-.3827$ 23) $\frac{1}{2}$ 25) .7880   27) $-.9088$

29) $-.9991$ 31) .9767   33) $s = -1, t = 0$ 35) False

37) False   39) True   41) False   43) True

45) .76 , 2.38   47) .34 , 5.94

49) 4.62 , 4.80   51) 2.96 , 3.32

SOLUTIONS TO ODD NUMBERED PROBLEMS

CHAPTER 3

Exercise Set 3-1

1) $\tan \theta = \frac{-2}{5}$, $\cot \theta = \frac{-5}{2}$

3) $\tan \theta = \frac{-5}{12}$, $\cot \theta = \frac{-12}{5}$

5) $\frac{-8}{5}$

7) .3778

9)

θ:	π	$\frac{3\pi}{2}$	2π
$\tan \theta$:	0	undefined	0
$\cot \theta$:	undefined	0	undefined

11) $\tan 45^O$

13) $\tan 45^O$

15) $- \tan 65^O$

17) $\tan 14^O$

19) $- \tan 60^O$

21) $\cot \frac{\pi}{4}$

23) $- \cot \frac{3\pi}{8}$

25) $\cot \frac{3\pi}{10}$

27) $\cot \frac{\pi}{3}$

29) $- \cot 1.01$

31) 1.402

33) -1.428

35) 1.280

37) -1.117

39) 0.3281

41) 9.188

43) -2.414

45) 0.3506

47) 4.502

49) -1.376

51) 6° 30'

53) 20° 40'

55) 64° 10'

57) .68

59) 1.48

61) 155° 20'

63) 248° 40'

65) 186° 40'

67) 102° 30'

Exercise Set 3-2.

	Sine	Cosine	Tangent	Cotangent
1)	$\frac{3}{\sqrt{13}}$	$\frac{2}{\sqrt{13}}$	$\frac{3}{2}$	$\frac{2}{3}$
3)	$\frac{-7}{25}$	$\frac{-24}{25}$	$\frac{7}{24}$	$\frac{24}{7}$
5)	$\frac{8}{17}$	$\frac{-15}{17}$	$\frac{-8}{15}$	$\frac{-15}{8}$
7)	$\frac{1}{\sqrt{5}}$	$\frac{2}{\sqrt{5}}$	$\frac{1}{2}$	$\frac{2}{1}$

9) 56°

11) 196°

13) 152°

15) 27°

17) $\tan\theta\cot\theta + \cos\theta\tan\theta = \dfrac{\sin\theta}{\cos\theta}\cdot\dfrac{\cos\theta}{\sin\theta} + \dfrac{\cos\theta}{1}\cdot\dfrac{\sin\theta}{\cos\theta}$

$$= 1 + \sin\theta$$

19) $\dfrac{1 - \sin^2\theta}{\cos^2\theta} = \dfrac{\cos^2\theta}{\cos^2\theta} = 1$

21) $\dfrac{\cos\theta}{1 - \sin\theta} - \tan\theta = \dfrac{\cos\theta}{1 - \sin\theta} - \dfrac{\sin\theta}{\cos\theta} = \dfrac{\cos^2\theta}{(1 - \sin\theta)(\cos\theta)} - \dfrac{(\sin\theta)(1 - \sin\theta)}{(\cos\theta)(1 - \sin\theta)}$

$$= \dfrac{\cos^2\theta - \sin\theta + \sin^2\theta}{\cos\theta\,(1 - \sin\theta)} = \dfrac{(\cos^2\theta + \sin^2\theta) - \sin\theta}{\cos\theta\,(1 - \sin\theta)}$$

$$= \dfrac{(1 - \sin\theta)}{\cos\theta\,(1 - \sin\theta)} = \dfrac{1}{\cos\theta}$$

23) $\dfrac{1 + \cot^2\theta}{1 + \tan^2\theta} = \dfrac{1 + \dfrac{\cos^2\theta}{\sin^2\theta}}{1 + \dfrac{\sin^2\theta}{\cos^2\theta}} = \dfrac{\dfrac{\sin^2\theta + \cos^2\theta}{\sin^2\theta}}{\dfrac{\cos^2\theta + \sin^2\theta}{\cos^2\theta}} = \dfrac{\dfrac{1}{\sin^2\theta}}{\dfrac{1}{\cos^2\theta}}$

$$= \dfrac{1}{\sin^2\theta}\cdot\dfrac{\cos^2\theta}{1} = \dfrac{\cos^2\theta}{\sin^2\theta} = \cot^2\theta$$

25) $\tan\theta$ increases from 0 to ∞

27) true   29) true   31) indeterminate   33) true

35) true

Exercise Set 3-3.

1) true   3) false   5) true   7) true   9) false

11) true   13) false   15) true   17) false   19) false

21) $\sin 36^{\circ}$ 23) $\cos 11^{\circ}$ 25) $\cos 22^{\circ}$ 27) $-\cot 5^{\circ}$

29) $-\sin 39^{\circ}$ 31) $-\cot\dfrac{\pi}{8}$ 33) $\sin\dfrac{\pi}{14}$ 35) $-\tan\dfrac{\pi}{10}$

37) $-\sin\left(2 - \dfrac{\pi}{2}\right)$ or $-\sin .43$ 39) $\cot .41$

Exercise Set 3-4.

1) 1.529 3) 1.036 5) -1.056 7) 3.599

9) 2.906 11) $-\sqrt{2}$ 13) -2 15) 1.131

17) Undefined 19) 3.790

21) $\cos \theta = \frac{-3}{5}$, $\tan \theta = \frac{-4}{3}$, $\cot \theta = \frac{-3}{4}$, $\sec \theta = \frac{-5}{3}$, $\csc \theta = \frac{5}{4}$

23) $\sin \theta = \frac{2}{\sqrt{5}}$, $\cos \theta = \frac{1}{\sqrt{5}}$, $\tan \theta = 2$, $\cot \theta = \frac{1}{2}$, $\sec \theta = \sqrt{5}$

25) $\sin \theta = -\frac{\sqrt{15}}{4}$, $\tan \theta = -\sqrt{15}$, $\cot \theta = \frac{-1}{\sqrt{15}}$, $\sec \theta = 4$, $\csc \theta = \frac{-4}{\sqrt{15}}$

27) $\sin \theta = \frac{1}{2\sqrt{2}} = \frac{\sqrt{2}}{4}$, $\cos \theta = \frac{-\sqrt{14}}{4}$, $\tan \theta = \frac{-1}{\sqrt{7}}$, $\cot \theta = -\sqrt{7}$, $\sec \theta = \frac{-4}{\sqrt{14}}$

29) .21 31) 2.23

33) $\cos^2 \theta - \sin^2 \theta = \cos^2 \theta - (1 - \cos^2 \theta) = \cos^2 \theta - 1 + \cos^2 \theta$
$$= 2 \cos^2 \theta - 1$$

35) $(1 + \sin \theta)(1 - \sin \theta) = 1 - \sin^2 \theta = \cos^2 \theta$

37) $\frac{\sin \theta}{1 - \cos \theta} + \frac{\sin \theta}{1 + \cos \theta} = \frac{\sin \theta (1 + \cos \theta) + \sin \theta (1 - \cos \theta)}{(1 - \cos \theta)(1 + \cos \theta)}$

$= \frac{\sin \theta + \sin \theta \cos \theta + \sin \theta - \sin \theta \cos \theta}{1 - \cos^2 \theta} = \frac{2 \sin \theta}{\sin^2 \theta} = \frac{2}{\sin \theta}$

$= 2 \cdot \frac{1}{\sin \theta}$

$= 2 \csc \theta$

39) $\cos^2 \theta \cot^2 \theta = \cos^2 \theta \cdot \frac{\cos^2 \theta}{\sin^2 \theta} = \frac{\cos^2 \theta (1 - \sin^2 \theta)}{\sin^2 \theta}$

$= \frac{\cos^2 \theta}{\sin^2 \theta} - \frac{\cos^2 \theta \sin^2 \theta}{\sin^2 \theta} = \cot^2 \theta - \cos^2 \theta$

41) $\frac{1 + \cos \theta}{1 + \sec \theta} = \frac{1 + \cos \theta}{1 + \frac{1}{\cos \theta}} = \frac{1 + \cos \theta}{\frac{\cos \theta + 1}{\cos \theta}} = \frac{(1 + \cos \theta)}{1} \cdot \frac{\cos \theta}{(\cos \theta + 1)} = \cos \theta$

43) $\frac{\sec^3 \theta}{\tan^2 \theta} = \frac{\sec^3 \theta}{\sec^2 \theta - 1}$ 45) $- \csc 33^0$ 47) $- \sec 24^0$

49) $- \csc 4^0 \, 50'$ 51) $- \csc \frac{\pi}{8}$ 53) $\frac{3}{\sqrt{2}}$ 55) 7

Exercise Set 3-5.

1) $\sin \theta = \frac{-4}{5}$, $\tan \theta = \frac{-4}{3}$, $\cot \theta = \frac{-3}{4}$, $\sec \theta = \frac{5}{3}$, $\csc \theta = \frac{-5}{4}$

3) $\sin \theta = \frac{1}{\sqrt{5}}$, $\cos \theta = \frac{2}{\sqrt{5}}$, $\tan \theta = \frac{1}{2}$, $\cot \theta = 2$, $\sec \theta = \frac{\sqrt{5}}{2}$

5) $\cos \theta = \frac{-1}{2}$, $\tan \theta = \sqrt{3}$, $\cot \theta = \frac{1}{\sqrt{3}}$, $\sec \theta = -2$, $\csc \theta = \frac{-2}{\sqrt{3}}$

7) $\cos \theta = 0$, $\tan \theta$ undefined, $\cot \theta = 0$, $\sec \theta$ undefined, $\csc \theta = -1$

9) true   11) indeterminate   13) false   15) true

17) true   19) false

21) $\cot \theta \sec \theta = \dfrac{\cos \theta}{\sin \theta} \cdot \dfrac{1}{\cos \theta} = \dfrac{1}{\sin \theta} = \csc \theta$

23) $\sin \theta \cot \theta + \cos \theta \tan \theta = \sin \theta \cdot \dfrac{\cos \theta}{\sin \theta} + \cos \theta \cdot \dfrac{\sin \theta}{\cos \theta}$

$$= \cos \theta + \sin \theta$$

$$= \sin \theta + \cos \theta$$

25) $\dfrac{1}{\tan \theta + \cot \theta} = \dfrac{1}{\dfrac{\sin \theta}{\cos \theta} + \dfrac{\cos \theta}{\sin \theta}} = \dfrac{1}{\dfrac{\sin^2 \theta + \cos^2 \theta}{\cos \theta \sin \theta}}$

$$= \dfrac{\cos \theta \sin \theta}{\sin^2 \theta + \cos^2 \theta} = \dfrac{\cos \theta \sin \theta}{1} = \sin \theta \cos \theta$$

27) $\dfrac{1 + \sec \theta}{\sec \theta} = \dfrac{1}{\sec \theta} + \dfrac{\sec \theta}{\sec \theta} = \cos \theta + 1 = \dfrac{(1 + \cos \theta)}{1} \dfrac{(1 - \cos \theta)}{(1 - \cos \theta)}$

$$= \dfrac{1 - \cos^2 \theta}{1 - \cos \theta} = \dfrac{\sin^2 \theta}{1 - \cos \theta}$$

29) 5.089   31) -3.078   33) -.6152   35) 0

37) $\frac{1}{4}$ 39) 1   41) $\frac{1}{\sqrt{2}} + \frac{1}{\sqrt{3}}$ 43) $\frac{-1}{\sqrt{2}}$

45) $\sec 22^{0}$ 47) $- \cot 19^{0}$ 49) $\cot 30^{0}$ 51) $- \cos 42.6^{0}$

SOLUTIONS TO ODD NUMBERED PROBLEMS

CHAPTER 4

Exercise Set 4-1.

1) Function one-to-one.   3) Not a function

5) Function many-to-one   7) Not a function

9) Not a function   11) $\cos \theta$, $-\pi \leq \theta \leq 3\pi$

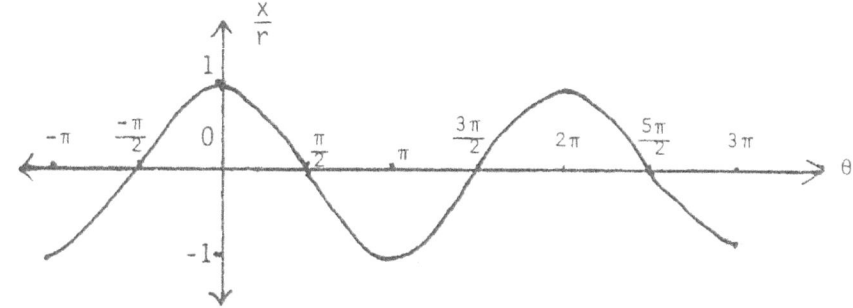

13) $\sin \theta$, $-2\pi \leq \theta \leq \frac{\pi}{2}$

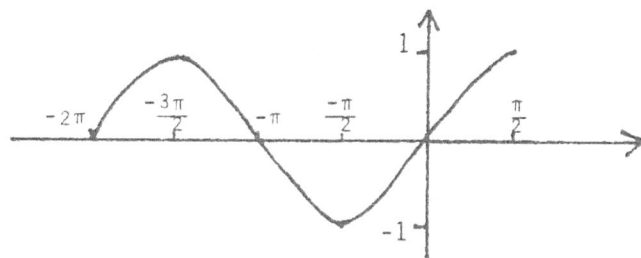

15) $\cos \theta$, $0 \leq \theta \leq 5$

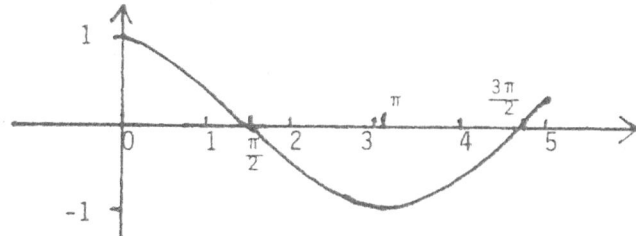

17) Neither   19) NA   21) Even   23) NA   25) NA

27) $\sin \theta = 0$, $\theta = -2\pi, -\pi, 0, \pi, 2\pi$ 29) 4

31)

θ	-5	-4	-3	-2	-1	0	1	2	3	4	5
$\cos \theta$	.28	-.65	-.99	-.42	.54	1	.54	-.42	-.99	-.65	.28

Exercise Set 4-2.

1) Decreasing   3) Decreasing   5) Increasing

7) Decreasing   9) Neither   11) $y = \sin x$, $\frac{-\pi}{4} \le x \le \frac{\pi}{4}$

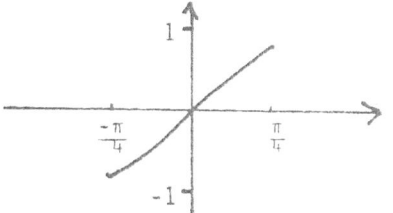

13) $y = \cos x$, $-\frac{\pi}{2} \le x \le \frac{\pi}{2}$

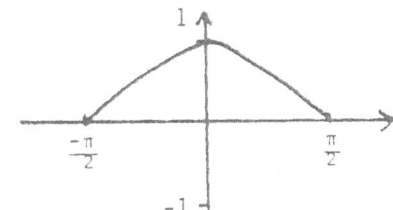

15) 3

17) $\sin x = \frac{1}{2}$, $x = \frac{\pi}{6}$

19) $\cos x = \frac{-1}{2}$, $x = \frac{2\pi}{3}$

21) $\sin x = \frac{\sqrt{2}}{2}$, $x = \frac{\pi}{4}$

23) $\cos x = -1$, $x = \pi$

25) $\sin x = -.4502$, $x = -.47$

27) $\sin x = .48$, $x = .50$

29) $\cos 2.35 = x$, $x = -.7027$

31) $\sin(-1) = x$, $x = -.8415$

Exercise Set 4-3.

1) True   3) True   5) No values   7) .67

9) $\tan(-.48) = -.5206$ 11) $\tan 74.3$ is undefined.

13) $\cot(-.043) = -23.24$ 15) $\tan x = .7083$, $x = .62$

17) Cot x = 1.954, x = .47   19) cot x = -1.594, $\frac{3\pi}{2} < x < 2\pi$. x = 5.72

21) 3   23) 2

25)

y = tan x

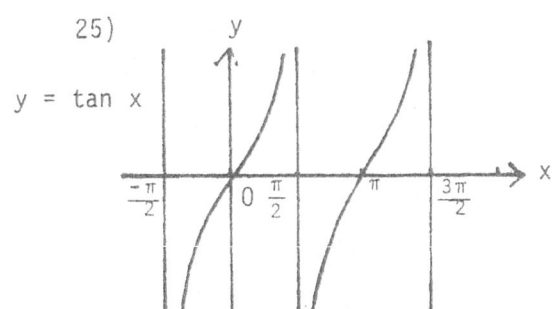

27)

y = cot x

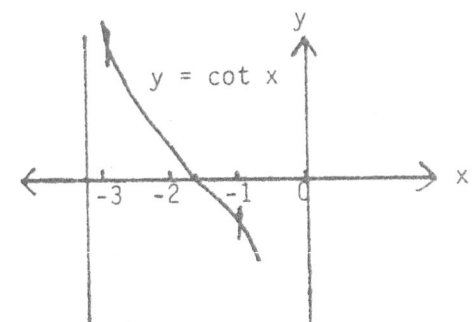

29)

y = |tan x|

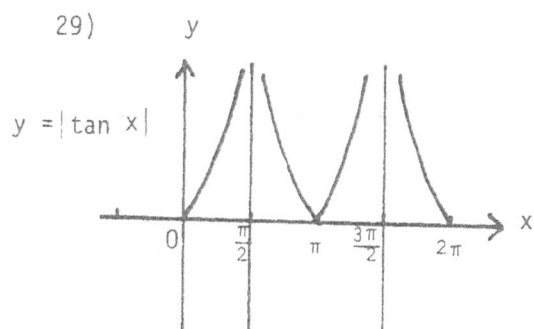

31)

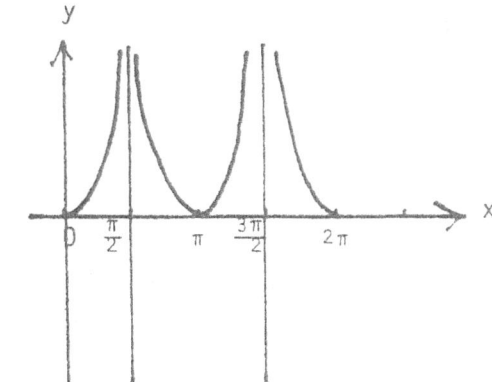

Exercise Set 4-4.

1) I, II   3) III, IV   5) II   7) III   9) 2

11) No   13) 3   15) No solution

17) x = kπ, k = ±1, ±3, ±5,...

19) Cosecant reaches a maximum value of -1 between III and IV.

  Cosecant reaches a minimum value of 1 between I and II

29) Sec x = 2.854, x = 1.21   31) Sec x = -4.327, x = -1.8

33) Csc x = -1.032, x = -1.82   35) sec x = 2.541, x = 1.17, -1.17

37) csc x = 1.798, x = .59, 2.55   39) sec x = -1.123, x = -2.67 or 2.67

21) 2.448   23) 7.714   25) -1.038   27) -1.180

Exercise Set 4-5.

1) "Amplitude" = 3, Period = π

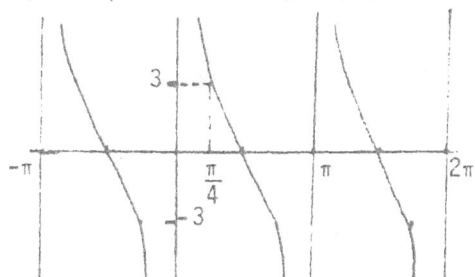

3) amplitude = $\frac{2}{5}$, period = 2π

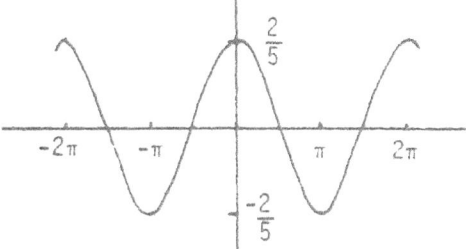

5) "Amplitude" = $\frac{3}{2}$ Period = 2π

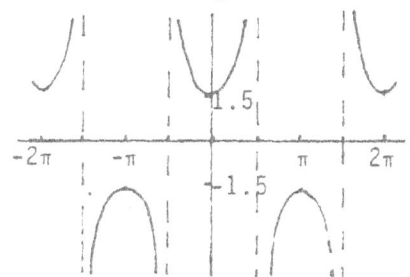

7) "Amplitude" = $\frac{8}{3}$, period = 3

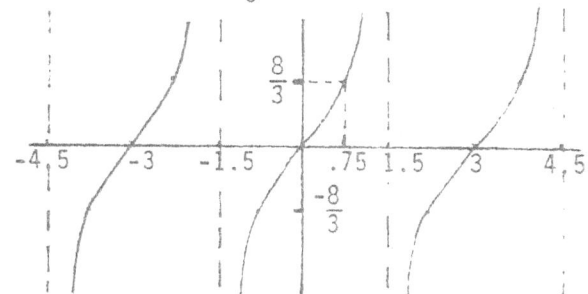

9) Amplitude = 2 , period = $\frac{2\pi}{3}$

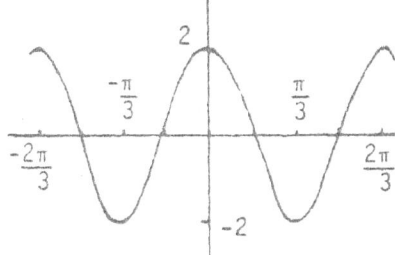

11) Domain = $\left[0,\frac{\pi}{5}\right]$, Range = $\left[-2,2\right]$

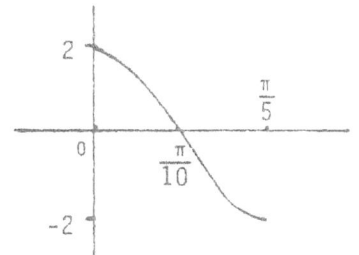

13) Domain = $\left[-\frac{\pi}{4},\frac{\pi}{4}\right]$, Range = $\left[-3,3\right]$ 15) Domain = $(0,\frac{4\pi}{3})$, Range = Reals

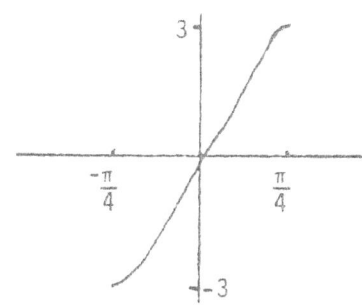

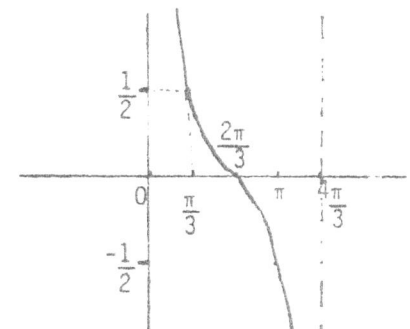

Exercise Set 4-6.

1) $f(x) = 2 \cos 3x$ amp = 2, period = $\frac{2\pi}{3}$

3) $g(x) = 5 \tan 3x$ amp = 5, period = $\frac{\pi}{3}$

5) $f(x) = -2 \csc \frac{4}{3}x$, amp = 2, period = $\frac{2\pi}{\frac{4}{3}}$ = $2\pi \cdot \frac{3}{4}$ = $\frac{3\pi}{2}$

7) $h(x) = \frac{3}{4} \sin (4x - \pi) = \frac{3}{4} \sin 4 (x - \frac{\pi}{4})$, amp = $\frac{3}{4}$, period = $\frac{\pi}{2}$, p.s. = $\frac{\pi}{4}$ lag

9) $t(x) = 10 \cos (\pi - 3x) = 10 \cos (3x - \pi) = 10 \cos 3 (x - \frac{\pi}{3})$ amp = 10,

period = $\frac{2\pi}{3}$, p.s. = $\frac{\pi}{3}$ lag.

11) $f(x) = 2 \cos 3x$

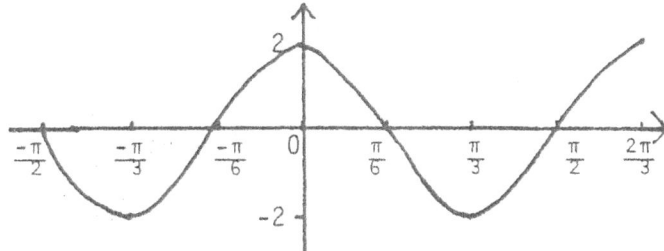

13) $h(x) = -\cot \frac{1}{2} x + 2 \sin x$

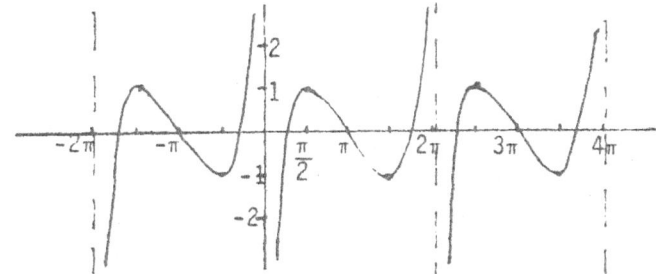

15) $t(x) = \frac{3}{2} \csc \frac{1}{3} x - \cos 3x$

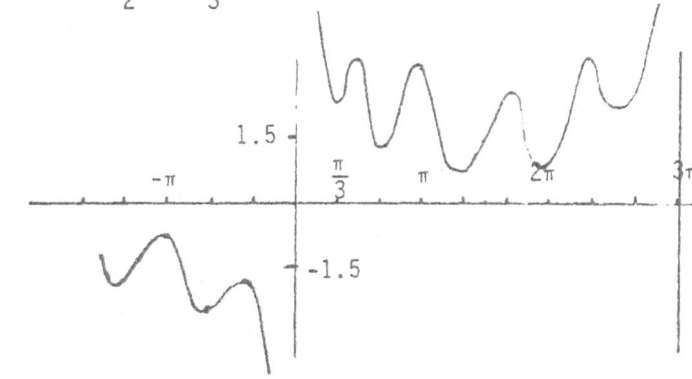

17) $f(x) = 2 \cos (x - \frac{\pi}{3})$ amp = 2, period = 2π, p.s. = $\frac{\pi}{3}$ lag

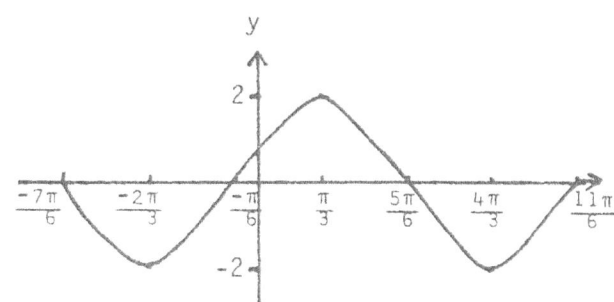

19) $g(x) = 3 \tan \pi x$ amp = 3, period = 1

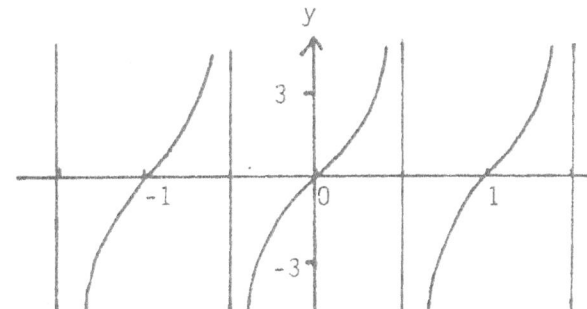

21) $p(x) = \frac{5}{4} \sin (2\pi x - \frac{\pi}{2}) = \frac{5}{4} \sin 2\pi (x - \frac{1}{4})$ amp = $\frac{5}{4}$, period = 1,
p.s. = $\frac{1}{4}$ lag

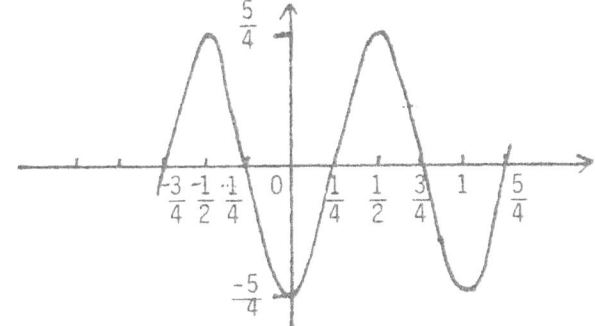

23) $t(x) = 4 \cot (x + \frac{\pi}{3})$ amp = 4, period = π, p.s. = $\frac{\pi}{3}$ lead

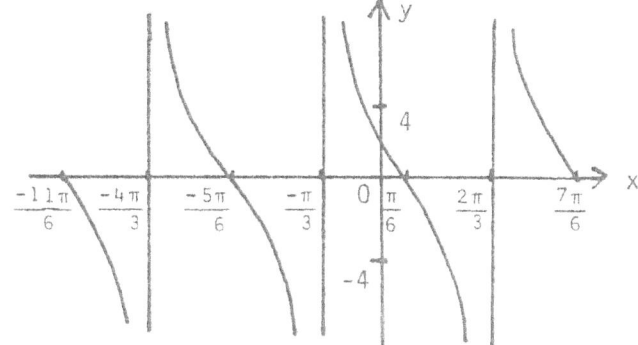

Exercise Set 4-7.

1) True   3) False   5) False   7) True   9) False

11) True   13) True   15) True   17) True   19) False

21) $f(x) = \frac{2}{3} \cos 5x$.   amp $= \frac{2}{3}$, period $= \frac{2\pi}{5}$

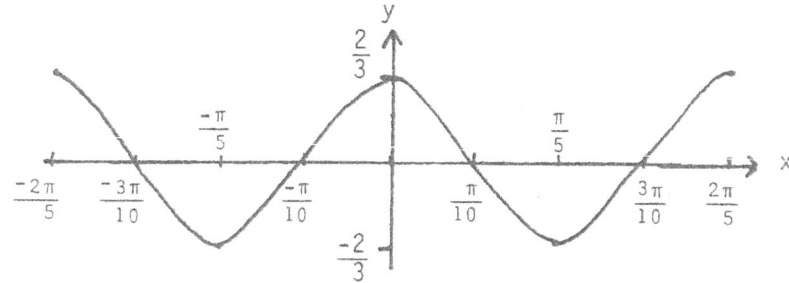

23) $h(x) = \frac{1}{2} \tan \frac{1}{2} x$ ·   amp $= \frac{1}{2}$, period $= 2\pi$

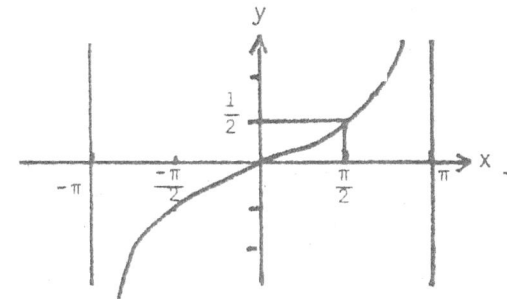

25) $5(x) = 4 \cot (2x - \frac{4\pi}{3}) = 4 \cot 2 (x - \frac{2\pi}{3})$, amp $= 4$, period $= \frac{\pi}{2}$

p.s. $= \frac{2\pi}{3}$ lag or $\frac{\pi}{6}$ lag

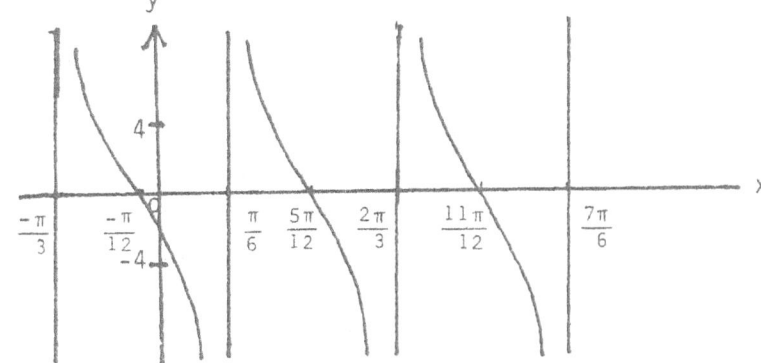

27) $n(x) = 10 \cos 3\pi x$ · amp = 10, period = $\frac{2}{3}$

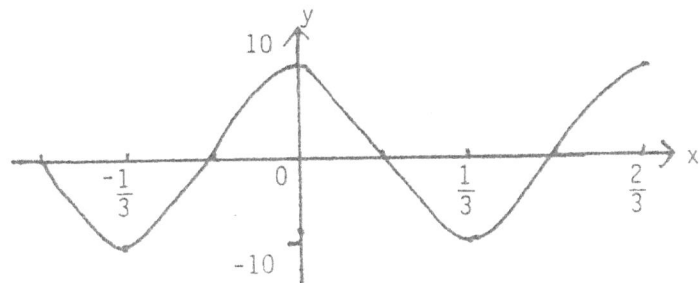

29) $\tan (\pi - x) = -\tan (x - \pi) = -\tan x$

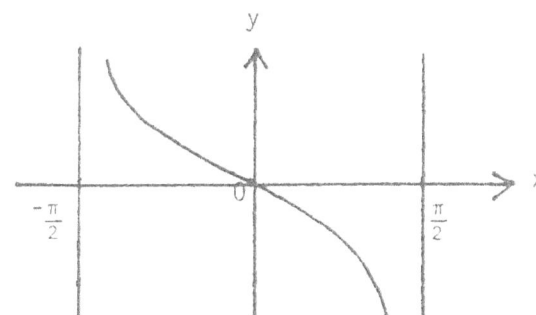

31) .33

33) no solution

35) .3946

37) .71

39) -1.671

41) -1.86

43) -2.613

45) -.0998

47) -.3085

49) 3.71

51) 1.85

53) 1.68

55) 2.09

Exercise Set 5-1.

1)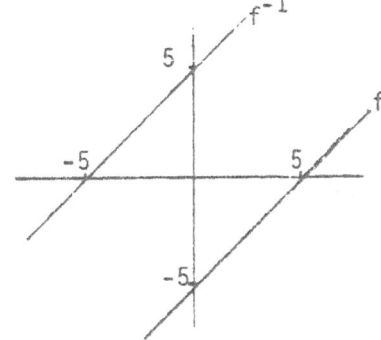

3)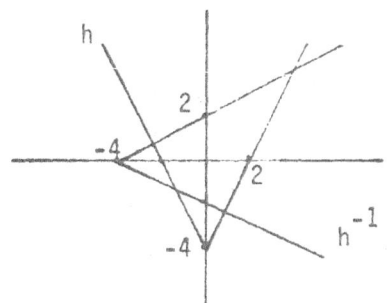

11) $y = x + 5$, function

13) $x = 2|y| - 4$, not a function

5)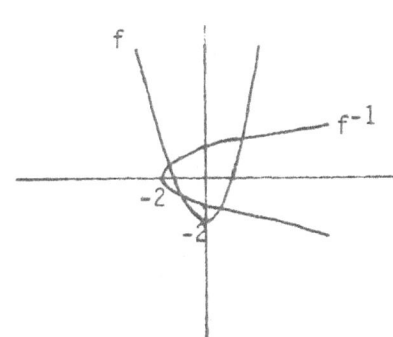

7)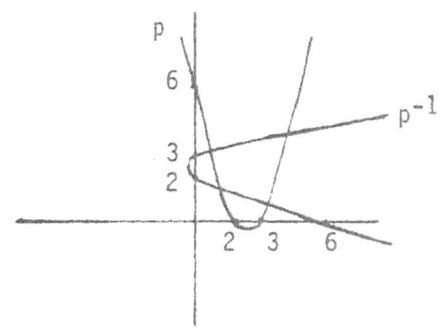

15) $x = y^2 - 2$, not a function

17) $x = y^2 - 5y + 6$, not a function

21) $-4/3$ 23) $1/2$ 25) 18

27) 3   29) $1/2$ 31)   4

33) $f_1^{-1}(x) = \dfrac{5 + \sqrt{4x + 9}}{2}$

   $f_2^{-1}(x) = \dfrac{5 - \sqrt{4x + 9}}{2}$

35) $f_1^{-1}(x) = \sqrt{1 - x}$

   $f_2^{-1}(x) = -\sqrt{1 - x}$

9)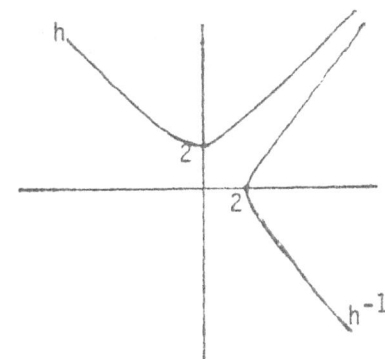

19) $x = \sqrt{4 + y^2}$, not a function

37) $2x - 6$ 39) $2x^2 - 1$ 41) 3^{x-1}

43) $|x|$ 45) $f(g(x)) = 7g(x) - 2 = ((7(x+2))/7) - 2 = x + 2 - 2 = x$

   $g(f(x)) = (f(x) + 2)/7 = (7x - 2 + 2)/7 = x$

47) $f(g(x)) = \sqrt{g(x) - 2} = \sqrt{x^2 + 2 - 2} = \sqrt{x^2} = x$

   $g(f(x)) = (f(x))^2 + 2 = \sqrt{(x - 2)^2} + 2 = x - 2 + 2 = x$

Exercise Set 5-2.

1) $y = \cos^{-1}x$

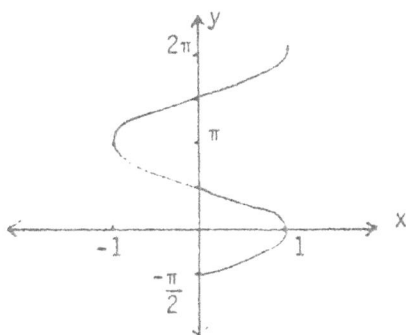

3) $y = 2\cos^{-1}x$

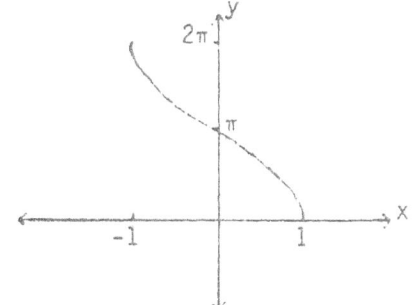

5) $y = \cos^{-1}2x$

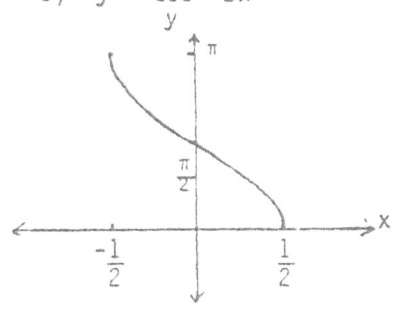

7) $y = \cos^{-1}(-x)$

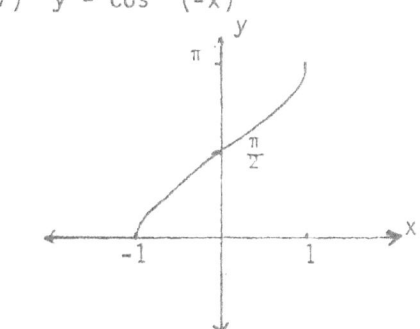

9) $-\dfrac{\pi}{6} + 2k\pi$ or $\dfrac{-5\pi}{6} + 2k\pi$, $k = 0, \pm1, \pm2, \ldots$ 11) not defined

13) $\pi + 2k\pi$, $k = 0, \pm1, \pm2, \ldots$ 15) $\dfrac{3\pi}{4}$ 17) $\dfrac{5\pi}{6}$

19) .57   21) .88   23) .7683   25) 1.98   27) -.95

29) 1/2   31) $\sqrt{3}/2$ 33) $3\pi/4$ 35) $-\pi/6$ 37) 3/5

39) $-\pi/4$ 41) 12/13   43) $3/\sqrt{13}$ 45) .67   47) -1.15

49) .6178   51) .9985   53) 1.02   55) -.91

Exercise Set 5-3.

1) $y = \text{Arctan } x$, $0 \le y < \frac{\pi}{2}$

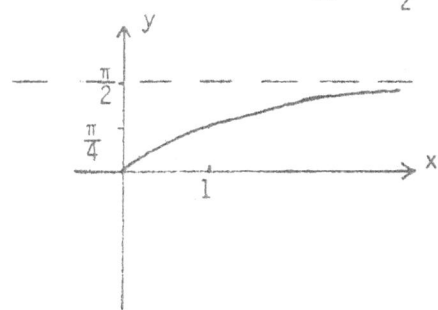

3) $y = \text{arccot } x$, $-\pi < y < 0$

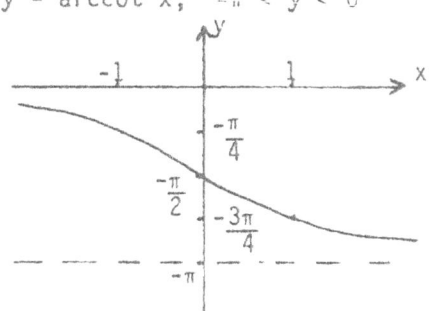

5) $y = \text{Tan}^{-1}x + \frac{\pi}{2}$

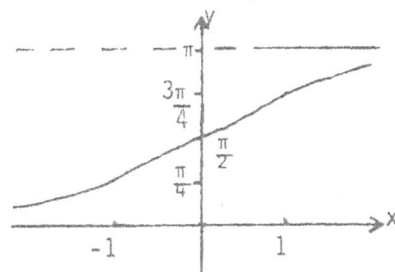

7) $\frac{-\pi}{4}$

9) $0 + k\pi$, $k = 0$, ± 1 , ± 2 , ...

11) $\frac{5\pi}{6} + k\pi$, $k = 0$, ± 1 , ± 2 , ...

13) 2

15) $\frac{-2}{3}$

17) $\frac{-3}{7}$

19) $\frac{4\pi}{5}$

21) $\frac{-\pi}{4}$

23) $\frac{\pi}{6}$

25) $\frac{\sqrt{3}}{2}$

27) $\frac{\pi}{2}$

29) $\frac{3}{\sqrt{13}}$

31) $\frac{-2\sqrt{10}}{3}$

33) $\frac{12}{13}$

35) $\frac{3}{5}$

37) .26

39) -.72

41) $.42 + k\pi$, $k = 0$, ± 1 , ± 2, ...

43) $2.36 + k\pi$, $k = 0$, ± 1 , ± 2 , ..

45) -.01

47) 2.70

49) -.88

51) -.37

53) .58

55) .53

57) .3318

59) .5184

Exercise Set 5-4.

1) $y = \text{arcsec } x,\ -\pi < y < \pi$

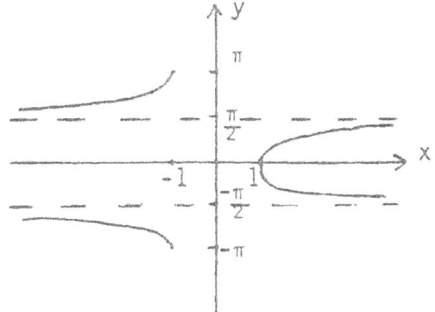

3) $y = \text{Arcsec } x,\ -\pi < y < \pi$

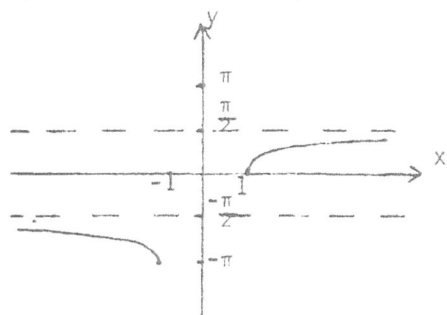

5) $y = \text{Sec}^{-1}x + \dfrac{\pi}{2}$

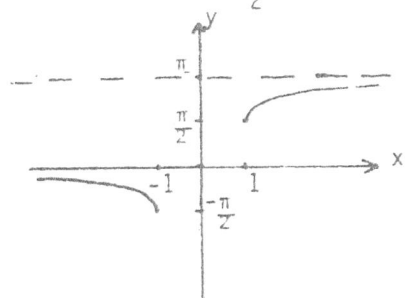

7) $\dfrac{\pi}{4}$

9) $\dfrac{-5\pi}{6}$

11) $\dfrac{\pi}{3}$

13) $\dfrac{1}{\sqrt{3}}$

15) undefined

17) 0

19) $\dfrac{\pi}{2}$

21) $\dfrac{\pi}{2}$

23) $\dfrac{-2\pi}{3}$

25) undefined   27) -2.19

29) .9322

31) .4999

33) undefined   35) -1.82

Exercise Set 5-5.

1) $y = \dfrac{1}{2} \text{Cos}^{-1} x$

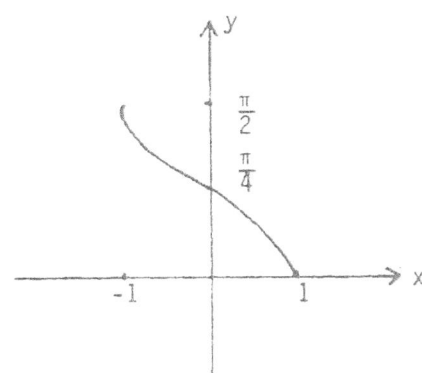

3) $y = -\text{Sin}^{-1} x$

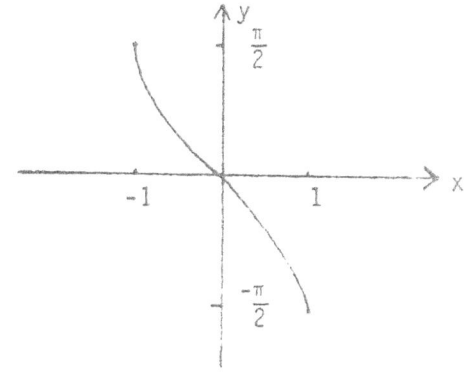

Exercise Set 5-5.(Continued)

5) $y = 2 \text{Tan}^{-1} 2x$

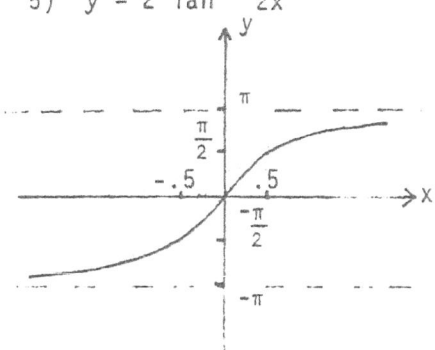

7) $y = \pi \text{Cot}^{-1} x$

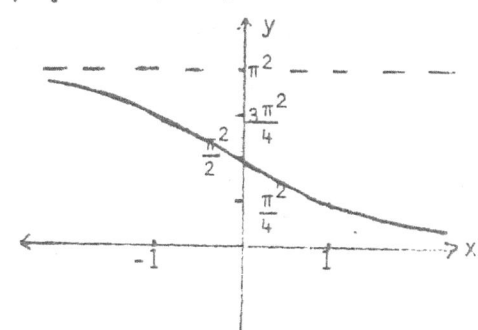

9) $y = \frac{2\pi}{3} \text{Arccos} (x - 1)$

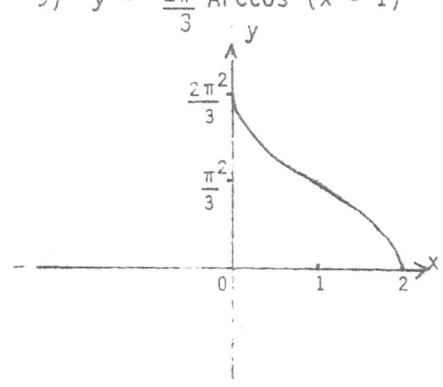

11) $y = 2 \text{Sin}^{-1} x - \frac{\pi}{3}$

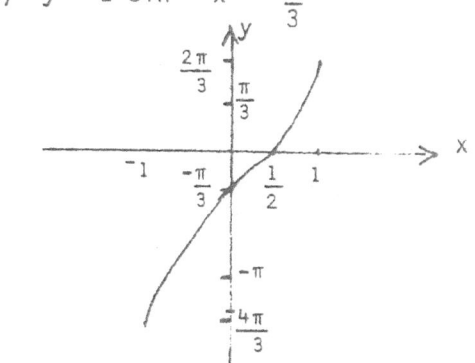

13) $y = \cos^{-1} x + \frac{\pi}{4}$ 15) $y = \tan^{-1} 2x + \frac{\pi}{6}$ 17) $y = 3 \text{Cos} \frac{1}{2} (x + \frac{\pi}{4})$

Exercise Set 5-6.

1) $\pi/3$ 3) $-\pi/3$ 5) .50 or 3.64   7) ± 1.11 9) $5\pi/6$ or $11\pi/6$

11) $-2\pi/3$ 13) $\pi/6$ or $7\pi/6$ 15) no solution   17) 0 , $\pi/3$, $-\pi/3$

19) $\pm \pi/6$ 21) $\pi/3$, $2\pi/3$, $4\pi/3$, $5\pi/3$ 23) $\pi/6$, $5\pi/6$

25) $\pi/3$ 27) $\pi/6$, $5\pi/6$, $7\pi/6$, $11\pi/6$ 29) $-.34$

31) $-.37$, 1.04 33) $-.58$, $.26$ 35) $\pi/2$, 1.35 37) $.93$

39) 1.88 41) 2.54 43) 1.38 45) $.07$ 47) 1.02 , $-.55$

49) $\pi/3$, $4\pi/3$ 51) $.37$ 53) $k\pi/9$, $k = 1,2,4,5,7,8,10,11,13,14,16,17$

55) 2.32 57) 1.32 , π , 4.96 59) $\pi/4$

Exercise Set 5-7.

1) $y = \dfrac{x - 7}{3}$, function

3) $y = x^2$, function

5) $y = \dfrac{1}{4} \cos^{-1} \dfrac{x}{2}$, not a function

7) $y = \sec^{-1} 2x - \dfrac{\pi}{3}$, function

9) $y = 3 \sin (2x + \dfrac{\pi}{2})$
function

11) $f(x) = 2 \arcsin \dfrac{x}{3}$

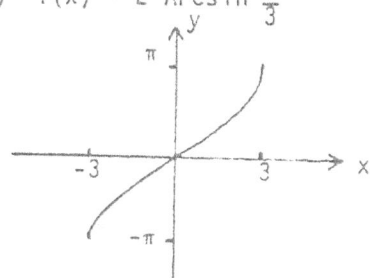

13) $t(x) = \dfrac{1}{3} \cot^{-1} x - \dfrac{\pi}{4}$

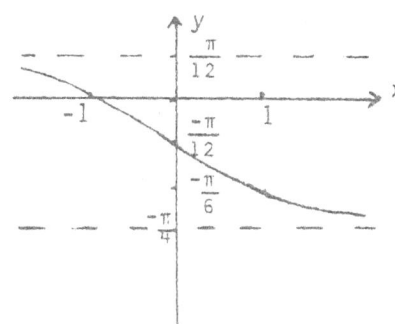

15) $p(x) = 2\sin^{-1}x + \cos^{-1}x$

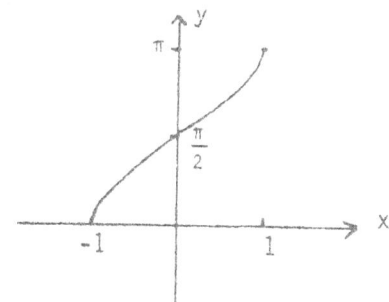

17) $-1/\sqrt{3}$

19) 3.96

21) $3/5$

23) 2.78

25) $.26$, 1.31

27) $f(g(x)) = 2 \log |x - 5|$

29) $f(g(x)) = e^x$

31) $\dfrac{k\pi}{9}$ where $k = 2, 4, 8, 10, 14, 16$

SOLUTIONS TO ODD NUMBERED EXERCISES

Chapter 6

Exercise Set 6-1.

1) $\sqrt{293}$ 3) 18.43   5) 7.76   7) 1.40

9) To derive the formula $\cos(x + y) = \cos x \cos y - \sin x \sin y$, note that $d(A,R) = d(Q,S)$ in the figure and apply the distance formula.

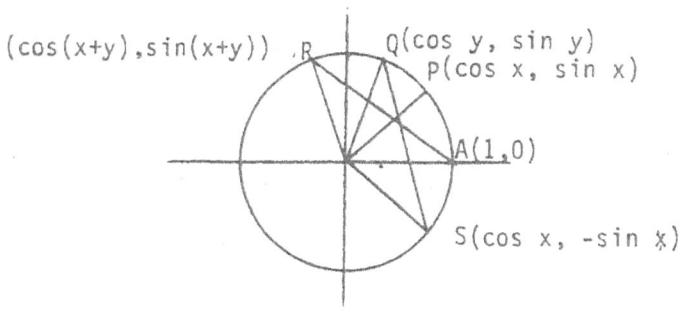

$(\cos(x+y),\sin(x+y))$ · R   Q$(\cos y, \sin y)$
P$(\cos x, \sin x)$
A$(1,0)$
S$(\cos x, -\sin x)$

11) $\dfrac{\sqrt{6} + \sqrt{2}}{4}$ 13) $\dfrac{\sqrt{2} - \sqrt{6}}{4}$ 15) $\dfrac{\sin(x+y)}{\sin(x-y)} = \dfrac{\dfrac{\sin x\cos y}{\cos x\cos y} + \dfrac{\cos x\sin y}{\cos x\cos y}}{\dfrac{\sin x\cos y}{\cos x\cos y} - \dfrac{\cos x\sin y}{\cos x\cos y}}$

$= \dfrac{\tan x + \tan y}{\tan x - \tan y}$.

17) $\tan x - \tan y = \dfrac{\sin x}{\cos x} - \dfrac{\sin y}{\cos y}$

$= \dfrac{\sin x \cos y - \cos x \sin y}{\cos x \cos y}$

$= \dfrac{\sin(x - y)}{\cos x \cos y}$

19) $\sin(x + y) + \sin(x - y)$

$= \sin x \cos y + \cos x \sin y$

$+ \sin x \cos y - \cos x \sin y$

$= 2 \sin x \cos y$

21) $\cos 118^{\circ}$ 23) $\sin 71^{\circ}$

25) $\dfrac{\sqrt{3} \cos x + \sin x}{2}$

27) $\dfrac{\sin x + \cos x}{\sqrt{2}}$ 29) $- \sin x$ 31) $\dfrac{63}{65}$ 33) $\dfrac{-4}{3}$

Exercise Set 6-2.

1) $\cos 2x = \cos^2 x - \sin^2 x = (1 - \sin^2 x) - \sin^2 x = 1 - 2 \sin^2 x$.

3) $\tan \frac{1}{2}x = \dfrac{\sin \frac{1}{2}x}{\cos \frac{1}{2}x} = \dfrac{\pm\sqrt{\dfrac{1 - \cos x}{2}}}{\pm\sqrt{\dfrac{1 + \cos x}{2}}} = \pm\sqrt{\dfrac{1 - \cos x}{1 + \cos x}}$

5) $\sin 270^O = 2 \sin 135^O \cos 135^O = 2 \cdot \dfrac{1}{\sqrt{2}} \cdot \dfrac{-1}{\sqrt{2}} = -1$

7) $\tan 90^O = \dfrac{2 \tan 45^O}{1 - \tan^2 45^O} = \dfrac{2 \cdot 1}{1 - 1} = \dfrac{2}{0}$ which is undefined.

9) $\frac{1}{2}\sqrt{2 + \sqrt{2}}$

11) $-\frac{1}{2}\sqrt{2 - \sqrt{3}}$

13) $-2 - \sqrt{3}$

15) $\frac{1}{2}\sqrt{2 + \sqrt{2}} + \frac{1}{2}\sqrt{2 - \sqrt{2}}$

17) $\dfrac{1 - \sqrt{5}}{2}$

19) $-.9569$

21) $-.8126$

23) $\pm\sqrt{\dfrac{1 + x}{2}}$

25) $1 + \sin 2x = 1 + 2 \sin x \cos x$

$= (\cos^2 x + \sin^2 x) + 2 \sin x \cos x$

$= \cos^2 x + 2 \sin x \cos x + \sin^2 x$

$= (\cos x + \sin x)^2$

27) $\sin(x+y)\sin(x-y) = (\sin x \cos y + \cos x \sin y)(\sin x \cos y - \cos x \sin y)$

$= \sin^2 x \cos^2 y - \cos^2 x \sin^2 y$

$= \sin^2 x(1 - \sin^2 y) - (1 - \sin^2 x) \sin^2 y$

$= \sin^2 x - \sin^2 x \sin^2 y - \sin^2 y + \sin^2 x \sin^2 y$

$= \sin^2 x - \sin^2 y$

29) $\cot x - \tan x = \dfrac{1}{\tan x} - \tan x = \dfrac{1 - \tan^2 x}{\tan x} = \dfrac{\dfrac{1}{\tan x}}{\dfrac{1 - \tan^2 x}{1}} = \dfrac{2}{\dfrac{2 \tan x}{1 - \tan^2 x}}$

$= \dfrac{2}{\tan 2x} = 2 \cdot \dfrac{1}{\tan 2x} = 2 \cot 2x$.

Exercise Set 6-3.

1) $2 \sin 40^0 \cos 7^0$ 3) $2 \cos 51^0 \sin 21^0$

5) $-2 \cos \frac{3\pi}{4} \sin \frac{\pi}{12}$ 7) $\sin 120^0 + \sin 80^0$

9) $\cos 47^0 + \cos 31^0$ 11) $\cos \frac{4\pi}{5} - \cos \frac{\pi}{7}$

13)
$$\begin{array}{ll}
\sin (x + y) & = \sin x \cos y + \cos x \sin y \\
\underline{\quad - \sin (x - y)} & = \underline{-\sin x \cos y + \cos x \sin y} \\
\sin (x + y) - \sin (x - y) & = \qquad 2 \cos x \sin y
\end{array}$$

Let $x = \dfrac{A + B}{2}$ $y = \dfrac{A - B}{2}$

$\sin (x + y) = \sin \left(\dfrac{A + B}{2} + \dfrac{A - B}{2}\right) = \sin A$

$\sin (x - y) = \sin \left(\dfrac{A + B}{2} - \dfrac{A - B}{2}\right) = \sin B$

Substituting, $\sin A - \sin B = 2 \cos \dfrac{A + B}{2} \sin \dfrac{A - B}{2}$, 6-3B.

15)
$$\begin{array}{ll}
\cos (x + y) & = \cos x \cos y - \sin x \sin y \\
\underline{\quad - \cos (x - y)} & = \underline{-\cos x \cos y - \sin x \sin y} \\
\cos (x + y) - \cos (x - y) & = \qquad -2 \sin x \sin y
\end{array}$$

Substituting as in Exercise 13 above,

$\cos A - \cos B = -2 \sin \dfrac{A + B}{2} \sin \dfrac{A - B}{2}$, 6-3D.

17) $\sin \frac{\pi}{5} + \sin \frac{2\pi}{15} = 2 \sin \dfrac{\frac{\pi}{5} + \frac{2\pi}{15}}{2} \cos \dfrac{\frac{\pi}{5} - \frac{2\pi}{15}}{2} = 2 \sin \frac{\pi}{6} \cos \frac{\pi}{30}$

$= 2 \cdot \frac{1}{2} \cos \frac{\pi}{30} = \cos \frac{\pi}{30}$

19) $-2\sin^2 \frac{\pi}{8}$ 21) $2 \cos 7x \sin 3x$ 23) $\tan^{-1} \frac{4}{3} \doteq 53.13^0$

25) $2 \sin x + 3 \cos x$ 27) $\sin x + 2 \cos x$

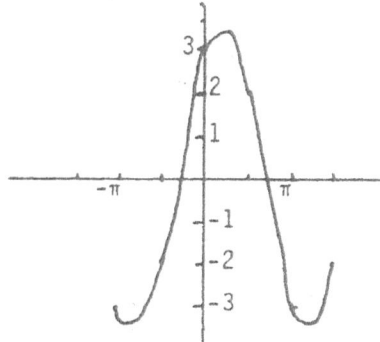

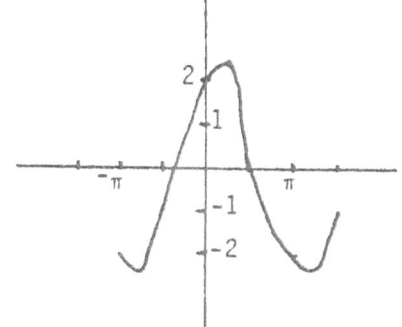

29) $\sin 2x - 2 \cos x$

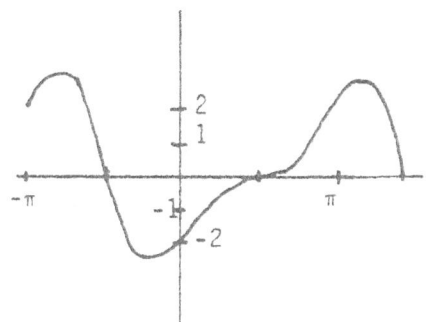

31) $\sin 3x - \sin x = 2 \cos 2x \sin x$

$$= 2(1 - 2 \sin^2 x) \sin x$$

$$= 2 \sin x - 4 \sin^3 x$$

33) $\dfrac{\cos 4x - \cos 2x}{\cos 4x + \cos 2x} = \dfrac{-2 \sin 3x \sin x}{2 \cos 3x \cos x}$

$$= - \tan 3x \tan x$$

32) $\dfrac{\sin x + \sin 3x}{\sin x - \sin 3x} = \dfrac{2 \sin 2x \cos(-x)}{2 \cos 2x \sin(-x)} = \tan 2x(-\cot x)$

$$= \left(\dfrac{2 \tan x}{1 - \tan^2 x}\right)\left(\dfrac{-1}{\tan x}\right) = \dfrac{2}{\tan^2 x - 1}$$

35) $\dfrac{\cos x + \cos 2x + \cos 3x}{\sin x + \sin 2x + \sin 3x} = \dfrac{2 \cos 2x \cos x + \cos 2x}{2 \sin 2x \cos x + \sin 2x}$

$$= \dfrac{\cos 2x (2 \cos x + 1)}{\sin 2x (2 \cos x + 1)} = \dfrac{\cos 2x}{\sin 2x} = \cot 2x$$

Exercise Set 6-4.

1) $\dfrac{\pi}{12}$ 3) $.88 , -.69$ 5) $1.32, 4.97, \pi$ 7) $\dfrac{\pi}{6}, \dfrac{5\pi}{6} , \dfrac{3\pi}{2}$

9) $.25, 1.30, 2.34, 3.39, 4.44, 5.48$ 11) $0 , \pi , 2\pi, \dfrac{\pi}{3} , \dfrac{5\pi}{3}$

13) $\sin^{-1} \dfrac{8}{\sqrt{89}} - \sin^{-1} \dfrac{2}{\sqrt{89}} \doteq .80 , 4.37$ 15) $\dfrac{\pi}{2}$ 17) $\dfrac{19\pi}{12} , \dfrac{23\pi}{12}$

19) $\dfrac{7\pi}{6} , \dfrac{11\pi}{6}$ 21) $.9955 (\text{or } 1.00)$ 23) $\dfrac{\pi}{12} , \dfrac{5\pi}{12}$

25) 4.07 27) $.16$ 29) ± 1.13 31) $(\sin x , \cos y) = \left(\dfrac{5}{7} , \dfrac{-3}{7}\right)$

33) $.9478$ 35) $.8927$ 37) $.2298$ $x = .80 , 2.35$
$y = 2.01, 4.27$

39) $\dfrac{-7\pi}{15}$

Exercise Set 6-5.

1) $\sqrt{136}$ 3) 1.88   5) $\dfrac{-56}{65}$ 7) $\dfrac{-63}{16}$

9) $\dfrac{1}{8}$ 11) $\dfrac{\pm\sqrt{2}}{4}$ 13) $(\sin x - \cos x)^2 + \sin 2x$

$$= \sin^2 x - 2 \sin x \cos x + \cos^2 x$$

$$+ 2 \sin x \cos x = \sin^2 x + \cos^2 x = 1$$

15) $\csc^2 \frac{1}{2}x + \cot^2 \frac{1}{2}x = (1 + \cot^2 \frac{1}{2}x) + \cot^2 \frac{1}{2}x = 1 + 2 \cot^2 \frac{1}{2}x$

$$= 1 + 2 \cdot \frac{1 + \cos x}{1 - \cos x} = \frac{(1 - \cos x) + (2 + 2 \cos x)}{1 - \cos x}$$

$$= \frac{3 + \cos x}{1 - \cos x}$$

17) $\tan 8x$ 19) $\cos 4x$ 21) $\sin y$ 23) $3 \sin 12x$

25) $\cos 4x - \cos 14x$ 27) $\dfrac{1}{2}(\sin 9x + \sin 5x)$ 29) $\dfrac{-\sqrt{2 - \sqrt{3}}}{2}$

31) $\sqrt{3} \cos 80^0$ 33) $\dfrac{\pi}{2}, \dfrac{3\pi}{2}, \dfrac{\pi}{3}, \dfrac{2\pi}{3}$ 35)

$$\dfrac{k\pi}{6}, \ k = 1,2,4,5,7,8,10,11$$

37) $0, \pi, \dfrac{3\pi}{4}$ $\dfrac{7\pi}{4}$

39) $0, \pi, 1.95, 4.34$

SOLUTIONS TO ODD NUMBERED EXERCISES

CHAPTER 7

Exercise Set 7-1.

1) a) $(2\sqrt{2} \cos \frac{\pi}{4} , 2\sqrt{2} \sin \frac{\pi}{4})$ b) $(2\sqrt{2} , \frac{\pi}{4})$

3) a) $(6 \cos \frac{11\pi}{6}, 6 \sin \frac{11\pi}{6})$ b) $(6 , \frac{11\pi}{6})$

5) a) $(13 \cos 2.75, 13 \sin 2.75)$ b) $(13, 2.75)$

7) a) $(5.47 \cos 5.25, 5.47 \sin 5.25)$ b) $(5.47, 5.25)$

9) $(-2 , 0)$ 11) $(-1, -\sqrt{3})$ 13) $(\sqrt{2} , \sqrt{2})$

15) $(2\sqrt{3} , 2)$ 17) $(\sqrt{3} , -1)$ 19) $(-1.71, 4.70)$

21) Both are equivalent to rectangular coordinates: $(1 , \sqrt{3})$.

23) Both are equivalent to rectangular coordinates: $(-\frac{3\sqrt{2}}{2} , \frac{3\sqrt{2}}{2})$

Exercise Set 7-2 .

1) $-1 \pm 2i$ 3) $\frac{1 \pm \sqrt{26} \ i}{9}$ 5) $\frac{1 \pm 3\sqrt{3} \ i}{4}$ 7) $\frac{2 \pm \sqrt{21} \ i}{5}$

9) -4 11) $28i$ 13) $6\sqrt{2} \ i$ 15) $3i$

17) $-i$ 19) $-i$ 21) $-5 - 8i$ 23) $6 + 7i$

25) 6 27) $\frac{16}{15} + \frac{1}{6} i$ 29) $\sqrt{34}$ 31) $\sqrt{170}$

33) $2\sqrt{10} + 0i$ 35) $2\sqrt{10} + 4i$ 37) $6.05 + 0i$

39) $v = -4 + i$ 41) $v = 3 + 7i$

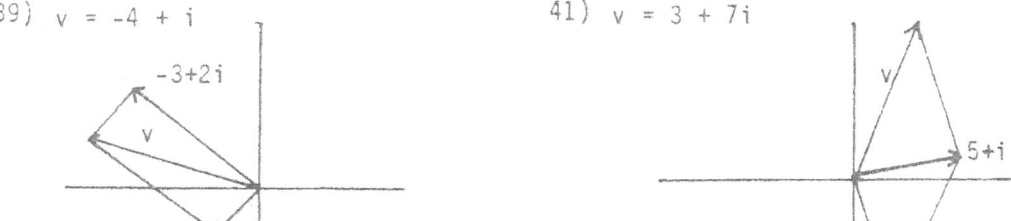

43) $v = 8 + 0i$

45) $\sqrt{82}$

47) 8

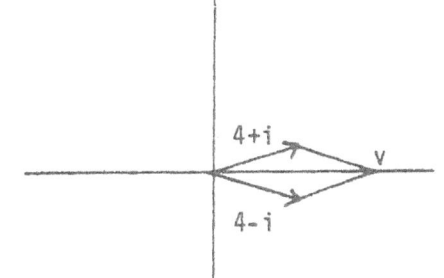

Exercise Set 7-3 .

1) $-24 + 30\,i$

3) $-12 - 14\,i$

5) $10 - 14\,i$

7) $(2\sqrt{3} - 2) - (4 + \sqrt{3})i$

9) $(2 + 2\sqrt{2}) + (2 - 2\sqrt{2})i$

11) i

13) $-7 + 26\,i$

15) -4

17) $|-24 + 30i| = 6\sqrt{41}$, $\Theta = 2.25$ radians

19) $|10 - 14i| = \sqrt{296}$, $\Theta = 5.33$ radians

21) $|(2 + 2\sqrt{2}) + (2 - 2\sqrt{2})| = 2\sqrt{6}$, $\Theta = 6.11$ radians

23) $|-7 + 26i| = 5\sqrt{29}$, $\Theta = 1.83$ radians

25) $\frac{3}{2} + \frac{3\sqrt{3}}{2}\,i$

27) $0 - 4i$

29) $\sqrt{2} + \sqrt{2}\,i$

31) $\frac{3\sqrt{3}}{2} - \frac{3}{2}\,i$

33) $\frac{-7}{2} - \frac{7\sqrt{3}}{2}\,i$

35) $3.53 + 1.88\,i$

37) a) $3\sqrt{2}(\cos 315^{\circ} + i \sin 315^{\circ})$

b) $(3\sqrt{2}, 315^{\circ})$

39) a) $4(\cos 120^{\circ} + i \sin 120^{\circ})$

b) $(4, 120^{\circ})$

41) a) $\sqrt{7}(\cos 40.89^{\circ} + i \sin 40.89^{\circ})$

b) $(\sqrt{7}, 40.89^{\circ})$

43) a) $\sqrt{5}(\cos 63.43^{\circ} + i \sin 63.43^{\circ})$, b) $(\sqrt{5}, 63.43^{\circ})$

45) $15(\cos 118^{\circ} + i \sin 118^{\circ})$

47) $8(\cos \frac{61\pi}{30} + i \sin \frac{61\pi}{30})$

49) $(-8)(\cos 131^{\circ} + i \sin 131^{\circ})$

51) $21(\cos \frac{7\pi}{9} + i \sin \frac{7\pi}{9})$

53) $\frac{-12}{5} + \frac{1}{5}\,i$

55) $\frac{-1}{5} + \frac{7}{5}\,i$

57) $9(\cos 144^{\circ} + i \sin 144^{\circ})$

59) $\frac{9}{2}(\cos\frac{13\pi}{12} + i\sin\frac{13\pi}{12})$ 61) $(-2)(\cos 180^0 + i\sin 180^0)$

63) $4(\cos\frac{\pi}{14} + i\sin\frac{\pi}{14})$

65) $\frac{12 \text{ cis } 240^0}{3 \text{ cis } 90^0} = 4 \text{ cis } 150^0 = 4(\cos 150^0 + i\sin 150^0) = 4(\frac{-\sqrt{3}}{2}) + 4i(\frac{1}{2})$

$$= -2\sqrt{3} + 2i \ .$$

$\frac{12 \text{ cis } 240^0}{3 \text{ cis } 90^0} = \frac{12(\cos 240^0 + i\sin 240^0)}{3(\cos 90^0 + i\sin 90^0)} = \frac{4(-\frac{1}{2}) + 4(-\frac{\sqrt{3}}{2})i}{i} \cdot \frac{i}{i}$

$$= -2\sqrt{3} + 2 i \ .$$

67) $\frac{-2\sqrt{3} - 2i}{\sqrt{3} + i} \ \frac{\sqrt{3} - i}{\sqrt{3} - i} = \frac{-6 - 2\sqrt{3}\, i + 2\sqrt{3}\, i - 2}{3 + 1} = \frac{-8}{4} = -2 \ .$

$\frac{(4 \ , \ 210^0)}{(2 \ , \ 30^0)} = (2 \ , \ 180^0) \backsim 2(\cos 180^0 + i\sin 180^0) = -2 \ .$

Exercise Set 7-4.

1) $8(\cos 144^0 + i\sin 144^0)$ 3) $25(\cos(-260^0) + i\sin(-260^0))$

5) $8(\cos 441^0 + i\sin 441^0)$ 7) $\frac{1}{16}(\cos\frac{4\pi}{3} + i\sin\frac{4\pi}{3})$

9) $32(\cos 180^0 + i\sin 180^0)$ 11) $64(\cos 360^0 + i\sin 360^0)$

13) $128\sqrt{2}(\cos 135^0 + i\sin 135^0)$ 15) $25(\cos 253.74^0 + i\sin 253.74^0)$

17) a) $2(\cos 90^0 + i\sin 90^0) \ , \ 2(\cos 270^0 + i\sin 270^0)$

   b) $0 + 2i \ , \ \ 0 - 2i$

19) a) $\cos 0^0 + i\sin 0^0 \ , \ \cos 120^0 + i\sin 120^0 \ , \ \cos 240^0 + i\sin 240^0$

   b) $1 + 0i \ , \ \frac{-1}{2} + \frac{\sqrt{3}}{2}i \ , \ \frac{-1}{2} - \frac{\sqrt{3}}{2}i$

21) a) $2(\cos 22.5^0 + i\sin 22.5^0) \ , \ 2(\cos 112.5^0 + i\sin 112.5^0)$

   $2(\cos 202.5^0 + i\sin 202.5^0) \ , \ 2(\cos 292.5^0 + i\sin 292.5^0)$

   b) $\sqrt{2 + \sqrt{2}} + \sqrt{2 - \sqrt{2}}\, i \ , \ -\sqrt{2 - \sqrt{2}} + \sqrt{2 + \sqrt{2}}\, i \ ,$

   $-\sqrt{2 + \sqrt{2}} - \sqrt{2 - \sqrt{2}}\, i \ , \ \sqrt{2 - \sqrt{2}} - \sqrt{2 + \sqrt{2}}\, i$ by 6-2 D, E

23) a) $8^{\frac{1}{4}} (\cos 67.5^0 + i \sin 67.5^0)$, $8^{\frac{1}{4}} (\cos 247.5^0 + i \sin 247.5^0)$

    b) $2^{-\frac{1}{4}} \sqrt{2 - \sqrt{2}} + 2^{-\frac{1}{4}} \sqrt{2 + \sqrt{2}} \, i$, $-2^{-\frac{1}{4}} \sqrt{2 - \sqrt{2}} - 2^{-\frac{1}{4}} \sqrt{2 + \sqrt{2}} \, i$

25) a) $29^{\frac{1}{4}} (\cos 10.9^0 + i \sin 10.9^0)$, $29^{\frac{1}{4}} (\cos 190.9^0 + i \sin 190.9^0)$

    b) $2.28 + .44 \, i$, $-2.28 - .44 \, i$

27) a) $4^{1/3} (\cos 49^0 + i \sin 49^0)$, $4^{1/3} (\cos 169^0 + i \sin 169^0)$,

        $4^{1/3} (\cos 289^0 + i \sin 289^0)$

    b) $1.04 + 1.20 \, i$, $-1.56 + .30 \, i$, $.52 - 1.5 \, i$

29) $0 + 4i$, $-2\sqrt{3} - 2i$, $2\sqrt{3} - 2i$

31) $-2.33 + .64 \, i$, $2.33 - .64 \, i$

<u>Exercise Set 7-5.</u>

1) $(\sqrt{2} , 135^0)$ 3) $(2 , 330^0)$ 5) $(\sqrt{58} , 66.80^0)$

7) $(5\sqrt{5} , 190.30^0)$ 9) $(3 , 20^0)$ 11) $\sqrt{3} + i$

13) $\dfrac{3 \sqrt{2 - \sqrt{3}}}{2} + \dfrac{3 \sqrt{2 + \sqrt{3}}}{2} \, i$ 15) $4 + 0i$ 17) $\dfrac{-1}{5} \pm \dfrac{2}{5} \, i$

19) $\dfrac{3}{4} \pm \dfrac{\sqrt{23}}{4} \, i$ 21) $-3 - 4i$ 23) $13 - 4i$

25) $10 + 0i$ 27) $-33 - 17i$ 29) $0 + i$

31) $\dfrac{27}{17} + \dfrac{11}{17} \, i$ 33) $-30 + 0i$ 35) $\dfrac{2}{3} + 0i$

37) $\sqrt{13}$ 39) $\sqrt{5}$

41)

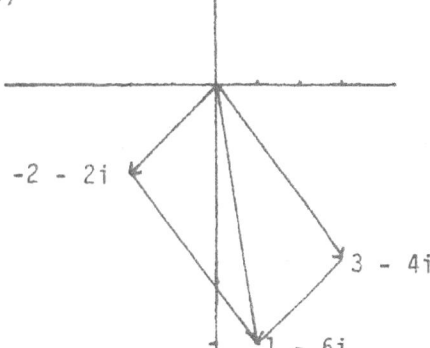

43)

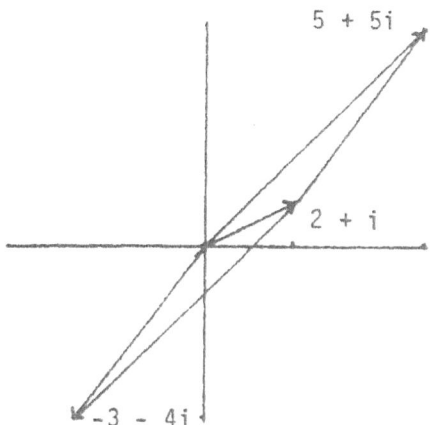

45) $2\sqrt{2}(\cos 135^\circ + i \sin 135^\circ)$ 47) $32(\cos 30^\circ + i \sin 30^\circ)$

49) $-8 + 8\sqrt{3}\, i$ 51) $-9\sqrt{2+\sqrt{3}} - 9\sqrt{2-\sqrt{3}}\, i$ 53) $-24.95 - 12.71\, i$

55) $r_1 = -\sqrt{\dfrac{2-\sqrt{3}}{2}} + \sqrt{\dfrac{2+\sqrt{3}}{2}}\, i$, $r_2 = \sqrt{\dfrac{2-\sqrt{3}}{2}} - \sqrt{\dfrac{2+\sqrt{3}}{2}}\, i$

57) $r_1 = 1.18 + .79\, i$ $\sim (\sqrt{2}\ ,\ 33.75^\circ)$

  $r_2 = -.79 + 1.18\, i$ $\sim (\sqrt{2}\ ,\ 123.75^\circ)$

  $r_3 = -1.18 - .79\, i$ $\sim (\sqrt{2}\ ,\ 213.75^\circ)$

  $r_4 = .79 - 1.18\, i$ $\sim (\sqrt{2}\ ,\ 303.75^\circ)$

59) $x_1 = 1 + \sqrt{3}\, i$ $\sim (2\ ,\ 60^\circ)$

  $x_2 = -1 + \sqrt{3}\, i$ $\sim (2\ ,\ 120^\circ)$

  $x_3 = -2 + 0\, i$ $\sim (2\ ,\ 180^\circ)$

  $x_4 = -1 - \sqrt{3}\, i$ $\sim (2\ ,\ 240^\circ)$

  $x_5 = 1 - \sqrt{3}\, i$ $\sim (2\ ,\ 300^\circ)$

  $x_6 = 2 + 0\, i$ $\sim (2\ ,\ 360^\circ)$

SOLUTIONS TO ODD NUMBERED EXERCISES

Exercise Set 8-1.

Answers to exercises 1 - 9 are contained in the following table:

Exercise	sin A	cos A	tan A	cot A	sec A	csc A
1)	5/13	12/13	5/12	12/5	13/12	13/5
3)	3/5	4/5	3/4	4/3	5/4	5/3
5)	3/5	4/5	3/4	4/3	5/4	5/3
7)	8/17	15/17	8/15	15/8	17/15	17/8
9)	1/2	$\sqrt{3}/2$	$1/\sqrt{3}$	$\sqrt{3}$	$2/\sqrt{3}$	2
	cos B	sin B	cot B	tan B	csc B	sec B

11) $E = 45^0$ 13) $E = 45^0$ 15) $E = 70^0$ 17) 15.1   19) 33.8^0

21) 49.22   23) 38.9^0 25) $\sqrt{1 - b^2}/b$ 27) $(2b^2 - c^2)/c^2$

29) $\sqrt{\dfrac{c - a}{2c}}$ 31) $\sin(A + B) = \sin A \cos B + \cos A \sin B$

$$= \frac{a}{c} \cdot \frac{a}{c} + \frac{b}{c} \cdot \frac{b}{c} = \frac{a^2}{c^2} + \frac{b^2}{c^2}$$

$$= \frac{a^2 + b^2}{c^2} = \frac{c^2}{c^2} = 1$$

33) 169 feet   35) 5 feet   37) 1454 feet   39) 294 feet (river)

184 feet (cliff)

Exercise Set 8-2.

1) $C = 97^0$, $a = 23.8$, $b = 27.3$ 3) $C = 67^0$, $b = 18$, $c = 88$

5) $A = 104^0 \, 08'$, $a = 37.66$, $b = 31.92$ 7) 2   9) 0

11) $B = 27^0$, $C = 118^0$, $c = 22$ 13) $R = 50.0^0$, $S = 92.2^0$, $s = 39$

$R = 130.0^0$, $S = 12.2^0$, $s = 8$

15) $Y = 52.4^0$, $Z = 43.3^0$, $z = 8.5$ 17) 89 feet   19) 130 feet

Exercise Set 8-3.

1) $c = 12$, $A = 53^O$, $B = 78^O$ 3) $b = 320$, $A = 52^O$, $C = 55.2^O$

5) $c = 38.8$, $A = 59.9^O$, $B = 27.3^O$ 7) $C = 83.5^O$, $A = 29.5^O$, $b = 3.62$

9) 148 pounds   11) 21.0^O 13) 46° 15) 1693 miles

17) $B = 89^O$, $A = 36^O$, $c = 229$ 19) $A = 28^O$, $B = 105^O$, $C = 47^O$

Exercise Set 8-4.

1) 36   3) no triangle   5) 689.1   7) 51.7

9) 1978   11) 53 or 37   13) 37.25   15) 115

17) 387   19) 3662 ft^2

Exercise Set 8-5.

1) $c = \sqrt{58}$, $A = 67^O$, $B = 23^O$ 3) $H = 67^O$, $i = 39$, $k = 25$

5) $o = 23.3$, $P = 48^O$, $Q = 29^O$

7) $d = 52$, $e = 25$, $f = 67$, $D = 43^O$, $E = 19^O$, $F = 118^O$

   $d = 113$, $e = 54$, $f = 67$, $D = 137^O$, $E = 19^O$, $F = 24^O$

9) $u = 78.6$, $v = 117.8$, $w = 98.5$, $U = 41^O$, $V = 83^O$, $W = 56^O$

   $u = 78.6$, $v = 117.8$, $w = 174.4$, $U = 22^O$, $V = 34^O$, $W = 124^O$

11) 26.83   13) no solution   15) 47.1   17) 87 feet

19) 57 feet   21) 87 feet   23) 27° , 242 pounds

TABLE 1: VALUES OF CIRCULAR FUNCTIONS

θ or RAD	SIN	COS	TAN	CSC	SEC	COT
0.00	0.0000	1.0000	0.0000	------	1.0000	------
0.01	0.0100	1.0000	0.0100	100.0	1.0000	100.0
0.02	0.0200	0.9998	0.0200	50.0	1.0002	50.0
0.03	0.0300	0.9996	0.0300	33.3	1.0005	33.3
0.04	0.0400	0.9992	0.0400	25.0	1.0008	25.0
0.05	0.0500	0.9988	0.0500	20.0	1.0013	20.0
0.06	0.0600	0.9982	0.0601	16.7	1.0018	16.6
0.07	0.0699	0.9976	0.0701	14.3	1.0025	14.3
0.08	0.0799	0.9968	0.0802	12.5	1.0032	12.5
0.09	0.0899	0.9960	0.0902	11.1	1.0041	11.1
0.10	0.0998	0.9950	0.1003	10.0	1.0050	9.9666
0.11	0.1098	0.9940	0.1104	9.1093	1.0061	9.0542
0.12	0.1197	0.9928	0.1206	8.3534	1.0072	8.2933
0.13	0.1296	0.9916	0.1307	7.7140	1.0085	7.6489
0.14	0.1395	0.9902	0.1409	7.1662	1.0099	7.0961
0.15	0.1494	0.9888	0.1511	6.6917	1.0114	6.6166
0.16	0.1593	0.9872	0.1614	6.2767	1.0129	6.1966
0.17	0.1692	0.9856	0.1717	5.9108	1.0146	5.8256
0.18	0.1790	0.9838	0.1820	5.5857	1.0164	5.4954
0.19	0.1889	0.9820	0.1923	5.2950	1.0183	5.1997
0.20	0.1987	0.9801	0.2027	5.0335	1.0203	4.9332
0.21	0.2085	0.9780	0.2131	4.7971	1.0225	4.6917
0.22	0.2182	0.9759	0.2236	4.5823	1.0247	4.4719
0.23	0.2280	0.9737	0.2341	4.3864	1.0270	4.2709
0.24	0.2377	0.9713	0.2447	4.2069	1.0295	4.0864
0.25	0.2474	0.9689	0.2553	4.0420	1.0321	3.9163
0.26	0.2571	0.9664	0.2660	3.8898	1.0348	3.7591
0.27	0.2667	0.9638	0.2768	3.7491	1.0376	3.6133
0.28	0.2764	0.9611	0.2876	3.6185	1.0405	3.4776
0.29	0.2860	0.9582	0.2984	3.4971	1.0436	3.3511
0.30	0.2955	0.9553	0.3093	3.3839	1.0468	3.2327
0.31	0.3051	0.9523	0.3203	3.2781	1.0501	3.1218
0.32	0.3146	0.9492	0.3314	3.1790	1.0535	3.0176
0.33	0.3240	0.9460	0.3425	3.0860	1.0570	2.9195
0.34	0.3335	0.9428	0.3537	2.9986	1.0607	2.8270
0.35	0.3429	0.9394	0.3650	2.9163	1.0645	2.7395
0.36	0.3523	0.9359	0.3764	2.8387	1.0685	2.6567
0.37	0.3616	0.9323	0.3879	2.7654	1.0726	2.5782
0.38	0.3709	0.9287	0.3994	2.6960	1.0768	2.5037
0.39	0.3802	0.9249	0.4111	2.6303	1.0812	2.4328
θ or RAD	SIN	COS	TAN	CSC	SEC	COT

TABLE 1: (CONTINUED)

θ or RAD	SIN	COS	TAN	CSC	SEC	COT
0.40	0.3894	0.9210	0.4228	2.5679	1.0857	2.3652
0.41	0.3986	0.9171	0.4346	2.5087	1.0904	2.3008
0.42	0.4078	0.9131	0.4466	2.4524	1.0952	2.2393
0.43	0.4169	0.9090	0.4586	2.3988	1.1002	2.1805
0.44	0.4259	0.9048	0.4708	2.3478	1.1053	2.1241
0.45	0.4350	0.9004	0.4831	2.2990	1.1106	2.0702
0.46	0.4439	0.8961	0.4954	2.2525	1.1160	2.0184
0.47	0.4529	0.8916	0.5080	2.2081	1.1216	1.9686
0.48	0.4618	0.8870	0.5206	2.1655	1.1274	1.9208
0.49	0.4706	0.8823	0.5334	2.1248	1.1334	1.8748
0.50	0.4794	0.8776	0.5463	2.0858	1.1395	1.8305
0.51	0.4882	0.8727	0.5594	2.0484	1.1458	1.7878
0.52	0.4969	0.8678	0.5726	2.0126	1.1523	1.7465
0.53	0.5055	0.8628	0.5859	1.9781	1.1590	1.7067
0.54	0.5141	0.8577	0.5994	1.9450	1.1659	1.6683
0.55	0.5227	0.8525	0.6131	1.9132	1.1730	1.6310
0.56	0.5312	0.8473	0.6269	1.8826	1.1803	1.5950
0.57	0.5396	0.8419	0.6410	1.8531	1.1878	1.5601
0.58	0.5480	0.8365	0.6552	1.8247	1.1955	1.5263
0.59	0.5564	0.8309	0.6696	1.7974	1.2035	1.4935
0.60	0.5646	0.8253	0.6841	1.7710	1.2116	1.4617
0.61	0.5729	0.8196	0.6989	1.7456	1.2200	1.4308
0.62	0.5810	0.8139	0.7139	1.7211	1.2287	1.4007
0.63	0.5891	0.8080	0.7291	1.6974	1.2376	1.3715
0.64	0.5972	0.8021	0.7445	1.6745	1.2467	1.3431
0.65	0.6052	0.7961	0.7602	1.6524	1.2561	1.3154
0.66	0.6131	0.7900	0.7761	1.6310	1.2658	1.2885
0.67	0.6210	0.7838	0.7923	1.6103	1.2758	1.2622
0.68	0.6288	0.7776	0.8087	1.5903	1.2861	1.2366
0.69	0.6365	0.7712	0.8253	1.5710	1.2966	1.2116
0.70	0.6442	0.7648	0.8423	1.5523	1.3075	1.1872
0.71	0.6518	0.7584	0.8595	1.5343	1.3186	1.1634
0.72	0.6594	0.7518	0.8771	1.5166	1.3301	1.1402
0.73	0.6669	0.7452	0.8949	1.4995	1.3420	1.1174
0.74	0.6743	0.7385	0.9131	1.4830	1.3542	1.0952
0.75	0.6816	0.7317	0.9316	1.4671	1.3667	1.0734
0.76	0.6889	0.7248	0.9505	1.4515	1.3796	1.0521
0.77	0.6961	0.7179	0.9697	1.4365	1.3929	1.0313
0.78	0.7033	0.7109	0.9893	1.4219	1.4066	1.0109
0.79	0.7104	0.7038	1.0092	1.4078	1.4208	0.9908
θ or RAD	SIN	COS	TAN	CSC	SEC	COT

TABLE 1 : (CONTINUED)

θ or RAD	SIN	COS	TAN	CSC	SEC	COT
0.80	0.7174	0.6967	1.0296	1.3940	1.4353	0.9712
0.81	0.7243	0.6895	1.0505	1.3807	1.4503	0.9520
0.82	0.7311	0.6822	1.0717	1.3677	1.4658	0.9331
0.83	0.7379	0.6749	1.0934	1.3551	1.4818	0.9146
0.84	0.7446	0.6675	1.1156	1.3429	1.4982	0.8964
0.85	0.7513	0.6600	1.1383	1.3311	1.5152	0.8785
0.86	0.7578	0.6524	1.1616	1.3195	1.5327	0.8609
0.87	0.7643	0.6448	1.1853	1.3083	1.5508	0.8437
0.88	0.7707	0.6372	1.2097	1.2975	1.5695	0.8267
0.89	0.7771	0.6294	1.2346	1.2869	1.5888	0.8100
0.90	0.7833	0.6216	1.2602	1.2766	1.6087	0.7936
0.91	0.7895	0.6137	1.2864	1.2666	1.6293	0.7774
0.92	0.7956	0.6058	1.3133	1.2569	1.6507	0.7615
0.93	0.8016	0.5978	1.3409	1.2475	1.6727	0.7458
0.94	0.8076	0.5898	1.3692	1.2383	1.6955	0.7303
0.95	0.8134	0.5817	1.3984	1.2294	1.7191	0.7151
0.96	0.8192	0.5735	1.4284	1.2207	1.7436	0.7001
0.97	0.8249	0.5653	1.4592	1.2123	1.7690	0.6853
0.98	0.8305	0.5570	1.4910	1.2041	1.7953	0.6707
0.99	0.8360	0.5487	1.5237	1.1961	1.8225	0.6563
1.00	0.8415	0.5403	1.5574	1.1884	1.8508	0.6421
1.01	0.8468	0.5319	1.5922	1.1809	1.8802	0.6281
1.02	0.8521	0.5234	1.6281	1.1736	1.9107	0.6142
1.03	0.8573	0.5148	1.6652	1.1665	1.9424	0.6005
1.04	0.8624	0.5062	1.7036	1.1595	1.9754	0.5870
1.05	0.8674	0.4976	1.7433	1.1528	2.0098	0.5736
1.06	0.8724	0.4889	1.7844	1.1463	2.0455	0.5604
1.07	0.8772	0.4801	1.8270	1.1400	2.0828	0.5473
1.08	0.8820	0.4713	1.8712	1.1338	2.1217	0.5344
1.09	0.8866	0.4625	1.9171	1.1279	2.1622	0.5216
1.10	0.8912	0.4536	1.9648	1.1221	2.2046	0.5090
1.11	0.8957	0.4447	2.0143	1.1164	2.2489	0.4964
1.12	0.9001	0.4357	2.0660	1.1110	2.2952	0.4840
1.13	0.9044	0.4267	2.1197	1.1057	2.3438	0.4718
1.14	0.9086	0.4176	2.1759	1.1006	2.3947	0.4596
1.15	0.9128	0.4085	2.2345	1.0956	2.4481	0.4475
1.16	0.9168	0.3993	2.2958	1.0907	2.5041	0.4356
1.17	0.9208	0.3902	2.3600	1.0861	2.5631	0.4237
1.18	0.9246	0.3809	2.4273	1.0815	2.6252	0.4120
1.19	0.9284	0.3717	2.4979	1.0772	2.6906	0.4003
θ or RAD	SIN	COS	TAN	CSC	SEC	COT

TABLE 1: (CONTINUED)

θ or RAD	SIN	COS	TAN	CSC	SEC	COT
1.20	0.9320	0.3624	2.5722	1.0729	2.7597	0.3888
1.21	0.9356	0.3530	2.6503	1.0688	2.8327	0.3773
1.22	0.9391	0.3436	2.7327	1.0649	2.9100	0.3659
1.23	0.9425	0.3342	2.8198	1.0610	2.9919	0.3546
1.24	0.9458	0.3248	2.9109	1.0573	3.0788	0.3434
1.25	0.9490	0.3153	3.0096	1.0538	3.1713	0.3323
1.26	0.9521	0.3058	3.1133	1.0503	3.2699	0.3212
1.27	0.9551	0.2963	3.2236	1.0470	3.3752	0.3102
1.28	0.9580	0.2867	3.3413	1.0438	3.4878	0.2993
1.29	0.9608	0.2771	3.4672	1.0408	3.6085	0.2884
1.30	0.9636	0.2675	3.6021	1.0378	3.7383	0.2776
1.31	0.9662	0.2579	3.7471	1.0350	3.8782	0.2669
1.32	0.9687	0.2482	3.9033	1.0323	4.0294	0.2562
1.33	0.9711	0.2385	4.0723	1.0297	4.1933	0.2456
1.34	0.9735	0.2288	4.2556	1.0272	4.3715	0.2350
1.35	0.9757	0.2190	4.4552	1.0249	4.5661	0.2245
1.36	0.9779	0.2092	4.6734	1.0226	4.7792	0.2140
1.37	0.9799	0.1995	4.9130	1.0205	5.0138	0.2035
1.38	0.9819	0.1896	5.1774	1.0185	5.2731	0.1931
1.39	0.9837	0.1798	5.4707	1.0166	5.5613	0.1828
1.40	0.9854	0.1700	5.7979	1.0148	5.8835	0.1725
1.41	0.9871	0.1601	6.1653	1.0133	6.2459	0.1622
1.42	0.9887	0.1502	6.5811	1.0115	6.6566	0.1520
1.43	0.9901	0.1403	7.0554	1.0100	7.1259	0.1417
1.44	0.9915	0.1304	7.6018	1.0086	7.6673	0.1315
1.45	0.9927	0.1205	8.2380	1.0073	8.2985	0.1214
1.46	0.9939	0.1106	8.9885	1.0062	9.0440	0.1113
1.47	0.9949	0.1006	9.8873	1.0051	9.9377	0.1011
1.48	0.9959	0.0907	11.0	1.0041	11.0	0.0910
1.49	0.9967	0.0807	12.3	1.0033	12.4	0.0810
1.50	0.9975	0.0707	14.1	1.0025	14.1	0.0709
1.51	0.9982	0.0608	16.4	1.0019	16.5	0.0609
1.52	0.9987	0.0508	19.7	1.0013	19.7	0.0508
1.53	0.9992	0.0408	24.5	1.0008	24.5	0.0408
1.54	0.9995	0.0308	32.5	1.0005	32.5	0.0308
1.55	0.9998	0.0208	48.1	1.0002	48.1	0.0208
1.56	0.9999	0.0108	92.6	1.0001	92.6	0.0108
1.57	1.0000	0.0008	1253.9	1.0000	1254.2	0.0008
θ or RAD	SIN	COS	TAN	CSC	SEC	COT

TABLE II : VALUES OF TRIGONOMETRIC FUNCTIONS

DEG		SIN	CSC	TAN	COT	SEC	COS		
00	00	0.0000	------	0.0000	------	1.0000	1.0000	90	00
	10	0.0029	343.8	0.0029	343.8	1.0000	1.0000		50
	20	0.0058	171.9	0.0058	171.9	1.0000	1.0000		40
	30	0.0087	114.6	0.0087	114.6	1.0000	1.0000		30
	40	0.0116	85.9	0.0116	85.9	1.0001	0.9999		20
	50	0.0145	68.8	0.0145	68.7	1.0001	0.9999		10
01	00	0.0175	57.3	0.0175	57.3	1.0002	0.9998	89	00
	10	0.0204	49.1	0.0204	49.1	1.0002	0.9998		50
	20	0.0233	43.0	0.0233	43.0	1.0003	0.9997		40
	30	0.0262	38.2	0.0262	38.2	1.0003	0.9997		30
	40	0.0291	34.4	0.0291	34.4	1.0004	0.9996		20
	50	0.0320	31.3	0.0320	31.2	1.0005	0.9995		10
02	00	0.0349	28.7	0.0349	28.6	1.0006	0.9994	88	00
	10	0.0378	26.5	0.0378	26.4	1.0007	0.9993		50
	20	0.0407	24.6	0.0407	24.5	1.0008	0.9992		40
	30	0.0436	22.9	0.0437	22.9	1.0010	0.9990		30
	40	0.0465	21.5	0.0466	21.5	1.0011	0.9989		20
	50	0.0494	20.2	0.0495	20.2	1.0012	0.9988		10
03	00	0.0523	19.1	0.0524	19.1	1.0014	0.9986	87	00
	10	0.0552	18.1	0.0553	18.1	1.0015	0.9985		50
	20	0.0581	17.2	0.0582	17.2	1.0017	0.9983		40
	30	0.0610	16.4	0.0612	16.3	1.0019	0.9981		30
	40	0.0640	15.6	0.0641	15.6	1.0021	0.9980		20
	50	0.0669	15.0	0.0670	14.9	1.0022	0.9978		10
04	00	0.0698	14.3	0.0699	14.3	1.0024	0.9976	86	00
	10	0.0727	13.8	0.0729	13.7	1.0027	0.9974		50
	20	0.0756	13.2	0.0758	13.2	1.0029	0.9971		40
	30	0.0785	12.7	0.0787	12.7	1.0031	0.9969		30
	40	0.0814	12.3	0.0816	12.3	1.0033	0.9967		20
	50	0.0843	11.9	0.0846	11.8	1.0036	0.9964		10
05	00	0.0872	11.5	0.0875	11.4	1.0038	0.9962	85	00
	10	0.0901	11.1	0.0904	11.1	1.0041	0.9959		50
	20	0.0930	10.8	0.0934	10.7	1.0043	0.9957		40
	30	0.0958	10.4	0.0963	10.4	1.0046	0.9954		30
	40	0.0987	10.1	0.0992	10.1	1.0049	0.9951		20
	50	0.1016	9.8391	0.1022	9.7882	1.0052	0.9948		10
06	00	0.1045	9.5667	0.1051	9.5143	1.0055	0.9945	84	00
	10	0.1074	9.3091	0.1080	9.2553	1.0058	0.9942		50
	20	0.1103	9.0651	0.1110	9.0098	1.0061	0.9939		40
	30	0.1132	8.8336	0.1139	8.7769	1.0065	0.9936		30
	40	0.1161	8.6138	0.1169	8.5555	1.0068	0.9932		20
	50	0.1190	8.4046	0.1198	8.3449	1.0072	0.9929		10
		COS	SEC	COT	TAN	CSC	SIN	DEG	

TABLE 2: (CONTINUED)

DEG	SIN	CSC	TAN	COT	SEC	COS		
07 00	0.1219	8.2055	0.1228	8.1443	1.0075	0.9925	83	00
10	0.1248	8.0156	0.1257	7.9530	1.0079	0.9922		50
20	0.1276	7.8344	0.1287	7.7703	1.0082	0.9918		40
30	0.1305	7.6613	0.1317	7.5957	1.0086	0.9914		30
40	0.1334	7.4957	0.1346	7.4287	1.0090	0.9911		20
50	0.1363	7.3372	0.1376	7.2687	1.0094	0.9907		10
08 00	0.1392	7.1853	0.1405	7.1154	1.0098	0.9903	82	00
10	0.1421	7.0396	0.1435	6.9682	1.0102	0.9899		50
20	0.1449	6.8998	0.1465	6.8269	1.0107	0.9894		40
30	0.1478	6.7655	0.1495	6.6911	1.0111	0.9890		30
40	0.1507	6.6363	0.1524	6.5605	1.0116	0.9886		20
50	0.1536	6.5121	0.1554	6.4348	1.0120	0.9881		10
09 00	0.1564	6.3924	0.1584	6.3137	1.0125	0.9877	81	00
10	0.1593	6.2772	0.1614	6.1970	1.0129	0.9872		50
20	0.1622	6.1661	0.1644	6.0844	1.0134	0.9868		40
30	0.1650	6.0588	0.1673	5.9758	1.0139	0.9863		30
40	0.1679	5.9553	0.1703	5.8708	1.0144	0.9858		20
50	0.1708	5.8554	0.1733	5.7694	1.0149	0.9853		10
10 00	0.1736	5.7588	0.1763	5.6713	1.0154	0.9848	80	00
10	0.1765	5.6653	0.1793	5.5764	1.0160	0.9843		50
20	0.1794	5.5749	0.1823	5.4845	1.0165	0.9838		40
30	0.1822	5.4874	0.1853	5.3955	1.0170	0.9833		30
40	0.1851	5.4026	0.1883	5.3093	1.0176	0.9827		20
50	0.1880	5.3205	0.1914	5.2257	1.0181	0.9822		10
11 00	0.1908	5.2408	0.1944	5.1445	1.0187	0.9816	79	00
10	0.1937	5.1636	0.1974	5.0658	1.0193	0.9811		50
20	0.1965	5.0886	0.2004	4.9894	1.0199	0.9805		40
30	0.1994	5.0158	0.2035	4.9151	1.0205	0.9799		30
40	0.2022	4.9452	0.2065	4.8430	1.0211	0.9793		20
50	0.2051	4.8765	0.2095	4.7728	1.0217	0.9787		10
12 00	0.2079	4.8097	0.2126	4.7046	1.0223	0.9781	78	00
10	0.2108	4.7448	0.2156	4.6382	1.0230	0.9775		50
20	0.2136	4.6817	0.2186	4.5736	1.0236	0.9769		40
30	0.2164	4.6202	0.2217	4.5107	1.0243	0.9763		30
40	0.2193	4.5604	0.2247	4.4494	1.0249	0.9757		20
50	0.2221	4.5021	0.2278	4.3897	1.0256	0.9750		10
13 00	0.2250	4.4454	0.2309	4.3315	1.0263	0.9744	77	00
10	0.2278	4.3901	0.2339	4.2747	1.0270	0.9737		50
20	0.2306	4.3362	0.2370	4.2193	1.0277	0.9730		40
30	0.2334	4.2836	0.2401	4.1653	1.0284	0.9724		30
40	0.2363	4.2324	0.2432	4.1126	1.0291	0.9717		20
50	0.2391	4.1824	0.2462	4.0611	1.0299	0.9710		10
	COS	SEC	COT	TAN	CSC	SIN	DEG	

TABLE 2: (CONTINUED)

DEG		SIN	CSC	TAN	COT	SEC	COS		
14	00	0.2419	4.1336	0.2493	4.0108	1.0306	0.9703	76	00
	10	0.2447	4.0859	0.2524	3.9616	1.0314	0.9696		50
	20	0.2476	4.0394	0.2555	3.9136	1.0321	0.9689		40
	30	0.2504	3.9939	0.2586	3.8667	1.0329	0.9681		30
	40	0.2532	3.9495	0.2617	3.8208	1.0337	0.9674		20
	50	0.2560	3.9061	0.2648	3.7759	1.0345	0.9667		10
15	00	0.2588	3.8637	0.2679	3.7320	1.0353	0.9659	75	00
	10	0.2616	3.8222	0.2711	3.6891	1.0361	0.9652		50
	20	0.2644	3.7816	0.2742	3.6470	1.0369	0.9644		40
	30	0.2672	3.7420	0.2773	3.6059	1.0377	0.9636		30
	40	0.2700	3.7031	0.2805	3.5656	1.0386	0.9628		20
	50	0.2728	3.6651	0.2836	3.5261	1.0394	0.9621		10
16	00	0.2756	3.6279	0.2867	3.4874	1.0403	0.9613	74	00
	10	0.2784	3.5915	0.2899	3.4495	1.0412	0.9605		50
	20	0.2812	3.5559	0.2931	3.4124	1.0421	0.9596		40
	30	0.2840	3.5209	0.2962	3.3759	1.0429	0.9588		30
	40	0.2868	3.4867	0.2994	3.3402	1.0439	0.9580		20
	50	0.2896	3.4532	0.3026	3.3052	1.0448	0.9572		10
17	00	0.2924	3.4203	0.3057	3.2708	1.0457	0.9563	73	00
	10	0.2952	3.3881	0.3089	3.2371	1.0466	0.9554		50
	20	0.2979	3.3565	0.3121	3.2041	1.0476	0.9546		40
	30	0.3007	3.3255	0.3153	3.1716	1.0485	0.9537		30
	40	0.3035	3.2951	0.3185	3.1397	1.0495	0.9528		20
	50	0.3062	3.2653	0.3217	3.1084	1.0505	0.9520		10
18	00	0.3090	3.2361	0.3249	3.0777	1.0515	0.9511	72	00
	10	0.3118	3.2074	0.3281	3.0475	1.0525	0.9502		50
	20	0.3145	3.1792	0.3314	3.0178	1.0535	0.9492		40
	30	0.3173	3.1515	0.3346	2.9887	1.0545	0.9483		30
	40	0.3201	3.1244	0.3378	2.9600	1.0555	0.9474		20
	50	0.3228	3.0977	0.3411	2.9319	1.0566	0.9465		10
19	00	0.3256	3.0715	0.3443	2.9042	1.0576	0.9455	71	00
	10	0.3283	3.0458	0.3476	2.8770	1.0587	0.9446		50
	20	0.3311	3.0206	0.3508	2.8502	1.0598	0.9436		40
	30	0.3338	2.9957	0.3541	2.8239	1.0608	0.9426		30
	40	0.3365	2.9713	0.3574	2.7980	1.0619	0.9417		20
	50	0.3393	2.9474	0.3607	2.7725	1.0631	0.9407		10
20	00	0.3420	2.9238	0.3640	2.7475	1.0642	0.9397	70	00
	10	0.3448	2.9006	0.3673	2.7228	1.0653	0.9387		50
	20	0.3475	2.8778	0.3706	2.6985	1.0665	0.9377		40
	30	0.3502	2.8554	0.3739	2.6746	1.0676	0.9367		30
	40	0.3529	2.8334	0.3772	2.6511	1.0688	0.9356		20
	50	0.3557	2.8117	0.3805	2.6279	1.0700	0.9346		10
		COS	SEC	COT	TAN	CSC	SIN	DEG	

TABLE 2: (CONTINUED)

DEG		SIN	CSC	TAN	COT	SEC	COS	RAD		
21	00	0.3584	2.7904	0.3839	2.6051	1.0711	0.9336	1.203	69	00
	10	0.3611	2.7694	0.3872	2.5826	1.0723	0.9325	1.201		50
	20	0.3638	2.7488	0.3906	2.5605	1.0736	0.9315	1.198		40
	30	0.3665	2.7285	0.3939	2.5386	1.0748	0.9304	1.195		30
	40	0.3692	2.7085	0.3973	2.5171	1.0760	0.9293	1.192		20
	50	0.3719	2.6888	0.4006	2.4960	1.0773	0.9283	1.189		10
22	00	0.3746	2.6695	0.4040	2.4751	1.0785	0.9272	1.186	68	00
	10	0.3773	2.6504	0.4074	2.4545	1.0798	0.9261	1.183		50
	20	0.3800	2.6316	0.4108	2.4342	1.0811	0.9250	1.180		40
	30	0.3827	2.6131	0.4142	2.4142	1.0824	0.9239	1.177		30
	40	0.3854	2.5949	0.4176	2.3945	1.0837	0.9228	1.174		20
	50	0.3881	2.5770	0.4210	2.3750	1.0850	0.9216	1.171		10
23	00	0.3907	2.5593	0.4245	2.3558	1.0864	0.9205	1.169	67	00
	10	0.3934	2.5419	0.4279	2.3369	1.0877	0.9194	1.166		50
	20	0.3961	2.5247	0.4314	2.3183	1.0891	0.9182	1.163		40
	30	0.3988	2.5078	0.4348	2.2998	1.0904	0.9171	1.160		30
	40	0.4014	2.4912	0.4383	2.2817	1.0918	0.9159	1.157		20
	50	0.4041	2.4748	0.4417	2.2637	1.0932	0.9147	1.154		10
24	00	0.4067	2.4586	0.4452	2.2460	1.0946	0.9135	1.151	66	00
	10	0.4094	2.4426	0.4487	2.2286	1.0961	0.9124	1.148		50
	20	0.4120	2.4269	0.4522	2.2113	1.0975	0.9112	1.145		40
	30	0.4147	2.4114	0.4557	2.1943	1.0989	0.9100	1.142		30
	40	0.4173	2.3961	0.4592	2.1775	1.1004	0.9088	1.139		20
	50	0.4200	2.3811	0.4628	2.1609	1.1019	0.9075	1.137		10
25	00	0.4226	2.3662	0.4663	2.1445	1.1034	0.9063	1.134	65	00
	10	0.4253	2.3515	0.4699	2.1283	1.1049	0.9051	1.131		50
	20	0.4279	2.3371	0.4734	2.1123	1.1064	0.9038	1.128		40
	30	0.4305	2.3228	0.4770	2.0965	1.1079	0.9026	1.125		30
	40	0.4331	2.3087	0.4806	2.0809	1.1095	0.9013	1.122		20
	50	0.4358	2.2949	0.4841	2.0655	1.1110	0.9001	1.119		10
26	00	0.4384	2.2812	0.4877	2.0503	1.1126	0.8988	1.116	64	00
	10	0.4410	2.2677	0.4913	2.0353	1.1142	0.8975	1.113		50
	20	0.4436	2.2543	0.4950	2.0204	1.1158	0.8962	1.110		40
	30	0.4462	2.2412	0.4986	2.0057	1.1174	0.8949	1.107		30
	40	0.4488	2.2282	0.5022	1.9912	1.1190	0.8936	1.105		20
	50	0.4514	2.2153	0.5059	1.9768	1.1207	0.8923	1.102		10
27	00	0.4540	2.2027	0.5095	1.9626	1.1223	0.8910	1.099	63	00
	10	0.4566	2.1902	0.5132	1.9486	1.1240	0.8897	1.096		50
	20	0.4592	2.1779	0.5169	1.9347	1.1257	0.8883	1.093		40
	30	0.4617	2.1657	0.5206	1.9210	1.1274	0.8870	1.090		30
	40	0.4643	2.1537	0.5243	1.9074	1.1291	0.8857	1.087		20
	50	0.4669	2.1418	0.5280	1.8940	1.1308	0.8843	1.084		10
		COS	SEC	COT	TAN	CSC	SIN	RAD	DEG	

TABLE 2: (CONTINUED)

DEG		SIN	CSC	TAN	COT	SEC	COS		
28	00	0.4695	2.1301	0.5317	1.8807	1.1326	0.8829	62	00
	10	0.4720	2.1185	0.5354	1.8676	1.1343	0.8816		50
	20	0.4746	2.1070	0.5392	1.8546	1.1361	0.8802		40
	30	0.4772	2.0957	0.5430	1.8418	1.1379	0.8788		30
	40	0.4797	2.0846	0.5467	1.8291	1.1397	0.8774		20
	50	0.4823	2.0736	0.5505	1.8165	1.1415	0.8760		10
29	00	0.4848	2.0627	0.5543	1.8040	1.1434	0.8746	61	00
	10	0.4874	2.0519	0.5581	1.7917	1.1452	0.8732		50
	20	0.4899	2.0413	0.5619	1.7795	1.1471	0.8718		40
	30	0.4924	2.0308	0.5658	1.7675	1.1490	0.8704		30
	40	0.4950	2.0204	0.5696	1.7556	1.1509	0.8689		20
	50	0.4975	2.0101	0.5735	1.7437	1.1528	0.8675		10
30	00	0.5000	2.0000	0.5774	1.7320	1.1547	0.8660	60	00
	10	0.5025	1.9900	0.5812	1.7205	1.1566	0.8646		50
	20	0.5050	1.9801	0.5851	1.7090	1.1586	0.8631		40
	30	0.5075	1.9703	0.5890	1.6977	1.1606	0.8616		30
	40	0.5100	1.9606	0.5930	1.6864	1.1626	0.8601		20
	50	0.5125	1.9511	0.5969	1.6753	1.1646	0.8587		10
31	00	0.5150	1.9416	0.6009	1.6643	1.1666	0.8572	59	00
	10	0.5175	1.9323	0.6048	1.6534	1.1687	0.8557		50
	20	0.5200	1.9230	0.6088	1.6426	1.1707	0.8542		40
	30	0.5225	1.9139	0.6128	1.6318	1.1728	0.8526		30
	40	0.5250	1.9048	0.6168	1.6212	1.1749	0.8511		20
	50	0.5275	1.8959	0.6208	1.6107	1.1770	0.8496		10
32	00	0.5299	1.8871	0.6249	1.6003	1.1792	0.8480	58	00
	10	0.5324	1.8783	0.6289	1.5900	1.1813	0.8465		50
	20	0.5348	1.8697	0.6330	1.5798	1.1835	0.8450		40
	30	0.5373	1.8612	0.6371	1.5697	1.1857	0.8434		30
	40	0.5398	1.8527	0.6412	1.5597	1.1879	0.8418		20
	50	0.5422	1.8443	0.6453	1.5497	1.1901	0.8403		10
33	00	0.5446	1.8361	0.6494	1.5399	1.1924	0.8387	57	00
	10	0.5471	1.8279	0.6536	1.5301	1.1946	0.8371		50
	20	0.5495	1.8198	0.6577	1.5204	1.1969	0.8355		40
	30	0.5519	1.8118	0.6619	1.5108	1.1992	0.8339		30
	40	0.5544	1.8039	0.6661	1.5013	1.2015	0.8323		20
	50	0.5568	1.7960	0.6703	1.4919	1.2039	0.8307		10
34	00	0.5592	1.7883	0.6745	1.4826	1.2062	0.8290	56	00
	10	0.5616	1.7806	0.6788	1.4733	1.2086	0.8274		50
	20	0.5640	1.7730	0.6830	1.4641	1.2110	0.8258		40
	30	0.5664	1.7655	0.6873	1.4550	1.2134	0.8241		30
	40	0.5688	1.7581	0.6916	1.4460	1.2158	0.8225		20
	50	0.5712	1.7507	0.6959	1.4370	1.2183	0.8208		10
		COS	SEC	COT	TAN	CSC	SIN	DEG	

TABLE 2: (CONTINUED)

DEG		SIN	CSC	TAN	COT	SEC	COS		
35	00	0.5736	1.7434	0.7002	1.4281	1.2208	0.8192	55	00
	10	0.5760	1.7362	0.7046	1.4193	1.2233	0.8175		50
	20	0.5783	1.7291	0.7089	1.4106	1.2258	0.8158		40
	30	0.5807	1.7220	0.7133	1.4019	1.2283	0.8141		30
	40	0.5831	1.7151	0.7177	1.3934	1.2309	0.8124		20
	50	0.5854	1.7081	0.7221	1.3848	1.2335	0.8107		10
36	00	0.5878	1.7013	0.7265	1.3764	1.2361	0.8090	54	00
	10	0.5901	1.6945	0.7310	1.3680	1.2387	0.8073		50
	20	0.5925	1.6878	0.7355	1.3597	1.2413	0.8056		40
	30	0.5948	1.6812	0.7400	1.3514	1.2440	0.8039		30
	40	0.5972	1.6746	0.7445	1.3432	1.2467	0.8021		20
	50	0.5995	1.6681	0.7490	1.3351	1.2494	0.8004		10
37	00	0.6018	1.6616	0.7536	1.3270	1.2521	0.7986	53	00
	10	0.6041	1.6553	0.7581	1.3190	1.2549	0.7969		50
	20	0.6065	1.6489	0.7627	1.3111	1.2577	0.7951		40
	30	0.6088	1.6427	0.7673	1.3032	1.2605	0.7934		30
	40	0.6111	1.6365	0.7720	1.2954	1.2633	0.7916		20
	50	0.6134	1.6303	0.7766	1.2876	1.2661	0.7898		10
38	00	0.6157	1.6243	0.7813	1.2799	1.2690	0.7880	52	00
	10	0.6180	1.6182	0.7860	1.2723	1.2719	0.7862		50
	20	0.6202	1.6123	0.7907	1.2647	1.2748	0.7844		40
	30	0.6225	1.6064	0.7954	1.2572	1.2778	0.7826		30
	40	0.6248	1.6005	0.8002	1.2497	1.2807	0.7808		20
	50	0.6271	1.5947	0.8050	1.2423	1.2837	0.7790		10
39	00	0.6293	1.5890	0.8098	1.2349	1.2868	0.7771	51	00
	10	0.6316	1.5833	0.8146	1.2276	1.2898	0.7753		50
	20	0.6338	1.5777	0.8195	1.2203	1.2929	0.7735		40
	30	0.6361	1.5721	0.8243	1.2131	1.2960	0.7716		30
	40	0.6383	1.5666	0.8292	1.2059	1.2991	0.7698		20
	50	0.6406	1.5611	0.8342	1.1988	1.3022	0.7679		10
40	00	0.6428	1.5557	0.8391	1.1917	1.3054	0.7660	50	00
	10	0.6450	1.5504	0.8441	1.1847	1.3086	0.7642		50
	20	0.6472	1.5450	0.8491	1.1778	1.3118	0.7623		40
	30	0.6494	1.5398	0.8541	1.1708	1.3151	0.7604		30
	40	0.6517	1.5345	0.8591	1.1640	1.3184	0.7585		20
	50	0.6539	1.5294	0.8642	1.1571	1.3217	0.7566		10
41	00	0.6561	1.5243	0.8693	1.1504	1.3250	0.7547	49	00
	10	0.6583	1.5192	0.8744	1.1436	1.3284	0.7528		50
	20	0.6604	1.5141	0.8796	1.1369	1.3318	0.7509		40
	30	0.6626	1.5092	0.8847	1.1303	1.3352	0.7490		30
	40	0.6648	1.5042	0.8899	1.1237	1.3386	0.7470		20
	50	0.6670	1.4993	0.8952	1.1171	1.3421	0.7451		10
		COS	SEC	COT	TAN	CSC	SIN	DEG	

TABLE 2: (CONTINUED)

DEG		SIN	CSC	TAN	COT	SEC	COS		
42	00	0.6691	1.4945	0.9004	1.1106	1.3456	0.7431	48	00
	10	0.6713	1.4897	0.9057	1.1041	1.3492	0.7412		50
	20	0.6734	1.4849	0.9110	1.0977	1.3527	0.7392		40
	30	0.6756	1.4802	0.9163	1.0913	1.3563	0.7373		30
	40	0.6777	1.4755	0.9217	1.0850	1.3600	0.7353		20
	50	0.6799	1.4709	0.9271	1.0786	1.3636	0.7333		10
43	00	0.6820	1.4663	0.9325	1.0724	1.3673	0.7314	47	00
	10	0.6841	1.4617	0.9380	1.0661	1.3711	0.7294		50
	20	0.6862	1.4572	0.9435	1.0599	1.3748	0.7274		40
	30	0.6884	1.4527	0.9490	1.0538	1.3786	0.7254		30
	40	0.6905	1.4483	0.9545	1.0477	1.3824	0.7234		20
	50	0.6926	1.4439	0.9601	1.0416	1.3863	0.7214		10
44	00	0.6947	1.4396	0.9657	1.0355	1.3902	0.7193	46	00
	10	0.6967	1.4352	0.9713	1.0295	1.3941	0.7173		50
	20	0.6988	1.4310	0.9770	1.0235	1.3980	0.7153		40
	30	0.7009	1.4267	0.9827	1.0176	1.4020	0.7132		30
	40	0.7030	1.4225	0.9884	1.0117	1.4061	0.7112		20
	50	0.7050	1.4183	0.9942	1.0058	1.4101	0.7092		10
45	00	0.7071	1.4142	1.0000	1.0000	1.4142	0.7071	45	00
		COS	SEC	COT	TAN	CSC	SIN	DEG	

www.ingramcontent.com/pod-product-compliance
Lightning Source LLC
Chambersburg PA
CBHW081103170526
45165CB00008B/2310